Power Quality Indices in
Liberalized Markets

Power Quality Indices in Liberalized Markets

Pierluigi Caramia
University of Napoli Parthenope, Italy

Guido Carpinelli
University of Napoli Federico II, Italy

Paola Verde
University of Cassino, Italy

A John Wiley and Sons, Ltd., Publication

This edition first published 2009
© 2009 John Wiley & Sons, Ltd

Registered office
John Wiley & Sons Ltd, The Atrium, Southern Gate, Chichester, West Sussex,
PO19 8SQ, United Kingdom

For details of our global editorial offices, for customer services and for information about how to apply for
permission to reuse the copyright material in this book please see our website at www.wiley.com.

Library of Congress Cataloging-in-Publication Data

Caramia, Pierluigi.
  Power quality indices in liberalized markets / by Pierluigi
  Caramia, Guido Carpinelli, Paola Verde.
    p. cm.
  Includes bibliographical references and index.
  ISBN 978-0-470-03395-1 (cloth)
  1. Electric power systems—Quality control. 2. Electric power system
  stability—Measurement. 3. Electric power failures—Economic aspects.
  4. Electric utilities. I. Carpinelli, Guido. II. Verde, Paola. III. Title.
  TK1010.C365 2009
  333.793′2—dc22
                                                    2009015956

A catalogue record for this book is available from the British Library.

ISBN: 978-0-470-03395-1 (H/B)

Set in 10/12pt Times by Integra Software Services Pvt. Ltd, Pondicherry, India
Printed in Great Britain by CPI Antony Rowe, Chippenham, Wiltshire

To my parents, and to Rita and Marcella
Pierluigi Caramia

To Paola, Pierluigi, Pietro and Angela
Guido Carpinelli

To my father Achille and to my mother Serà
Paola Verde

Contents

About the authors

Pierluigi Caramia is associate professor of Electrical Power Systems at the University of Napoli 'Parthenope' (Italy). He received his MS degree in Electrical Engineering from the University of Cassino in 1991. His research interests concern power system analysis and power quality. In addition, he is a co-author of Chapter 8 and Chapter 11 of the book *Time-Varying Waveform Distortions in Power Systems*, published by John Wiley & Sons. He participates in the CIGRE and IEEE Working Group activities on power quality.

Guido Carpinelli is a full professor of Electrical Power Systems at the University of Napoli 'Federico II' (Italy). He received his MS degree in Electrical Engineering from the University of Napoli in 1978. His research interests concern electrical power quality and power system analysis. He is the author of several papers in journals published by IEEE, IEE, IET and Elsevier. In addition, he is a co-author of Chapter 8 and Chapter 11 of the book *Time-Varying Waveform Distortions in Power Systems*, published by John Wiley & Sons. He participates in the CIGRE and IEEE Working Group activities on power quality.

Paola Verde is a full professor of Electrical Power Systems at the University of Cassino (Italy). She received her MS degree in Electrical Engineering from the University of Napoli 'Federico II' in 1988. Her research interests concern power quality and power electronics in power systems. She is the author of several papers in journals published by IEEE, IEE, IET and Elsevier. In addition, she is a co-author of Chapter 8 and Chapter 11 of the book *Time-Varying Waveform Distortions in Power Systems*, published by John Wiley & Sons. She participates in the CIGRE and IEEE Working Group activities on power quality.

Preface

Power quality (PQ) has recently become a pressing concern in electrical power systems due to the increasing number of perturbing loads and the susceptibility of loads to power quality problems. Obviously, electrical disturbances can have significant economic consequences for customers, but they can also have serious economic impacts on the utility companies, because the new, liberalized competitive markets allow customers the flexibility of choosing which utility serves them. In practice, the new, liberalized markets throughout the world are changing the framework in which power quality is addressed, and power quality objectives are now of great importance to all power system operators.

In particular, this book deals with power quality indices, which are a powerful tool for quickly quantifying power quality disturbances. They also serve as the basis for illustrating the negative impacts of electrical disturbances on power system components and for assessing compliance with the required standards or recommendations within a given regulatory framework. Both traditional indices, which are currently in use, and new indices, which are likely to be useful in the future, are considered. The selected indices represent a tradeoff among three main characteristics: their ability to capture complex phenomena, the simplicity of the calculations they require and their mathematical and physical validity. Interruptions, which have been addressed extensively in several books and publications, are not considered.

Chapter 1 begins with some background concepts concerning the main classifications, causes and effects of power quality disturbances and some notes about the link between electromagnetic compatibility and power quality disturbances. Subsequently, the traditional indices are introduced, i.e. the indices that are most frequently used in international standards and recommendations or in the relevant literature.

Chapter 2 illustrates new indices for assessing whether the utility or customers are responsible for electrical disturbances. The problems associated with identifying the sources of power quality disturbances are analyzed, including locating the source of the PQ disturbance, assigning responsibility for the disturbance and identifying the 'prevailing source' of the disturbance, which is the side (utility or customer) that contributes more to the power quality disturbance level.

In Chapter 3, indices for nonstationary waveforms are analyzed. This topic is particularly important because, for such waveforms, the necessary information cannot be deduced easily from the analyses of their spectral components, since their location in time must also be known. Some advanced methods which are useful for overcoming difficulties that can arise in the analysis of power system waveforms, such as spectral leakage problems, are also addressed; in particular, this chapter deals with the theoretical background of these methods and their application to the calculation and proposal of power quality indices.

Global indices are presented in Chapter 4. The goal of global indices is to quantify the quality of the whole supply voltage at a monitored site by means of just one figure or, at most, two figures. The chapter outlines global indices that are based on a comparison between ideal and actual voltages, the proper treatment of traditional indices and the economic impact on the customers.

Chapter 5 deals with the problem of the definition of an adequate assessment of power quality levels in an electrical distribution system in the presence of dispersed generation (DG); in particular, probabilistic indices are illustrated that take into account the variation in power quality disturbance levels in the presence of DG.

Finally, Chapter 6 addresses some of the economic aspects of power quality disturbances. First, the focus is on the costs associated with some PQ disturbances and on the indices that are more sensitive in terms of making good cost estimates. Then, economic mechanisms for PQ regulation, based on financial penalties, incentives or both, are discussed.

Comprehensive coverage of the topics in the six chapters would have required a much more extensive book than this. In the interest of brevity, the authors have attempted to make the reader aware of the various issues as succinctly as possible. The reader can access more comprehensive treatment of these areas by consulting the references listed at the end of each chapter.

The book is based on the analysis and synthesis of a very large number of papers that have been published over the years on the subject of PQ indices, and the authors encourage and welcome input from anyone who reads this material and wishes to point out mistakes, make suggestions or ask questions to clarify any issue.

Pierluigi Caramia, Guido Carpinelli
and Paola Verde

Acknowledgements

The authors wish to thank Professor Alfredo Testa (Second University of Naples, Italy) for the invaluable discussions and suggestions related to the content of Chapter 3. Also, the authors acknowledge Antonio Bracale (Parthenope University of Naples, Italy) for the valuable contributions he made to Chapters 2, 3 and 4 and Daniela Proto (University of Naples Federico II, Italy) and Angela Russo (Polytechnic of Turin, Italy) for their contributions.

We also deeply thank all of the bright, young university students who we are lucky to encounter every day. Their curiosity and enthusiasm definitely were contributing factors in motivating us to write this book.

A further thanks goes to Nicky Skinner at John Wiley & Sons for helping us through the entire process of writing the book.

We would very much like to acknowledge the IEEE for allowing us to include material taken from previously published papers.

We are grateful for the permission to adapt sections of text from the following sources in Chapter 2:

1. Parsons, A.C., Mack Grady, W., Powers, E.J. and Soward, J.C. (1999) Rules for Locating the Source of Capacitor Switching Disturbances, *PES Summer Meeting, Edmonton (Canada)*, July © 1999 IEEE.

2. Parsons, A.C., Mack Grady, W., Powers, E.J. and Soward, J.C. (2000) 'A Direction Finder for Power Quality Disturbances Based upon Disturbance Power and Energy', *IEEE Trans. on Power Delivery*, **15**(3), 1081–1086 © 2000 IEEE.

3. Li, C., Xu, W. and Tayjasanant, T. (2004) 'A Critical Impedance Based Method for Identifying Harmonic Sources', *IEEE Trans. on Power Delivery*, **19**(2), 671–678 © 2004 IEEE.

4. Leborgne, R.C., Karlsson, D. and Daalder, J. (2006) Voltage Sag Source Location Methods Performance under Symmetrical and Asymmetrical Fault Conditions, *Transmission & Distribution Conference and Exposition: Latin America, 2006. TDC '06. IEEE/PES*, Latin America, Caracas (Venezuela), 1–6 August © 2006 IEEE.

5. Axelberg, P.G.V., Bollen, M.H.J. and Gu, I.Y. (2008) 'Trace of Flicker Sources Using the Quantity of Flicker Power', *IEEE Trans. on Power Delivery*, **23**(1), 465–471 © 2008 IEEE.

We are grateful for the permission to adapt sections of text from the following sources in Chapter 3:

1. Bracale, A., Carpinelli, G., Leonowicz, Z., Lobos, T. and Rezmer, J. (2008) 'Measurement of IEC Groups and Subgroups using Advanced Spectrum Estimation Methods', *IEEE Transactions on Instrumentation and Measurement*, **57**(4), 672–681 © 2008 IEEE.

2. Andreotti, A., Bracale, A., Caramia, P. and Carpinelli, G. (2009) 'Adaptive Prony Method for the Calculation of Power-Quality Indices in the Presence of Nonstationary Disturbance Waveforms', *IEEE Transaction on Power Delivery*, **24**(2), 874–883 © 2009 IEEE.

3. Carpinelli, G., Chiodo, E. and Lauria, D. (2007) 'Indices for the Characterization of Bursts of Sort-duration Waveform Distortion', *IET Generation, Transmission & Distribution*, **1**(1), 170–179 © 2007 IEEE.

We are grateful for the permission to adapt sections of text from the below source in Chapter 4:

1. Herat, H.M.S.C., Gosbell, V.J. and Perera, S. (2005) 'Power Quality (PQ) Survey Reporting: Discrete Disturbance Limits', *IEEE Transaction on Power Delivery*, **20**(2), 851–858 © 2005 IEEE.

We are grateful for the permission to adapt sections of text from the following sources in Chapter 6:

1. McGranaghan, M.F. (2007) 'Quantifying Reliability and Service Quality for Distribution System', *IEEE Transaction on Industry Applications*, **43**(1), 188–195 © 2007 IEEE.

2. McEachern, A., Grady, W.M., Moncrief, W. A., Heydt, G. T. and McGranaghan, M. 'Revenue and Harmonics: An Evaluation of Some Proposed Rate Structure', *IEEE Transaction on Power Delivery*, **10**(1), 474–482 © 1995 IEEE.

1

Traditional power quality indices

1.1 Introduction

The term *power quality* should take into account different aspects of the behaviour of a power system, the fundamental function of which is to supply loads economically and with adequate levels of continuity and quality. Continuity usually refers to an uninterrupted electricity supply service, while quality concerns a variety of disturbances that arise in the power system and influence the waveforms' characteristics. A recent definition of power quality (PQ), adopted throughout this book, is related to [1]:

- the ability of a power system to operate loads without disturbing or damaging them – this property is mainly concerned with voltage quality at points of common coupling between customers and utilities;

- the ability of loads to operate without disturbing or reducing the efficiency of the power system – this property is mainly, but not exclusively, concerned with the current quality of the load's waveforms.

Within this framework, power quality issues include short-term events, lasting a few cycles to seconds, and transients, or disturbances present for longer intervals (sometimes continuously) in the waveforms.

This chapter deals with the power quality indices used to quantify the voltage quality and the current quality, without considering interruptions, which are widely treated in several books and papers covering reliability issues.[1] In particular, this chapter considers the 'traditional' power quality indices. Within our definition of 'traditional indices' we include both site indices, largely used in international standards and recommendations, and the indices most frequently used in the relevant literature for power system analysis or for sizing the electrical components in the presence of PQ disturbances; we also include the system indices

---

[1] The reliability is, in a general sense, a measure of the overall ability of the system to perform its intended function [2,3].

which, even though not affirmed in standards or recommendations, are of particular interest in the new liberalized market framework, because they can be used as a benchmark against which index values for different distribution systems, or for various parts of the same distribution system, can be compared.[2]

First, some background concepts concerning the main classifications and the main causes and effects of PQ disturbances are provided. The basic concepts of electromagnetic compatibility are recalled, and the link with PQ disturbances is discussed. Then, for each disturbance, the existing site and system indices and the objectives to be complied with are introduced.

We acknowledge that an exhaustive overview of all traditional indices and objectives would be an arduous task, and one which is beyond the scope of this book.

1.2 Background concepts

Deviations of voltage or current from the ideal waveforms are generally called *power quality disturbances*. The ideal waveforms in a three-phase power system are sinusoidal waveforms, characterized by proper fixed values of frequency and amplitude; each phase waveform has the same amplitude and the angular phase difference between them is $2/3\,\pi$ radians.

Section 1.2.1 deals with the classification of PQ disturbances and briefly recalls their main causes and some effects.

The presence of PQ disturbances degrades the electrical characteristics of the power systems where they occur and can lead to degradation of the performance of the equipment connected to the system. The study of the interactions between the loads causing disturbances and the performance of equipment operating in the same environment belongs within the framework of electromagnetic compatibility (EMC). Section 1.2.2 deals with the link between PQ disturbances and the EMC levels and limits; this section introduces some basic concepts (for example, planning levels and voltage characteristics) that are of great importance in the assessment of traditional site and system indices, which will be the subjects of Section 1.3.

1.2.1 Power quality disturbances

There are many classifications of PQ disturbances. Two classifications widely referenced are reported in the IEC's (International Electrotechnical Commission's) EMC series and in IEEE (Institute of Electrical and Electronic Engineers) 1159–1995.

The IEC's EMC series classifies the environmental phenomena that describe all disturbances into low-frequency phenomena (up to 9 kHz), high-frequency phenomena, electrostatic discharge phenomena and nuclear electromagnetic pulses.

Low- and high-frequency phenomena contain radiated and conducted disturbances, depending on the medium within which they occur. Radiated disturbances occur in the medium surrounding the equipment while conducted disturbances occur in various metallic media.

---

[2] Site indices refer to the points of common coupling between customers and utilities, while system indices refer to a segment of the distribution system or, more generally, to the utility's entire electrical system.

Conducted low-frequency phenomena include the following disturbances: harmonics, interharmonics, signalling voltages, voltage fluctuations, voltage dips, short and long interruptions, voltage unbalances, power frequency variations, inducted low-frequency voltages and direct current (DC) in alternative current (AC) networks. Conducted high-frequency phenomena include the following disturbances: induced continuous wave voltages or currents, unidirectional and oscillatory transients.

Radiated low- and high-frequency phenomena include magnetic fields, electric fields, electromagnetic fields, continuous waves and transients.

Electrostatic discharge is the sudden transfer of charge between bodies of differing electrostatic potential [4].

Nuclear electromagnetic pulses are produced by high-altitude nuclear explosions [5, 6]. Any conductor within the area of influence of this disturbance will act as an antenna, picking up the electromagnetic pulses. The low-frequency effects of nuclear electromagnetic pulses can also induce large currents and voltages in long-distance communications and telephone links, while the high-frequency components may be picked up by circuits within electronic and electrical apparatus.

Although it is arguable as to whether all of these phenomena may be considered power quality issues, it is generally accepted that only conducted phenomena qualify as such. Thus, most of the disturbances included in the conducted category will be analyzed extensively in this section.

IEEE 1159–1995 classifies the disturbances into seven categories: transients, short-duration variations, long-duration variations, voltage unbalances, waveform distortions, voltage fluctuations and power frequency variations. In each category, the disturbances are diversified as a function of their spectral content, duration and magnitude [7]. The category *short-duration variations* includes both short interruptions and IEC voltage dips (labelled voltage sags in IEEE 1159–1995); moreover, this category also covers voltage swells (the inverse phenomena to voltage dips). The category *long-duration variations* is added to deal with ANSI C84.1–1989 limits [8] and includes long interruptions, undervoltages and overvoltages. The category *waveform distortions* is used as a catch-all category for the IEC harmonics, interharmonics and DC in AC networks phenomena, as well as for an additional phenomenon called *notching*. In the last category the phenomenon *noise* is also introduced to deal with broadband-conducted phenomena.

A third useful and simple classification of PQ disturbances that was used in [9, 10] separates the disturbances into 'events' and 'variations'. Events are occasional but significant deviations of voltage or current from their nominal or ideal waveforms. Variations are small deviations of the voltage or current from their nominal or ideal waveforms. In addition, variations are practically characterized by a value at any moment in time (or over a sufficiently long interval of time) and they have to be monitored continuously. This classification is similar to another classification reported in the literature that separates the disturbances into 'discrete disturbances' and 'continuous disturbances'.

In order to briefly analyze some characteristics of the main PQ disturbances, in this section we refer to the classifications as 'events' and 'variations'.

1.2.1.1   Events

Events include the following PQ disturbances: interruptions, voltage dips (sags), voltage swells, transient overvoltages and phase-angle jumps.

Interruptions

Even though interruptions will not be a subject covered in any depth in this book, the definition of this type of PQ disturbance is reported for completeness of the classification of events.

Supply interruption is a condition in which the voltage at the supply terminals is lower than a pre-fixed threshold. EN 50160–2000 fixes the voltage interruption threshold at 1% of the declared voltage[3] [11]; IEEE 1159–1995 considers the interruption threshold to be 10% of the nominal voltage. IEC documents do not fix a particular value for the interruption threshold. IEC 61000-4-30–2003 recommends that, for the evaluation of a voltage interruption, the user of the power quality monitoring device sets an appropriate voltage interruption threshold [12].

EN 50160–2000 classifies interruptions as:

- pre-arranged (or planned), meaning that customers are informed in advance;

- accidental (or unplanned), meaning that the interruptions are caused by permanent or transient faults mostly related to external events, equipment failure or interference. Accidental interruptions are further classified as long interruptions (longer than three minutes) caused by a permanent fault, and short interruptions (up to three minutes) caused by a transient fault.

IEEE documents introduce the terms *instantaneous*, *momentary*, *temporary* and *sustained* interruptions, and different documents give different definitions. In particular, IEEE 1159–1995 distinguishes among momentary (between 0.5 cycles and 3 seconds), temporary (between 3 seconds and 1 minute) and sustained (longer than 1 minute with zero voltage) interruptions. IEEE 1250–1995, on the other hand, distinguishes among instantaneous (between 0.5 cycles and 30 cycles), momentary (between 30 cycles and 2 seconds), temporary (between 2 seconds and 2 minutes) and sustained (longer than 2 minutes) interruptions [13].

IEC 61000-2-8–2002 considers short interruptions to be sudden reductions in the voltage on all phases below a specified interruption threshold followed by its restoration after a brief interval [14]. Interruptions having durations up to 1 minute (or, in the case of some reclosing schemes, up to 3 minutes) are classified conventionally as short interruptions.

Voltage dips (sags)

A voltage dip (sag) is a reduction in the voltage at a point in the electrical system below a threshold, followed by a voltage recovery after a short period of time. The voltage dip threshold is different to the interruption threshold and, in particular, it is assumed to be equal to 90% of the declared voltage by EN 50160–2000 and IEC 61000-2-8–2002 and equal to 90% of the nominal voltage by IEEE 1159–1995. In IEEE 1159–1995 the term 'sag' is used instead of the IEC term 'dip'; this recommendation distinguishes between instantaneous (from 0.5 cycles to 30 cycles), momentary (from 30 cycles to 3 seconds) and temporary (from 3 seconds to

---

[3] The declared supply voltage U_c is normally the nominal voltage U_n of the system. The nominal voltage is the voltage by which a system is designated or identified and to which certain operating characteristics are referred. If, by agreement between the supplier and the customer, a voltage different from the nominal voltage is applied to the terminal, then this voltage is the declared supply voltage U_c.

1 minute) sags. We note that in IEEE 1159–1995, those voltages characterized by a magnitude between 80% and 90% of the nominal voltage with duration greater than 1 minute are classified as *undervoltages*.

A voltage dip is characterized by a pair of data: the residual voltage, or depth, and the duration. The residual voltage is the lowest value of the voltage during the event. The depth is the difference between the reference voltage (either a declared voltage or a sliding voltage reference used when the pre-event voltage is considered as a reference voltage) and the residual voltage. Duration is the time that the root mean square (RMS) stays below the threshold; generally, the duration of a voltage dip is between 10 ms and 1 minute.

A voltage dip can be caused by short circuits and subsequent fault clearing by protection equipment or by a sudden change of load, such as a motor starting. After a short circuit, the duration of the voltage dip depends on the protection system.

There are many factors that result in a wide variety of dip severities, e.g. arc characteristics, earthing impedance, feeder R/X ratio and motor and load characteristics.

An example of a voltage dip caused by a single-line-to-ground fault is shown in Figures 1.1 and 1.2; in particular, Figure 1.1 shows the waveform and Figure 1.2 shows the RMS of the voltage as a function of time. The RMS is obtained over a 10 ms (one half cycle) rectangular window which shifts through the waveform.

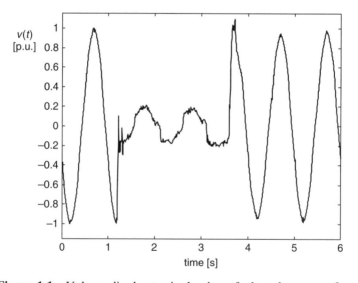

Figure 1.1   Voltage dip due to single-phase fault: voltage waveform

The interest in voltage dips is mainly due to the problems they cause on several types of equipment. The most relevant problem associated with dips is equipment shutdown. In many industries with critical process loads, this disturbance can cause process shutdowns which require hours to restart. Switch-mode power supplies, which are common at the front end of electronic equipment such as computers and PLCs, AC variable speed drives, relays, contactors and directly connected induction motors are among the most sensitive loads to voltage dips.

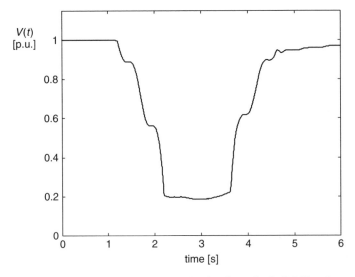

Figure 1.2   Voltage dip due to single-phase fault: RMS value

Voltage swells

A voltage swell is an increase in the RMS of the supply voltage to a value between 110% and 180% of the declared voltage, followed by a voltage recovery after a short period of time. Generally, the duration of a voltage swell is between 10 ms and 1 minute. IEEE 1159–1995 distinguishes swells as being instantaneous (from 0.5 cycles to 30 cycles with an amplitude between 110% and 180%), momentary (from 30 cycles to 3 seconds with an amplitude between 110% and 140%) and temporary (from 3 seconds to 1 minute with an amplitude between 110% and 120%). We note that in IEEE 1159–1995 voltages characterized by magnitudes between 110% and 120% of the nominal voltage with durations greater than 1 minute are classified as *overvoltages*.

Voltage swells may be caused by system faults, in the same way as voltage dips, but they are much less common than voltage dips. For example, they can appear in the non-faulty phases of a three-phase circuit that has developed a single-phase short circuit. As a further example, swells can be caused by switching off a large load. The severity of a voltage swell during a fault condition is a function of the fault location, system impedance and grounding.

Depending upon the frequency of occurrence, swells can cause the failure of components. Electronic devices, including adjustable speed drives, computers and electronic controllers, may show failure modes in the presence of swells. Transformers, switchgear, cables, rotating machinery and voltage and current transformers may suffer reduced equipment life. A temporary increase in voltage may also result in undesired intervention of some protective relays.

Rapid voltage changes

A rapid voltage change is a quick transition in RMS voltage between two steady-state conditions. The voltage during a rapid voltage change must not exceed the voltage dip and/ or the voltage swell threshold, otherwise the phenomenon is considered a voltage dip or swell. In addition, before and after the step variation the voltage should be characterized by a normal value (typically 90% to 110% of the nominal voltage).

This phenomenon can be caused by transformer tap-changers, sudden increases or decreases in loads (sometimes associated with fault switching), the in-rush current of a motor or the switching action in an electrical system. Rapid voltage changes will mainly lead to visual annoyance and generally do not cause any damage or malfunction of electrical equipment (although this is still a field for further research).

Transient overvoltages
A transient overvoltage is a short-duration oscillatory or nonoscillatory overvoltage, usually highly damped and with a duration of a few milliseconds or less.

Transient overvoltages can be divided into impulsive transient and oscillatory transient overvoltages. Impulsive transients are characterized by the fact that they are unidirectional in polarity and result from, for example, direct and indirect lightning strikes, arcing or insulation breakdowns. Figure 1.3 (a) shows an example of an impulsive transient occurring on a voltage waveform; the magnitude of the transient can be many times larger than the peak value of the voltage waveform. The impulsive transient is characterized by rise time, decay time and peak value, as shown in Figure 1.3 (b). The rise time T_1 is measured between the instant when the front edge rises from 10% to 90% of its peak value; it is usually calculated as 1.67 times the time for the transient to rise from 30% to 90% of the peak value. The decay time T_2 is measured from the start of the waveform to the time when the tail value is half the peak. The waveform shown in Figure 1.3 (b) is called a T_1/T_2 waveform, where T_1 and T_2 are expressed in microseconds.

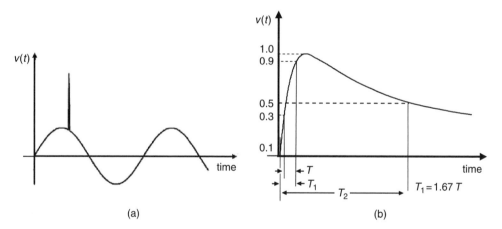

Figure 1.3   (a) Example of an impulsive transient overvoltage occurring on a voltage waveform; (b) characteristics of impulsive transients

Oscillatory transient overvoltages consist of voltages that oscillate both positively and negatively with respect to the voltage waveform; the frequencies of these oscillations can be quite high. Switching operations are the major cause of these oscillations. Oscillatory transients are described by their peak voltage, predominant frequency and decay time (duration).

These parameters are useful quantities for evaluating the potential impact of these transients on power system equipment. The absolute peak voltage, which is dependent on the transient magnitude and the point on the fundamental frequency voltage waveform at which the event occurs, is important for dielectric breakdown evaluation. Some equipment and types of insulation, however, may also be sensitive to rates of change in voltage or current. The predominant frequency, combined with the peak magnitude, can be used to estimate the rate of change.

Oscillatory transients can be subdivided into low-, medium- and high-frequency categories according to the classifications given in IEEE 1159–1995.

Low-frequency transients (a predominant frequency component less than 5 kHz and duration from 0.3 to 50 ms) are frequently caused by switching events; the most frequent is capacitor bank energization, which typically results in an oscillatory voltage transient with a predominant frequency between 300 and 900 Hz and a peak value magnitude between 1.3 and 1.5 times the peak voltage of the 50/60 Hz waveform, depending on the system damping. Oscillatory transients associated with ferroresonance and transformer energization also fall into this category. Figure 1.4 shows an example of an oscillatory transient overvoltage event due to capacitor switching.

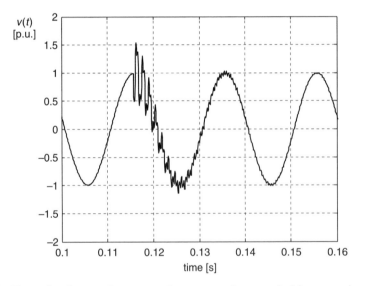

Figure 1.4   Example of a transient overvoltage event due to switching capacitor transients

Medium-frequency transients are characterized by a predominant frequency component between 5 and 500 kHz and durations measured in tens of microseconds. A typical example is the transient generated by back-to-back capacitor energization.[4]

---

[4] When a capacitor is energized in close proximity to one already in service, the energized bank sees the de-energized bank as low impedance, resulting in a current transient oscillating between them. These oscillations can last up to tens of microseconds, depending on the size of the capacitors and the damping (resistive losses) between them.

Finally, oscillatory transients with predominant frequency components greater than 500 kHz and typical durations in microseconds are considered to be high-frequency transients. These often occur when an impulsive transient excites the natural frequency of the local power system network.

Transient overvoltages can result in degradation or immediate dielectric failure of electrical equipment, such as rotating machines, transformers, capacitors and cables. In electronic equipment, power supply component failures can result from a single transient of relatively modest magnitude, and transients can cause nuisance tripping of adjustable speed drives.

Phase-angle jumps
A change in the electrical power system, like a short circuit, causes a change in voltage. Voltage is a complex quantity characterized by amplitude and phase angle. Sometimes, changes in voltage produced by changes in system state are not limited to the magnitude but include a change in phase angle as well. The phase-angle jump manifests itself as a shift in zero crossing of the instantaneous voltage. Power electronic converters that use phase angle information to determine their firing instants may be affected by this type of disturbance.

1.2.1.2   Variations

Within the category of variations fall the following PQ disturbances: waveform distortions (including both harmonics and interharmonics), slow voltage variations, unbalances, voltage fluctuations, mains signalling voltages, power frequency variations, voltage notches and noise.

Waveform distortions
'Waveform distortion' is usually discussed in terms of harmonic and interharmonic components (for brevity, simply referred to as harmonics and interharmonics), which are sine waves obtained by performing a Fourier analysis on the original waveform.[5] Strictly speaking, harmonics are sinusoidal waveforms with a frequency equal to an integer multiple of the fundamental frequency, assumed frequently to be the same as the power system frequency (50 or 60 Hz). Interharmonics are sinusoidal waveforms at frequencies that are not integer multiples of the fundamental frequency. The ratio of the harmonic (interharmonic) frequency to the fundamental frequency is the harmonic (interharmonic) order. As an example, for a fundamental frequency of 50 Hz, the frequency of the harmonic of order 5 is 250 Hz and the frequency of the interharmonic of order 1.2 is 60 Hz.

As an example of waveform distortion, Figure 1.5 shows the current waveform absorbed by a personal computer.

Examples of sources of waveform distortion are the saturation of a transformer core, static power converters and other non linear and time-varying loads (such as arc furnaces).

Harmonics have undesirable effects on power system components. For example, induction motor windings are overheated, accelerating the degradation of insulation and reducing the useful life of the motor. In a three-phase, four-wire system, the sum of the three phase currents returns through the neutral conductor, which can be overloaded. In fact, positive- and negative-sequence components add to zero at the neutral point, but zero-sequence components are

---

[5] For details about Fourier analysis and its problems, see Section 1.3.1 of Chapter 1 and Section 3.2 of Chapter 3.

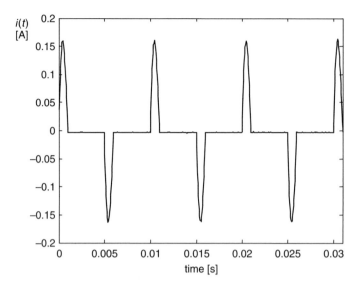

Figure 1.5   Current waveform absorbed by a personal computer

additive to neutral. The zero-sequence components sum in the neutral wire and can overload the neutral conductor, with consequent loss of life of the component and possible risk of fault.[6]

Transformers and capacitors can also be affected by harmonics. Distorted currents can cause transformers to overheat, reducing their useful life. Capacitors are affected by the applied voltage waveform; harmonic voltage, in fact, produces excessive harmonic currents in capacitors because of the inverse relationship between capacitor impedance and frequency. In turn, capacitors can cause excessive harmonic voltages in the presence of resonance conditions at frequencies where their capacitive impedances and the inductive impedance of the power system can combine to give a very high impedance; in this case a small harmonic current within this frequency range can give very high and undesirable harmonic voltages. As a further example, harmonics can also degrade meter accuracy. In addition, the operating condition of some equipment depends on an accurate voltage waveform, and such equipment can malfunction in the presence of harmonics. Examples are light dimmers and some computer-controlled loads.

Besides the typical problems caused by harmonics, such as overheating and the reduction of useful life, interharmonics create some new problems, such as subsynchronous oscillations and light flicker, even for low amplitude levels.

Slow voltage variations
In normal operating conditions, bus voltage magnitudes can be characterized by small deviations around their nominal values. Slow voltage variations are increases or decreases in voltage amplitude due to various causes, e.g. variations in the load demands with time. These variations are characterized by daily, weekly and seasonal cycles.

---

[6] In three-phase balanced systems, the harmonic orders $h = 3, 6, 9, \ldots$ are of zero sequence; the harmonic orders $h = 4, 7, 10, \ldots$ are of positive sequence and the harmonic orders $h = 2, 5, 8, \ldots$ are of negative sequence. Theoretically, in unbalanced systems, all three-phase harmonic currents (but also voltages) can be decomposed into positive-, negative- and zero-sequence components.

Unbalances

An unbalance is a condition of a poly-phase system in which the RMS values of the line voltages (fundamental components) and/or the phase angles between consecutive line voltages are not all equal and/or $2/3\pi$ displaced. Large single-phase loads, such as railway traction systems or arc furnaces, and untransposed overhead lines are the main sources of unbalances in transmission systems. Also, distribution systems with unbalanced lines and line sections carrying a mixture of single, double or three-phase loads are characterized by unbalances.

Voltage unbalances on electrical machines lead to increased losses; moreover, the net torque and speed, in the presence of an unbalanced voltage supply, are reduced, and torque pulsations and acoustic noise may occur. For power electronic converters in which the firing angle is derived from the voltages, unbalances can cause the presence of noncharacteristic harmonics, on both the DC and AC sides. Noncharacteristic harmonics are harmonics that are not produced by semiconductor converter equipment in the course of normal operation.

Figure 1.6 shows a plot of the three phase-unbalanced voltages in a bus of a distribution system.

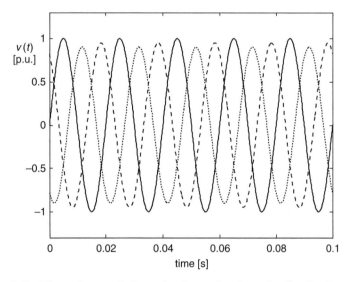

Figure 1.6   Three phase-unbalanced voltages in a bus of a distribution system

Obviously, the phase currents in poly-phase systems can also be unbalanced.

Voltage fluctuations

A series of voltage changes or a continuous variation in the RMS voltage are denoted voltage fluctuations (a qualitative example is shown in Figure 1.7). If fluctuations occur within proper frequency ranges, they cause changes in the luminance of lamps, which can create the visual phenomenon called 'flicker.' Above a certain threshold, the flicker becomes annoying, and annoyance grows very rapidly with the amplitude of the fluctuation. At certain repetition rates, even very small amplitudes can be annoying.

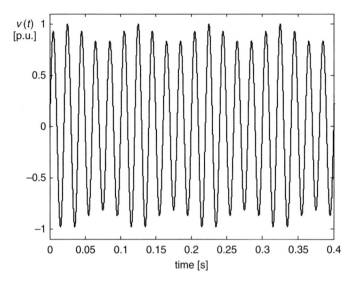

Figure 1.7   Example of voltage fluctuations

The main sources of severe voltage fluctuations are industrial loads, which are particularly troublesome in the steelmaking industry, where there is significant use of arc furnaces, rolling mills and multiple-welder loads. In some cases, large fluctuating motor loads, such as mine winders, can also cause voltage fluctuation problems. Domestic loads with repeated switching, such as cooker hobs, electric showers, heat pumps and air conditioning, have the potential to cause local flicker, but the problem is contained by imposing statutory product design constraints in terms of the size and frequency of load switching.

It should be noted that voltage fluctuations and interharmonics have an inherent relationship. In fact, the magnitude of a voltage waveform can fluctuate if it contains interharmonics, and conversely, the voltage fluctuation can cause the presence of interharmonics. Experience has shown that even a small amount of interharmonics can result in perceptible light flicker in incandescent and fluorescent lamps.

In addition to flicker, an effect produced by voltage fluctuations is the braking or acceleration of motors connected directly to the system.

Mains signalling voltages
Some utilities intentionally superimpose small signals on the supply voltage for the purpose of transmission of information in the public distribution system and to customers' premises. Signals in the public distribution system can be classified into three types:

- **ripple control signals**: superimposed sinusoidal voltage signals in the range of 0.11 to 3 kHz;

- **power-line-carrier signals**: superimposed sinusoidal voltage signals in the range of 3 to 148.5 kHz;

- **mains marking signals**: superimposed short-time alterations (transients) at selected points of the voltage waveform.

Power frequency variations

A power frequency variation is a deviation of the frequency from the nominal value (equal to 50 or 60 Hz and denoted the 'power supply frequency,' 'power system frequency' or 'power frequency'). A power frequency variation is linked directly to the variation in rotation speed of the generators supplying the electrical power system. There are slight variations in frequency due to imbalances between load and generation. Slight deviations in frequency can cause severe damage to generators and turbine shafts due to the large torque developed. In addition, cascading system separations can result with even slight deviations in frequency, since electric systems are closely connected and depend on synchronous operation. Actually, significant frequency variations are rare in modern interconnected power systems.

Voltage notches

Notching is a periodic transient occurring within each cycle as a result of the phase-to-phase short circuit caused by the commutation process in AC–DC converters (Figure 1.8). Being periodic, this disturbance is characterized by the presence of the harmonic spectrum of the voltage waveform; in particular, notching mainly results in high-order harmonics. Most problems caused by notches are confined to a customer's own installation. In fact, the high frequency content of notches not under a utility's control is 'filtered' by the power transformer at the service entrance. Voltage notches are characterized by their depth and duration in combination with the point on the sine wave at which the notching starts.

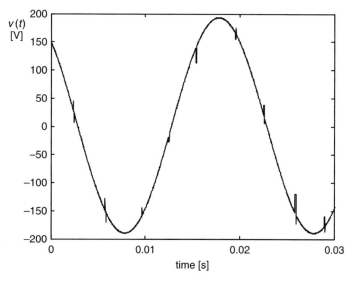

Figure 1.8   Example of voltage notching caused by converter operation

Noise

In IEEE 1159–1995, *noise* is indicated as an additional PQ disturbance. Noise is an unwanted electrical signal with broadband content lower than 200 kHz superimposed upon the power system voltage or current in phase conductors or found on neutral conductors or signal lines. Basically, noise consists of any unwanted distortion that cannot be classified as a waveform distortion or a transient. Noise problems are often exacerbated by improper grounding.

1.2.2   Power quality disturbances and electromagnetic compatibility

The study of the interaction between the emissions of disturbing loads and the performance of equipment operating in the same environment falls within the framework of electromagnetic compatibility (EMC).

In particular, EMC is defined as 'the ability of a piece of equipment or system to function satisfactorily in its electromagnetic environment without introducing intolerable electromagnetic disturbances to anything in that environment.' [15]. The principle of EMC can be explained considering two devices operating in the same environment: the first device produces disturbances ('disturbing load') and the second device can be affected by these disturbances ('susceptible load'). If the operating performance of the susceptible load is degraded by the disturbances emitted by the disturbing load, an EMC problem exists.

To achieve electromagnetic compatibility, it is necessary both that the emission of disturbances by the disturbing loads into the electromagnetic environment is below a level that would produce an undesirable degradation of the performance of equipment operating in the same environment, and that the susceptible loads operating in the electromagnetic environment have sufficient immunity from all disturbances at the levels existing in the environment [46].

The emission limit is the maximum amount of electromagnetic disturbance that a device is allowed to produce.

The immunity level is the minimum level of electromagnetic disturbance that a device can withstand. To take into account the different sensibilities of devices with respect to the disturbances, a probability density function (pdf) of the equipment immunity level is introduced. For the applied disturbance, this curve gives the probability of malfunction or of damage to the device. Practically, considering the pdf curve of the equipment immunity level, the immunity level of the device is chosen equal to the value of the disturbance, allowing for a small probability (typically 5%) of susceptibility[7] of the device.

To coordinate the emission of the disturbing loads and the immunity of the susceptible loads operating in the same environment, the concept of compatibility levels is introduced. In order to ensure EMC in the whole system, the immunity limit must be equal to or greater than the compatibility level and the emission limit must be equal to or less than the compatibility level (Figure 1.9). Compatibility levels are generally based on 95% non-exceeding probability levels of entire systems, using distributions that represent both temporal and spatial variations in disturbances.

Together with compatibility levels, standards and recommendations consider other objectives for disturbance levels useful in achieving EMC in the whole electrical system: planning levels and voltage characteristics.

Planning levels are levels that can be used for planning purposes in evaluating the impact of all customers on the supply system. Planning levels are specified by the utilities and can be considered internal quality objectives of the utilities. These levels are used to define the emission limits for large loads and installations that are to be connected to the system. Although each planning level mainly considers large equipment and installations, many other

---

[7] Susceptibility is the inability of a device, piece of equipment or system to resist an electromagnetic disturbance. Susceptibility is the lack of immunity.

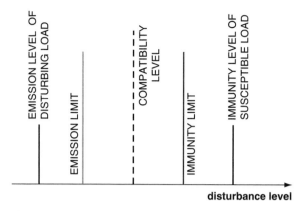

Figure 1.9   Illustration of EMC level and limits

sources of disturbance must be accounted for, notably numerous low-power devices supplied by low voltages. In order to ensure EMC in the system, planning levels are chosen equal to or lower than compatibility levels.

Voltage characteristics are the main characteristics of the voltage at the customers' supply terminals in electrical distribution systems under normal operating conditions. These are quasi-guaranteed limits covering any location of the power system and consequently are equal to or slightly greater than the compatibility levels.

1.3   Power quality disturbances: indices and objectives

Compatibility levels, planning levels, voltage characteristics and emission limits are indicated in standards, recommendations or guidelines in terms of reference values of one or more index [16]. In practice, the PQ indices represent, for compactness and practicality, the quickest and most useful way to describe the characteristics of PQ disturbances. They are convenient for condensing complex time and frequency domain waveform phenomena into a single number. Several international standards, recommendations and guidelines, as well as many papers published in the relevant literature, propose indices to characterize the PQ in general and, more particularly, the voltage and current quality.

An important classification of PQ indices, also used in this section, is based on the portion of the electrical system to which the index refers. In particular, we distinguish between site indices, which refer to a single customer point of common coupling, and system indices, which refer to a segment of the distribution system or, more generally, to the utility's entire electrical system.

While the site index treatment requires the collection of different observations in time, for system indices observations at different sites, usually a representative set of the system under study, must also be collected. Weighting factors can be introduced to take into account the sites not monitored and the difference in importance between different monitored sites; for example, weighting factors may be based on the number of substations/customers or the rated power of each site. It is evident that system indices do not give an exhaustive indication of the PQ referred to each customer, but as evidenced in Section 1.1, they serve as a metric only and are of particular interest in the new liberalized market framework.

The values of site and system indices have to be compared with objectives. These objectives can be defined in bilateral agreements between a distributor and a customer, set as self-imposed quality objectives by a network operator or set by a regulatory body. Usually, only objectives for site indices have been introduced in standards and recommendations.

In this section, an overview of the main traditional site and system indices together with objectives is provided for the following disturbances [7–56]:

- waveform distortions;

- slow voltage variations;

- unbalances;

- voltage fluctuations;

- mains signalling voltages;

- voltage dips (sags);

- transient overvoltages;

- rapid voltage changes.

The same analysis has not been carried out for phase-angle jumps or notches because, to the best of our knowledge, accepted specific site or system indices to quantify these power quality disturbances do not presently exist. For voltage swells, some site and system indices which are used for voltage dips can be used if they are properly modified, as will be highlighted in the voltage dip section.

However, before recalling the indices and objectives, we consider it useful to briefly recall some PQ data aggregation techniques that are frequently used. In fact, site and system indices require the aggregation of several values, obtained by monitoring, for long time intervals, a bus (site indices) or several busbars (system indices). So, adequate data aggregation techniques are needed and are useful both in reducing the data to be stored and managed and in facilitating the interpretation of these data.

With reference to the PQ voltage and current variations, a time aggregation technique frequently used consists of the combination of several values of a given parameter, each determined over an identical time interval, to provide a value for a longer time interval. For example, IEC 61000-4-30–2003 suggests, as a basic time interval for parameter magnitudes (RMS voltages, harmonics, interharmonics and unbalances), a time equivalent to 10 cycles for a power system frequency of 50 Hz or 12 cycles for a 60-Hz power system. This time interval is referred to as the 200 ms value, even though the interval is not exactly 200 ms, because, in the actual operating conditions, the power system frequency changes around the nominal value. Successive aggregations of the magnitudes determined using this time interval are performed using the square root of the arithmetic mean of the squared values calculated at the basic measurement time intervals. Three categories of time aggregation are frequently considered:

- very short interval (time): 3 seconds;

- short interval (time): 10 minutes;

- long interval (time): 2 hours.

The very short interval value is obtained by the aggregation of fifteen basic measurement interval values calculated in the considered 3-second time interval:

$$Q_{vsh} = \sqrt{\frac{1}{15} \sum_{i=1}^{15} Q_i^2},$$ (1.1)

where Q_{vsh} is the very short interval value (also called the 3-s value) and Q_i is the value of the parameter calculated at the ith basic time interval (200-ms value).

The short interval value is obtained by the aggregation of the two hundred 3-s values calculated in the considered 10-minute time interval:

$$Q_{sh} = \sqrt{\frac{1}{200} \sum_{j=1}^{200} Q_{vsh,j}^2},$$ (1.2)

where Q_{sh} is the short interval value (also called the 10-min value) and $Q_{vsh,j}$ is the jth very short interval value in the considered 10-minute time interval.

The long interval value is obtained by the aggregation of the twelve 10-min values calculated in the considered 2-hour time interval:

$$Q_{lt} = \sqrt{\frac{1}{12} \sum_{m=1}^{12} Q_{sh,m}^2},$$ (1.3)

where Q_{lt} is the long interval value (also called the 2-h value) and $Q_{sh,m}$ is the mth short interval value in the considered 2-hour time interval.

The interval values defined above are usually further aggregated to obtain the statistical characterization for a longer observation time. The results of this further data collection are the corresponding daily and weekly probability density functions (pdfs). These pdfs are the basis for calculating the site indices, often reported in standards and recommendations, for quantifying the level of a specific disturbance. Standards and recommendations, in fact, usually refer to the 95th or the 99th daily or weekly percentile of PQ indices. A *percentile* is the variable value below which a given percentage of occurrences can be expected.

With reference to voltage events, it is important to evidence that these disturbances are less frequent than variations (e.g., harmonics and unbalances can always be present), so the observation period has to be longer than the period considered for variations. The observation period for monitoring voltage dips or swells, for example, is usually assumed to be at least one year. Also for these types of PQ disturbances, an appropriate grouping of event parameters permits a reduction in the amount of data that must be reported and managed.

Site and system performance, with respect to a voltage event (mainly in the case of voltage dips) are often described in the form of an event table: the columns of the table represent ranges of event duration d; the rows represent ranges of voltage amplitude A, characterizing the event. In particular, in the table, the numbers of events falling into a predetermined range of duration and amplitude are reported. The choice of tables for the magnitude and duration ranges is a point of discussion. Different publications use different values. Table 1.1 shows the voltage-dip table recommended by a group of experts (UNIPEDE DISDIP group) that arranged a coordinated series of measurements in nine countries.

Table 1.1   Example of voltage event aggregation: UNIPEDE voltage-dip table

Residual voltage[%]	Duration[s]							
	< 0.02	0.02–0.1	0.1–0.5	0.5–1	1–3	3–20	20–60	60–180
85–90	1.2	6.8	3.6	0	0	0	0	0
70–85	0.17	9.33	2.33	0.17	0	0	0	0
40–70	0	4.83	2.67	0	0	0	0	0
10–40	0	0.5	0.5	0	0	0	0	0
< 10	0	0	0	0	0	0	0	0

1.3.1   Waveform distortions

1.3.1.1   Site indices

Waveform distortions (voltage or current) can be characterized by several indices and the most frequently used are:

- the individual harmonic;
- the total harmonic distortion factor;
- the individual interharmonic;
- the total interharmonic distortion factor.

The individual voltage or current harmonic A_h is the ratio between the RMS value of harmonic component of order h, X_h, and the RMS value of the fundamental component, X_1, of the voltage or current waveform:

$$A_h = \frac{X_h}{X_1}. \tag{1.4}$$

The *total harmonic distortion factor* (THD) is defined as the RMS of the harmonic content divided by the RMS value of the fundamental component, usually multiplied by 100:

$$\text{THD} = \frac{\sqrt{\sum_{h=2}^{H_{max}} X_h^2}}{X_1} \, 100, \tag{1.5}$$

where H_{max} is the order of the highest harmonic that is taken into account.

The THD value applied to current can be misleading when the fundamental component is low. A high THD value for input current cannot be of significant concern if the load is light, since the magnitude of the harmonic current is low, even though its relative distortion to the fundamental is high. To avoid such ambiguity, IEEE 519–1992 defined the *total demand distortion factor* (TDD). This term is similar to THD except that the distortion is expressed as a percentage of the rated or maximum load current magnitude rather than as a percentage of the current fundamental component.

The individual voltage or current interharmonic is defined as the ratio between the RMS value of the spectral interharmonic component and the RMS value of the fundamental component.

A total distortion index equivalent to that for harmonics can be defined for interharmonics. The *total interharmonic distortion factor* (TIHD) can be applied once again to both current and voltage and is defined as the RMS of the interharmonic content divided by the RMS value of the fundamental component, usually multiplied by 100.

The spectral components to be included in the above indices are usually obtained by performing a Fourier analysis of the current or voltage waveforms and using adequate data aggregation techniques.

In particular, in IEC 61000-4-7–2002 and IEC 61000-4-30–2003, the procedure for the measurements of waveform distortion indices refers to the discrete Fourier transform (DFT) of the current or voltage waveforms effected on the waveform inside a rectangular window with a length T_W equal to 10 cycles for 50-Hz systems and 12 cycles for 60-Hz systems of fundamental [12, 30]. The window width T_W determines the frequency resolution $\Delta f = 1/T_W$ for the spectral analysis (i.e., the frequency separation of the spectral lines).

It is important to highlight that the frequency resolution Δf is called the *frequency basis*, while the term 'fundamental' is currently used to indicate the power system frequency (50 or 60 Hz).[8] So the term 'fundamental frequency' is usually a synonym for 'power supply frequency' or 'power system frequency.'

As previously shown, the power system frequency can change around the nominal value, so the length of the basic window is about 200 ms for both 50- and 60-Hz systems, but it is not 'exactly' 200 ms. However, some literature references refer to this basic window as the 200-ms window and to the resulting spectrum as the 200-ms values with 5 Hz of resolution frequency.

The spectral components obtained by applying the DFT to the actual signal inside the window are aggregated in different configurations, called groups or subgroups, depending on what kind of measurement is to be performed (harmonics, interharmonics or both).

The RMS value of the harmonic group order associated with harmonic order n, G_{gn}, is obtained by applying Equation (1.6) for the 50-Hz system and Equation (1.7) for the 60-Hz system:

$$G_{gn} = \sqrt{\frac{X^2_{(10n-5)\Delta f}}{2} + \sum_{i=-4}^{4} X^2_{(10n+i)\Delta f} + \frac{X^2_{(10n+5)\Delta f}}{2}}, \qquad (1.6)$$

$$G_{gn} = \sqrt{\frac{X^2_{(12n-6)\Delta f}}{2} + \sum_{i=-5}^{5} X^2_{(12n+i)\Delta f} + \frac{X^2_{(12n+6)\Delta f}}{2}}, \qquad (1.7)$$

where $X_{(10n+i)\Delta f}$ and $X_{(12n+i)\Delta f}$ are the RMS values of the spectral components at $(10n + i)\Delta f$ and $(12n + i)\Delta f$ frequencies, respectively.[9] Figure 1.10 (a) illustrates the harmonic grouping for the 50-Hz system.

---

[8] Theoretically speaking, when a waveform contains both harmonics and interharmonics, the fundamental frequency should be the greatest common divisor of all the frequency components contained in the signal.
[9] It is useful to highlight that to define the harmonic (interharmonic) groups and subgroups, the subscripts of the harmonic component are used to indicate directly the harmonic frequency.

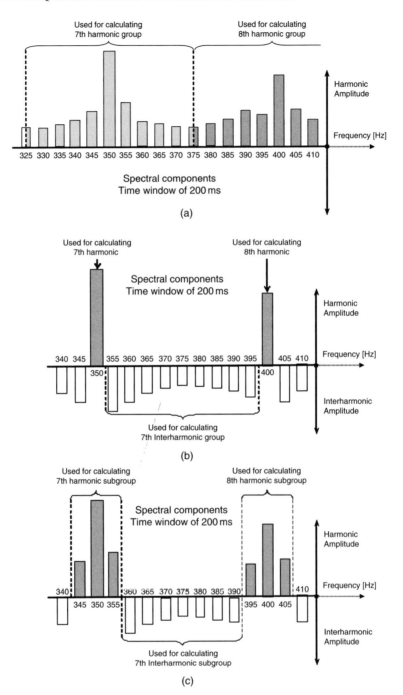

Figure 1.10 Illustration of IEC harmonic and interharmonic groupings: (a) harmonic groups; (b) interharmonic groups; (c) harmonic and interharmonic subgroups (for 50-Hz supply)

A similar evaluation is used for interharmonics. A grouping of the spectral components in the interval between two consecutive harmonic components forms an interharmonic group that can be evaluated according to Equation (1.8) for the 50-Hz-system and Equation (1.9) for the 60-Hz system:

$$C_{ign} = \sqrt{\sum_{i=1}^{9} X_{(10n+i)\Delta f}^2} \, , \tag{1.8}$$

$$C_{ign} = \sqrt{\sum_{i=1}^{11} X_{(12n+i)\Delta f}^2} \, . \tag{1.9}$$

In these relationships, the subscript 'ign' indicates the interharmonic group of order n, and the interharmonic group between the harmonic order n and $n+1$ is designated C_{ign}.

Figure 1.10 (b) illustrates the interharmonic grouping for the 50-Hz system. This grouping provides an overall value for the interharmonic components between two discrete harmonics.

The voltage magnitude of the power system may fluctuate, spreading out the energy of harmonic components to adjacent interharmonic frequencies. The effects of fluctuations that cause sidebands close to the harmonics can be partially reduced by excluding the lines immediately adjacent to the harmonic frequencies from the interharmonic groups, thereby introducing so-called *subgroups*. The harmonic subgroups permit a reduction in the error in the harmonic evaluation when spectral leakage is present, while in the presence of real interharmonics adjacent to harmonic components these, incorrectly, increase the harmonic distortion level.

The interharmonic components directly adjacent to a harmonic are grouped to form a harmonic subgroup according to Equation (1.10):

$$\begin{aligned} G_{sgn} &= \sqrt{\sum_{i=-1}^{1} X_{(10n+i)\Delta f}^2} \\ G_{sgn} &= \sqrt{\sum_{i=-1}^{1} X_{(12n+i)\Delta f}^2} \end{aligned} \tag{1.10}$$

for the 50-Hz system and the 60-Hz system, respectively.

The remaining interharmonic components form the centred interharmonic subgroup according to Equations (1.11) and (1.12) for the 50-Hz and 60-Hz systems respectively:

$$C_{isgn} = \sqrt{\sum_{i=2}^{8} X_{(10n+i)\Delta f}^2}, \tag{1.11}$$

$$C_{isgn} = \sqrt{\sum_{i=2}^{10} X_{(12n+i)\Delta f}^2}. \tag{1.12}$$

Figure 1.10 (c) gives a clear idea of the different lines that are grouped to form harmonic and interharmonic subgroups.

From the analysis of the definitions of groups and subgroups (Equations 1.6 to 1.12), it appears that each harmonic/interharmonic group contains more spectral components than the corresponding harmonic/interharmonic subgroup. Moreover, it is evident that, due to the overlapping between harmonic and interharmonic groups, some spectral components are considered twice. Inaccuracies in the spectral content characterization can also arise, as will be shown in Chapter 3.

IEC 61000-4-7–2002 defines two new distortion factors deriving from the definition of groups and subgroups: the *group total harmonic distortion* (THDG) and the *subgroup total harmonic distortion* (THDS).

The THDG is defined as the ratio of the RMS value of the harmonic groups to the RMS value of the group associated with the fundamental group:

$$\text{THDG} = \sqrt{\sum_{n=2}^{H_{\max}} \left(\frac{G_{gn}}{G_{g1}} \right)^2}. \tag{1.13}$$

The THDS is defined as the ratio of the RMS value of the harmonic subgroups to the RMS value of the subgroup associated with the fundamental group:

$$\text{THDS} = \sqrt{\sum_{n=2}^{H_{\max}} \left(\frac{G_{sgn}}{G_{sg1}} \right)^2}. \tag{1.14}$$

Table 1.2 provides a comparison of site indices adopted for voltage distortions in various standards and guidelines [11, 12, 16, 18, 20, 27, 38, 45, 46]. From the analysis in Table 1.2, it becomes clear that the site indices most frequently used are:

1. the 99th daily percentile (or other percentile) of the very short interval (3 s) RMS value of individual harmonic $V_{h,\text{vsh}}$;

2. the 95th weekly percentile (or other percentile) of the short interval (10 min) RMS value of individual harmonics $V_{h,\text{sh}}$ or of the total harmonic distortion factor THD_{sh};

3. the maximum weekly value of the short-interval RMS value of individual harmonics $V_{h,\text{sh}}$ or of the total harmonic distortion factor THD_{sh}.

The very short and short harmonic components are obtained by applying the aggregation time procedure illustrated at the beginning of this section, and then the daily or weekly probability density function is obtained by collecting the 3-s or 10-min values (depending on the requirements) over a one-day or one-week period. Moreover, Table 1.2 shows that, in most cases, the reference standard for performing harmonic measurements is IEC 61000-4-7–2002.

IEC 61000-3-6–2008 defines the planning level for voltage distortion which is the basis for determining emission limits for a single customer in medium voltage (MV) and high voltage (HV) networks. In fact, while observation of the emission limits for low-voltage (LV) customers is demanded by the equipment manufacturer, for customers connected to MV and HV networks, the emission limits should be set by the network operator in order to guarantee observation of the voltage distortion limits.

IEC 61000-4-30–2003 defines the PQ measurement methods for the main PQ disturbances and, in particular, refer to IEC 61000-4-7–2002 for waveform distortions; in Annex 6 guidelines for contractual applications are also suggested.

Table 1.2 Harmonic site indices proposed in different standards and guidelines

Standard & Document		International				National or regional		
		IEC 61000-3-6	IEC 61000-4-30	EN 50160	ANSI/IEEE 519	Norwegian directive	EDF Emeraude contract	NRS048-2-2007	Hydro Quebec
Purpose		Planning levels	Power quality measurement methods	Voltage characteristics	Recommended practice for emission limits and system design	Standard used by regulator	Voltage characteristics	Standard used by regulator	Voltage characteristics
Where it applies		International	International	Some European countries	Some countries, mostly USA	Norway	France	South Africa	Quebec, CA
Indices & assessment	Very short time (3-s)	$V_{h,vsh}$ 99th daily percentile	$V_{h,vsh}$ X daily percentile as agreed						
	Short time (10-min)	$V_{h,sh}$ 95th weekly percentile	$V_{h,sh}$ X weekly percentile as agreed	$V_{h,sh} + THD_{sh}$ 95th weekly percentile		$V_{h,sh} + THD_{sh}$ Max	$V_{h,sh} + THD_{sh}$ Max	$V_{h,sh} + THD_{sh}$ 95th weekly percentile	$V_{h,sh} + THD_{sh}$ 95th weekly percentile
	Other				$V_h + THD$	THD Week-average			
Period for statistical assessment		One week minimum	At least one week	One week	Undefined	One week minimum	One week	At least one week	One week or more
Measurement method		IEC 61000-4-7	IEC 61000-4-7	IEC 61000-4-7	No specific measurement method	In accordance with IEC or CENELEC [10] standards	IEC 61000-4-7	Specific measurement method	IEC 61000-4-7

[10] CENELEC: European committee for electrotechnical standardization

It is useful to highlight also that the version of IEEE 519–1992 currently in force does not consider the probabilistic nature of harmonic indices, but the harmonic recommendation is undergoing revision and, in the draft document (2005), the following probabilistic harmonic indices were introduced:

- the 99th percentile of the very short time (3-s) value over one day;

- the 95th percentile of the short time (10-min) value over one week.

Some regional or national standards and guidelines, such as NRS048-2-2007 (Norway), Emeraude contract (EDF France) and Hydro Quebec voltage characteristic (Canada), recommend indices similar to those considered earlier.

Site indices adopted in some countries or suggested by working groups and not reported in Table 1.2 are detailed in the following.

In Spain, the 95th percentile of the 10-min value of THD is obtained over each one-week time period and then an annual index is calculated as the average of the weekly indices [10].

In Argentina, the national authority (ENRE: Ente Nacional Regulador de la Electricidad) uses the 95th percentile of the 10-min value of THD over one week [50].

The CIGRE JWG C4.07/Cired recommends the following site harmonic indices: the 95th percentile of the 3-s values over one day, the 99th percentile of the 10-min values over one week, the 99th percentile of the 3-s values over one week and the 95th percentile of the 10-min values over one week. The first three indices are used for planning purposes, while the fourth index is used for voltage characteristics [16].

With reference to emission levels, in IEC 61000-3-6–2008 the following indices are proposed to compare the actual emission level with the customer's emission limit:

1. the 95th percentile of the 10-min values of individual harmonics over one week;

2. the greatest 99th percentile of the 3-s values of individual harmonic components over one day.

Where higher emission levels are allowed for short periods of time, such as during bursts or start-up conditions, the use of more than one index may be needed to assess the impact.

In the draft document of the revised IEEE 519 (2005), probabilistic limits on harmonic current are also introduced and indices for assessing emission levels are indicated. As an example, with reference to a distribution system with rated voltage up to 69 kV, the following indices are considered: the 99th percentile of the very short time values over one day, the 99th percentile of the short time values over one week and the 95th percentile of the short time values of the individual harmonic current components and of the total demand distortion factor over one week.

Before concluding this section on the waveform distortion site indices, for completeness we list some other harmonic indices that are frequently applied in the relevant literature for power system analysis or for the sizing of electrical components in the presence of waveform distortion. These indices are:

- the telephone influence factor (TIF) $= \dfrac{\sqrt{\sum\limits_{h=1}^{\infty} w_h^2 I_h^2}}{I_{\mathrm{RMS}}}$;

- the C-message index $= \dfrac{\sqrt{\sum\limits_{h=1}^{\infty} c_h^2 I_h^2}}{I_{RMS}}$;

- the IT $= \left(\sqrt{\sum\limits_{h=1}^{\infty} w_h^2 I_h^2}\right)$ and VT $= \left(\sqrt{\sum\limits_{h=1}^{\infty} w_h^2 V_h^2}\right)$ products;

- the K-factor $= \dfrac{\sum\limits_{h=1}^{\infty} (h I_h)^2}{I_{RMS}^2}$;

- the peak factor $= k_{pY} = \dfrac{Y_p}{Y_{p1n}}$;
- the true power factor: pf_{true}.

The TIF and C-message are measures of audio circuit interference produced by the harmonics I_h in electrical power systems. The weighting factors, w_h and c_h, take into account the sensitivity of the human ear to noise at different frequencies.

IT and VT products are used as another measure of harmonic interference in audio circuits.

The K-factor is a weighting of the harmonic load currents according to their effects on transformer heating. A K-factor of 1.0 indicates a linear load. The higher the K-factor, the greater the harmonic heating effects. When a nonlinear load is supplied from a transformer, it is sometimes necessary to de-rate the transformer capacity to avoid overheating and subsequent insulation failure. The K-factor is used by transformer manufacturers and their customers to adjust the load rating as a function of the harmonic currents caused by the load.

The peak factor k_{pY} is defined as the ratio between the peak value of the distorted waveform, Y_p, and the rated sinusoidal waveform, Y_{p1n}, and has been demonstrated to be an adequate index to quantify the severity of the electrical stress in the presence of harmonics with respect to that to which the insulation components are subjected in the nominal operating condition. This index can be split into two contributions: k_{p1Y} and k_{phY}. The first is related to the effects of fundamental variations and the second is related to the effects of harmonics superimposed on the fundamental [55].

The true power factor is the ratio of the active power (P) consumed in watts to the apparent power (S) drawn in volt-amperes [52]:

$$pf_{true} = \frac{P}{S}. \tag{1.15}$$

1.3.1.2 System indices

In general, the system indices can be obtained directly from the site indices. For example, the system index can be chosen to be coincident with the value of the site index not exceeded for a fixed percentage (50, 90, 95 or 99%) of sites, and the choice of the percentage of sites can be a matter of agreement between the system operator and the regulator. An alternative approach would be to define a system index equal to the percentage of sites that exceeds the objective in the reporting period.

Other system indices reported in the literature are [29]:

- the system total harmonic distortion 95th percentile, STHD95;

- the system average total harmonic distortion, SATHD;

- the system average excessive total harmonic distortion ratio index, SAETHDRI$_{\mathrm{THD}*}$.

To define these indices we consider a distribution system with N busbars and M monitoring sites. For the sake of simplicity, in the following we address only indices referring to the entire distribution system.

The STHD95 index is defined as the 95th percentile value of a weighted distribution; this weighted distribution is obtained by collecting the 95th percentile values of the M individual index distributions, with each distribution obtained from the measurements recorded at a monitoring site. The weights can be linked to the connected powers or the number of customers served from the area represented by the monitoring data. The system index STHD95 allows us to summarize the measurements both temporally and spatially by handling measurements at M sites of the system in a defined time period, assumed to be significant for the characterization of the system service condition. Due to the introduced weights, this system index allows the assignment of different levels of importance to the various sections of the entire distribution system.

The system average total harmonic distortion, SATHD, is based on the mean value of the distributions rather than the 95th percentile value. The SATHD index gives average indications on the system voltage quality.

The system average excessive total harmonic distortion ratio index, SAETHDRI$_{\mathrm{THD}*}$ quantifies the number of measurements that exhibit a THD value exceeding the THD* threshold. This index is computed by counting, for each monitoring site in the system, the measurements that exceed the THD* value and normalizing this number to the total number of the measurements conducted at the site s; finally, the weighted sum of these normalized numbers is calculated:

$$\mathrm{SAETHDRI}_{\mathrm{THD}*} = \frac{\sum_{s=1}^{M} L_s \left(\frac{N_{\mathrm{THD}*s}}{N_{\mathrm{Tot},s}} \right)}{L_{\mathrm{T}}}, \tag{1.16}$$

where $N_{\mathrm{THD}*s}$ is the number of measurements at monitoring site s that exhibit a THD value exceeding the specified threshold THD*, $N_{\mathrm{Tot},s}$ is the total number of measurements conducted at monitoring site s over the assessed period of time, L_s is the connected kVA served by the system segment at monitoring site s and L_{t} is the total connected kVA served by the system.

An alternative method of computing the aforementioned index concerns the statistical analysis of all the measurements conducted at the M sites of the system.

Analogous system indices have been proposed in the relevant literature for the peak factor [56].

1.3.1.3   Existing objectives

A comparison of some harmonic voltage objectives is given in Tables 1.3 and 1.4. The objectives reported in standards and guidelines refer to site indices [11, 16, 18, 20, 27, 33, 38, 45, 46].

Table 1.3 Harmonic voltage objectives proposed in different standards and guidelines

Standard & Document	International				National or regional			
	IEC 61000-2-12	IEC 61000-3-6	EN 50160	ANSI/IEEE 519	EDF Emeraude contract	Norwegian directive	NRS048-2-2007	Hydro Quebec
Purpose	Compatibility level	Planning levels	Voltage characteristics	Recommended practice for emission limits and system design	Voltage characteristics	Standard used by regulator	Standard used by regulator	Voltage characteristics
Where it applies	International	International	Some European countries	Some countries, mostly USA	France	Norway	South Africa	Quebec, CA
MV — Voltage level	1 to 35 kV	1 to 35 kV	1 to 35 kV	1 to 69 kV	1 to 50 kV	≤35 kV	1 to 33 kV	0.75 to 34.5 kV
Order	$h \leq 50$	$h \leq 50$	$h \leq 25$	All orders	$h \leq 25$	All orders	$h \leq 50$	$h \leq 25$
h	(e.g.: 6% at $h=5$) See Table 1.4	(e.g.: 5% at $h=5$) See Table 1.4	(e.g.: 6% at $h=5$) See Table 1.4	3% all orders	(e.g.: 6% at $h=5$) See Table 1.4	(e.g.: 6% at $h=5$) See Table 1.4	(e.g.: 6% at $h=5$) See Table 1.4	(e.g.: 6% at $h=5$) See Table 1.4
THD	8%	6.5%	8%	5%	8%	8% (10 min-value) 5% (week-average)	8%	8%

(continued overleaf)

Table 1.3 (Continued)

Standard & Document	International				National or regional			
	IEC 61000-2-12	IEC 61000-3-6	EN 50160	ANSI/IEEE 519	EDF Emeraude contract	Norwegian directive	NRS048-2-2007	Hydro Quebec
HV EHV — Voltage level / Order	Not applicable	>35 kV; $h \leq 50$	Not applicable[11]	>69 to 161 kV; >161 kV All orders	>50 kV; $h \leq 25$	>35 to 245 kV; >245 kV All orders	33 to 440 kV; $h \leq 25$	≥44 to <315 kV; $h \leq 50$ + THD
h		(e.g.: 2% at $h = 5$) See Table 1.4		1.5% all orders; 1% all orders	(e.g.: 2% at $h = 5$) See Table 1.4	(e.g.: 3% at $h = 5$) See Table 1.4; (e.g.: 2% at $h = 5$) See Table 1.4	(e.g.: 3% at $h = 5$) See Table 1.4	(e.g.: 2% at $h = 5$) See Table 1.4
THD		3%		2.5%; 1.5%	3%	3%; 2%	4%	3%

[11] The current draft of EN 50160 (March 2008) also considers HV supply characteristics.

Table 1.4 Voltage harmonic limits proposed in different standards and guidelines

MV – Harmonic voltages [% of nominal or declared voltage]

Order h	IEC 61000-2-12	IEC 61000-3-6	EN 50160 EDF Norwegian directive	NRS048-2-2007	Hydro Quebec
2	2	1.8	2	2	2
3	5	4	5	5	6
4	1	1	1	1	1.5
5	6	5	6	6	6
6	0.5	0.5	0.5	0.5	0.75
7	5	4	5	5	5
8	0.5	0.5	0.5	0.5	0.6
9	1.5	1.2	1.5	1.5	3.5
10	0.5	0.47	0.5	0.5	0.6
11	3.5	3	3.5	3.5	3.5
12	0.46	0.43	0.5	0.46	0.5
13	3	2.5	3	3	3
14	0.43	0.4	0.5	0.43	0.5
15	0.4	0.3	0.5	0.5	2
16	0.41	0.38	0.5	0.41	0.5
17	2	1.7	2	2	2
18	0.39	0.36	0.5	0.39	0.5
19	1.76	1.5	1.5	1.76	1.5
20	0.38	0.35	0.5	0.38	0.5
21	0.3	0.2	0.5	0.3	1.5
22	0.36	0.33	0.5	0.36	0.5
23	1.41	1.2	1.5	1.41	1.5
24	0.35	0.32	0.5	0.35	0.5
25	1.27	1.09	1.5	1.27	1.5

HV & EHV – Harmonic voltages [% of nominal or declared voltage]

Order h	IEC 61000-3-6	EDF Emeraude contract	Norwegian directive 35 to 245 kV	Norwegian directive > 245 kV	NRS048-2-2007	Hydro Quebec
2	1.4	1.5	1.5	1	—	1.5
3	2	2	3	2	2.5	2
4	0.8	1	1	0.5	—	1
5	2	2	3	2	3	2
6	0.4	0.5	0.5	0.5	—	0.5
7	2	2	2.5	2	2.5	2
8	0.4	0.5	0.3	0.3	—	0.4
9	1	1	1.5	1	—	1
10	0.35	0.5	0.3	0.3	—	0.4
11	1.5	1.5	2.5	1.5	1.7	1.5
12	0.32	0.5	0.3	0.3	—	0.3
13	1.5	1.5	2	1.5	1.7	1.5
14	0.3	0.5	0.3	0.3	—	0.3
15	0.3	0.5	0.5	0.5	—	0.75
16	0.28	0.5	0.3	0.3	—	0.3
17	1.2	1	2	1.5	1.2	1
18	0.27	0.5	0.3	0.3	—	0.3
19	1.07	1	1.5	1.5	1.2	1
20	0.26	0.5	0.3	0.3	—	0.3
21	0.2	0.5	0.5	0.5	—	0.5
22	0.25	0.5	0.3	0.3	—	0.3
23	0.89	0.7	1.5	1	0.8	0.7
24	0.24	0.5	0.3	0.3	—	0.3
25	0.82	0.7	1	1	0.8	0.7

The tables report the compatibility levels for the medium-voltage power supply system defined in IEC 61000-2-12–2003. These limits are close to the limits reported in EN 50160–2000, where the supply voltage characteristics for public networks are given.

The same tables report the planning levels for MV, HV and EHV power systems, according to IEC 61000-3-6–2008. The planning levels are assessed by using the measured 3-s and 10-min values over at least a one-week period. In particular:

1. the 95th weekly percentile of the 10-min values should not exceed the value reported in Table 1.4.

2. the greatest 99th daily percentile of the 3-s values should not exceed the planning levels reported in Table 1.4 multiplied by a factor $k_{hvs} = 1.3 + \frac{0.7}{45}(h - 5)$.

Most national standards or regional guidelines provide indicative values for harmonic voltages at HV–EHV. In most cases, these values correspond to the planning levels for HV–EHV published in IEC 61000-3-6–2008, locally adapted to account for specific system configurations or circumstances. A standard such as EN 50160 does not exist for defining voltage characteristics for transmission systems at HV–EHV. However, in the current draft of EN 50160 (March 2008), HV supply characteristics are also considered.

IEEE 519–1992 also recommends harmonic voltage limits at HV–EHV for system design purposes. This recommendation allows the same level of harmonic voltage for any harmonic order; this may change in the future, because this recommendation is currently under revision. In particular, in the draft document of IEEE 519 (2005), probabilistic limits on harmonics are introduced: the 99th percentile of the very short time values over one day should be less than 1.5 times the values given in Table 1.3, and the 95th percentile of the short time values over one week should be less than the values given in Table 1.3.

Other objectives adopted in some countries or suggested by working groups and standards not reported in Tables 1.3 and 1.4 are outlined in the following.

Compatibility levels for low-frequency conducted disturbances in public LV systems are given in IEC 61000-2-2–2002; with some exceptions, these limits are the same as in EN 50160–2000. The CIGRE WG C4-07 recommends keeping the planning levels as they are given in IEC 61000-3-6. With reference to the objectives for voltage characteristics, the CIGRE WG suggests the harmonic limits reported in EN 50160–2000 for MV systems and indicates the objectives (but only for dominant odd harmonics) for HV–EHV systems.

In Argentina, the national authority ENRE indicates limit levels of THD equal to 3% and 8% for HV and MV/LV power systems, respectively [17].

Australia has several standards addressing the problem of harmonic distortion. In AS 2279.2–1991 the limit on voltage total harmonic distortion in the 415 V supply system is fixed at 5%; limits on odd and even harmonics are equal to 4% and 2%, respectively.

In China the limit on voltage total harmonic distortion is fixed at 3%; limits on odd and even harmonics are equal to 2.4% and 1.2%, respectively. These limits are referred to the mean values for ten minutes [49].

With reference to the limits on the harmonic current injected into the network by polluting loads, the IEC standards considered here are 61000-3-2–2005, 61000-3-4–1998 and 61000-3-6–2008 [25, 41, 46] we also consider IEEE 519–1992 [18].

Emission limits for small LV equipment, classified into four classes, are defined in IEC 61000-3-2–2005. This standard document gives emission limits as absolute values that are related to the power consumption of the device; this document also contains a detailed description of the test equipment and the circumstances of the test.

For equipment with a rated current exceeding 16 A per phase, IEC 61000-3-4–1998 applies. In this document, the emission limits are reported as a function of the ratio between the rated power of the load and the short circuit capacity of the source.

For customers connected to the MV and HV networks, IEC 61000-3-6–2008 indicates the procedure to calculate the emission limits of disturbing loads as a function of the power of the customer, the power of the polluting equipment and the system characteristics. The procedure is based on the actual source impedance, the contribution from different voltage levels and the share of the emission limit among the different voltage levels. The objective is to limit the harmonic current injected by all distorting installations to levels that will not result in voltage distortion levels that exceed the planning level. So, when the planning levels are fixed and the procedure reported in IEC 61000-3-6–2008 is applied, it is possible to derive the coordinated emission limits. In particular, the following objectives are proposed:

1. the 95th weekly percentile of the 10-min values of individual harmonics should not exceed the emission limit;

2. the greatest 99th daily percentile of the 3-s values of individual harmonic components should not exceed the emission limit multiplied by the factor $k_{hvs} = 1.3 + \frac{0.7}{45}(h - 5)$. With reference to very short time effects of harmonics,[12] use of a very short time index for assessing emission is only needed for loads having a significant impact on the system, so the use of this index could depend on the ratio between the agreed power of the connecting customer and the fault level at the point of connection.

IEEE 519–1992 also gives limits for the harmonic current distortion for individual customers. The difference with respect to the IEC standards is that the limits are given at the point of common coupling (PCC), they are then referenced to the total emission of disturbing equipment and not to each individual piece of equipment belonging to the customer.

IEEE 519–1992 gives current limits for different voltage levels, different load sizes and different harmonic orders. As an example, Table 1.5 reports the current harmonic limits for voltage levels up to 69 kV. In the table, the current ratio is the ratio between the fault current and the load current, and the limits are expressed as a percentage of the maximum load current.

In the draft document of IEEE 519 (2005), analogously to the voltages, probabilistic limits on harmonic current are also introduced. As an example, with reference to a distribution system with rated voltage up to 69 kV, users should limit their harmonic current at the PCC as follows: the 99th percentile of the very short time values over one day should be less than 2.0 times the values given in Table 1.5, the 99th percentile of the short time values over one week should be less than 1.5 times the values given in Table 1.5 and the 95th percentile of the short time values over one week should be less than the values given in Table 1.5.

---

[12] The long-term effects of harmonics relate mainly to thermal effects on cables, transformers, motors, capacitors, etc; these effects arise from harmonic levels that are sustained for ten minutes or more. The very short-term effects relate mainly to disturbing effects on electronic devices that may be susceptible to harmonic levels sustained for three seconds or less; transients are not included [46].

Table 1.5 Current harmonic limits according to IEEE 519–1992 [18]

Current ratio	Harmonic order (Odd harmonics [a])					
	<11	11–16	17–22	23–34	>34	TDD
<20	4.0	2.0	1.5	0.6	0.3	5.0
20–49.9	7.0	3.5	2.5	1.0	0.5	8.0
50–99.9	10.0	4.5	4.0	1.5	0.7	12.0
100–999	12.0	5.5	5.0	2.0	1.0	15.0
>1000	15.0	7.0	6.0	2.5	1.4	20.0

[a] Even harmonics are limited to 25% of the odd harmonics.

Information related to the electromagnetic disturbance involved in interharmonics is still being developed. The first proposal made in standards was to fix a very low value (i.e. 0.2%) for the 95th percentile of the short time values over one week of the interharmonic voltages at low frequencies (less than twice the fundamental frequency).

Due to measurement difficulties, alternative solutions are still under discussion [44]; these are:

1. To limit individual interharmonic component voltage distortion to less than 1%, 3% or 5% (depending on voltage level) from 0 Hz up to 3 kHz, as has been done for harmonics.

2. To adopt limits correlated with the short-term flicker severity index P_{st} (see Section 1.3.4) equal to 1.0, to be checked by an IEC flickermeter for frequencies at which these limits are more restrictive than those of point 1. IEC 61000-2-2–2002, in fact, gives compatibility levels for the case of an interharmonic voltage occurring at a frequency close to the fundamental frequency (50 Hz or 60 Hz), resulting in amplitude modulation of the supply voltage. In these conditions, certain loads, especially lighting devices, exhibit a beat effect, resulting in flicker. The compatibility level for a single interharmonic voltage in the above case, expressed as the ratio of its amplitude to that of the fundamental, is shown in Figure 1.11 as a function of the beat frequency. The beat frequency is the difference between the frequencies of the interharmonic and fundamental frequencies. The curves in Figure 1.11 are based on a flicker level of $P_{st} = 1$ for lamps operated at 120 V and 230 V and it is clearly evident that the flicker is due to interharmonic amplitude which is less than 1%.

3. To develop appropriate limits for equipment and system effects, such as generator mechanical systems, signalling and communication systems and filters on a case-by-case basis using specific knowledge of the supply system and connected user loads.

Therefore, different limits are necessary for different ranges of frequency and two kinds of measurement (interharmonic components and light flicker) are simultaneously needed.

Most of the national or regional standards do not specify compatibility levels or limits for interharmonic voltage. Sometimes these standards refer to the characteristic values of interharmonic voltages on LV networks corresponding to the compatibility level with respect to the flicker effect reported in IEC 61000-2-2–2002 [28].

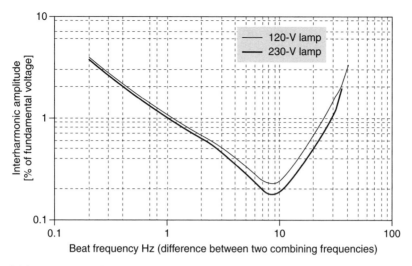

Figure 1.11 Compatibility level for the interharmonic voltage [28]. © 2002 IEC Geneva, Switzerland. www.iec.ch

1.3.2 Slow voltage variations

1.3.2.1 Site indices

Slow voltage variations are usually quantified by calculation of the RMS value of the supply voltage. In assessing RMS supply voltage, measurement has to take place over a relatively long period of time to avoid the instantaneous effect on the measurement caused by individual load switching (e.g. motor starting, inrush current) and faults.

EN 50160–2000 quantifies slow voltage variations using the 10-min mean RMS value and considering a week as the minimum measurement period; in particular, the 95th percentile of the 10-min mean RMS values over one week is considered the site index. In the current draft of EN 50160 (March 2008), instead of the 95th percentile, the 99th percentile and the maximum value of all 10-min mean RMS values over one week are considered.

Several utilities characterize slow voltage variations by using the index of the current EN 50160–2000. In some countries different site indices have been introduced. In Norway, the variations in voltage RMS values are measured as a mean value over one minute, and the maximum value in the monitored period is designated the site index. Other countries also use the 1-minute RMS voltage value, e.g. Hungary.

IEC 61000-4-30–2003 defines the procedure for the RMS voltage measurements and, as in EN 50160–2000, assigns the 10-min RMS value of the supply voltage (short interval value) to quantify slow voltage variations and considers a week as the minimum measurement period. Annex 6 of the same standard suggests guidelines for contractual applications.

The PQ directive of the Hydro-Quebec on MV and LV systems considers both the 95th percentile and the 99.9th percentile of the 10-min mean RMS voltage variations with respect to nominal voltage over one week [27].

The South African Standard NRS048-2–2007 considers the 95th percentile of the 10-min measurement value over a period of a week and the number of times that consecutive 10-min values have been outside the higher or lower permissible value [45].

In Argentina, the 95th percentile of the 10-min measurement value over a period of a week is considered the site index [50].

In Australia, many energy companies refer to the 10-min measurement values as the index to quantify slow voltage variations.

In China, the maximum value of the 10-min mean RMS values is considered the site index [49].

1.3.2.2   System indices

Starting from the site indices, the following system indices can be introduced:

- the percentage of sites that exceeds the objectives in the reporting period;

- the average or median value of the site indices;

- the value of the site index not exceeded for a fixed percentage (90, 95 or 99%) of sites; the percentage of sites can be a matter of agreement between the system operator and the regulator.

Recently, in Spain, a new system index was proposed. The procedure for calculating the index is as follows: for each measurement point, the number of 10-min RMS voltage measurements exceeding the limit is determined and a weighted average of this number is taken over all measurement points for the considered system. Weighting is based on the amount of the load connected to the substations that are coincident with the measurement points.

Other new system indices could be introduced, extending the definitions of system indices that are known to characterize other aspects of PQ disturbances, as happened with the system harmonic indices.

1.3.2.3   Existing objectives

The existing objectives refer only to site indices.

EN 50160–2000 considers the following condition to indicate acceptable performance of the system: during each period of one week, under normal operating conditions, the 95th percentile of the 10-min mean RMS values of the supply voltage shall be within the ranges $U_n \pm 10\%$ for the LV system and $U_C \pm 10\%$ for the MV system, where U_n is the nominal voltage of the system and U_C is the declared supply voltage.

For the Norwegian PQ directives, variations in the voltage RMS value, measured as a mean value over one minute, shall be within an interval of $\pm 10\%$ of the nominal voltage [38].

In France, for MV customers, the supply contracts contain the voltage variation limit $U_C \pm 5\%$ of the nominal voltage for 100% of the time, where U_C must be in the range $\pm 5\%$ around the nominal voltage [20].

The regulation in Hungary contains three different objectives [48]:

- 100% of the 10-min RMS voltage shall be between 85% and 110% of the nominal voltage;

- 95% of the 10-min RMS voltage shall be between 92.5% and 107.5% of the nominal voltage;

- 100% of the 1-min RMS voltage shall be less than 115% of the nominal voltage.

The regulation in Spain fixes the following limit: the 95th percentile of the 10-min mean RMS value over one week shall be between 93% and 107% of the declared voltage [42].

In the current draft of EN 50160 (March 2008), the following limits apply:

- 100% of all 10-min mean RMS values of the supply voltage shall be below +10 % of the nominal voltage; and

- at least 99% of the 10-min mean RMS values of the supply voltage shall be above −10% of the nominal voltage, with not more than two consecutive 10-min mean RMS values below the lower limit of −10%.

The PQ directive of the Hydro-Quebec specifies different limits for LV and MV systems. With reference to LV systems, for every one-week period, under normal operating conditions, excluding interruptions, 95% of RMS value voltage variations evaluated over ten minutes must be within the range –11.7% to +5.8% of nominal voltage, and 99.9% of RMS value variations evaluated over ten minutes must be within the range –15% to +10% of nominal voltage. With reference to MV systems, 95% of RMS value voltage variations evaluated over ten minutes must be within the range ± 6% of nominal voltage, and 99.9% of RMS value variations evaluated over ten minutes must be within the range ± 10% of nominal voltage [27].

In South Africa, the limits of voltage RMS variations are fixed at ±10% of the nominal voltage for LV systems (<500 V) and at ±5% of the declared voltage at other voltage levels (≥500 V). Moreover, there is the additional criterion that no more than two consecutive 10-min values can exceed the higher and the lower permissible levels [45].

In Argentina, the limits of voltage RMS variations expressed as a percentage of the nominal voltage are: ± 8% for LV and MV systems with overhead lines; ±5% for LV and MV systems with cable lines; ±5% for HV systems; and ±10% for rural systems [50].

In Australia, distribution companies fix different limits for RMS voltage variations. In particular, Integral Energy fixes the following limits: +14% and –6% of the nominal voltage for LV systems; ±10% for MV systems. EnergyAustralia's objective, on the other hand, is to maintain the 10-min RMS value of the phase-to-neutral supply voltage within the following range: between 253 V and 264 V (upper limit) and between 226 V and 216 V (lower limit) for systems with nominal phase voltage equal to 240 V.

In China, the limits of voltage RMS variations are fixed at ±10% of the declared supply voltage [49].

1.3.3   Unbalances

1.3.3.1   Site indices

The severity of voltage unbalances is often quantified by means of the (negative sequence) voltage unbalance factor, which is defined as the ratio between the negative-sequence voltage component V_{-1} and the positive-sequence voltage component V_{+1}, usually expressed as a percentage:

$$K_d = \frac{V_{-1}}{V_{+1}} 100. \tag{1.17}$$

Equation (1.17) can be substituted by the relationship:

$$K_d = \sqrt{\frac{1 - \sqrt{3 - 6\beta}}{1 + \sqrt{3 - 6\beta}}} 100 \tag{1.18}$$

where:

$$\beta = \frac{U_{12\text{fund}}^4 + U_{23\text{fund}}^4 + U_{31\text{fund}}^4}{\left(U_{12\text{fund}}^2 + U_{23\text{fund}}^2 + U_{31\text{fund}}^2\right)^2} \tag{1.19}$$

and $U_{12\text{fund}}$, $U_{23\text{fund}}$ and $U_{31\text{fund}}$ are the fundamental components of the phase-to-phase voltages.

Sometimes, local distributors quantify voltage unbalances by means of the differences between the highest and lowest phase-to-neutral or phase-to-phase steady-state voltages.

Table 1.6 provides a summary of unbalance site indices proposed in various standards and guidelines [11, 12, 16, 20, 27, 38, 45]; the site indices most frequently used are:

1. the 95th percentile of the very short interval value of the negative-sequence voltage unbalance factor ($K_{d,\text{vsh}}$) over one day;

2. the maximum of the short interval value of the negative-sequence voltage unbalance factor ($K_{d,\text{sh}}$) over one week;

3. the 95th percentile of the short interval value of the negative-sequence voltage unbalance factor ($K_{d,\text{sh}}$) over one week;

4. the 95th percentile of the long interval value (two hours) of the negative-sequence voltage unbalance factor ($K_{d,\text{lt}}$) over one week.

Annex 6 of IEC 61000-4-30–2003 also suggests guidelines for contractual applications.

EN 50160–2000 considers the 95th percentile of the short interval value of the unbalance factor over one week and it does not specify any limit for 5% of the time. Some national directives (for example, in Norway) choose to specify limits for 100% of the time [38].

All the standards and guidelines usually consider a week to be the minimum measurement period. The whole measurement and evaluation procedure for the short and long interval values of the negative-sequence voltage unbalance factor is defined in detail in IEC 61000-4-30–2003.

Table 1.6 Voltage unbalance site indices proposed in different standards and reference document

Standard & Document		International			National or Regional		
		IEC 61000-4-30	CIGRE WG C4.07 2004	EN 50160	Norwegian directive	EDF Emeraude contract	NRS048-2-2007	Hydro Quebec
Purpose		Power quality measurement methods	Planning level	Voltage characteristics	Standard used by regulator	Voltage characteristics	Standard used by regulator	Voltage characteristics
Where it applies		International	—	Some European countries	Norway	France	South Africa	Quebec, CA
Indices & assessment time	Very-short time (3-sec)	—	$K_{d,vsh}$ *95th daily percentile 99th weekly percentile*	—	—	—	—	—
	Short time (10-min)	$K_{d,sh}$ *95th weekly percentile (or as agreed)*	$K_{d,sh}$ *99th weekly percentile*	$K_{d,sh}$ *95th weekly percentile*	$K_{d,sh}$ *Max*	$K_{d,sh}$ *(no further specification)*	$K_{d,sh}$ *95th weekly percentile*	—
	Long time (2-hour)	$K_{d,lt}$ *95th weekly percentile (or as agreed)*	—	—	—	—	—	$K_{d,lt}$ *95th weekly percentile*
Period for statistical assessment		At least one week or more, as agreed	One week	One week	One week minimum	At least one week or more	At least one week	One week

Recently, indices have been proposed to measure the unbalance in power systems that contain nonsinusoidal waveforms. These indices, called the *equivalent voltage total harmonic distortion factor* (VTHD$_e$) and the *voltage total unbalance distortion factor* (VTUD), are obtained using an unbalance decomposition method developed by the authors in [34]. From the three-phase voltage phasors of fundamental and each harmonic component, the balanced, the first unbalanced and the second unbalanced components are obtained by means of three transformational matrices. Using these components, the equivalent RMS value of the three-phase voltage is calculated and then decomposed into the balanced fundamental component, the unbalanced fundamental component, the balanced harmonic component and the unbalanced harmonic component. Using these four components, the equivalent total harmonic distortion factor and the total unbalance distortion factor of the voltage are calculated. The voltage total unbalance distortion factor not only takes into account the fundamental frequency component, but also considers the influence of unbalanced harmonics; it can quantify the severity of the unbalance even in the presence of harmonic distortion. For more details see [34].

1.3.3.2   System indices

Once again, the system index may be the value of the site index not exceeded for a high percentage (90, 95 or 99%) of sites, or an alternative approach would be to define, as a system index, the percentage of sites that exceeds the objective in the reporting period.

Other possible unbalance system indices can be obtained by substituting, in the relationships that define the system indices for waveform distortions, the index THD by the index K_d; so, the following system indices can be introduced:

- the system unbalance factor 95th percentile, SK_d95;
- the system average unbalance factor, SAK_d;
- the system average excessive unbalance factor ratio index, $SAEK_dRI_{Kd*}$.

1.3.3.3   Existing objectives

Table 1.7 summarizes the objectives relevant to voltage unbalance among different standards and reference documents [11, 16, 20, 27, 33, 38, 45]. For MV systems, the objective is usually a 2% negative sequence voltage unbalance factor; in some areas the voltage unbalance may be up to 3% (usually in cases where the networks are predominantly single-phase, i.e. single-phase traction and single-phase distribution). In HV and EHV systems, objectives vary from 1% to 2%.

CIGRE WG C4.07 2004 suggests that the planning level for the 99% probability weekly value of $K_{d,vsh}$ may exceed the planning level of the 95% probability daily value, reported in Table 1.7, by a factor (for example, 1.25 to 2.0 times) to be specified by the system operator.

In Australia (not reported in Table 1.7), the Integral Energy distributor aims to achieve differences of no greater than 6% on the LV network and 3% on the HV network between the highest and the lowest phase-to-neutral or phase-to-phase steady-state voltages. The limits refer to the 10-min measurement values and the percentages refer to the nominal voltage.

Table 1.7 Voltage unbalance objectives proposed in different standards and reference documents

Standard & Document	International			National or Regional			
	IEC 61000-2-12	CIGRE WG C4.07 2004	EN50160	Norwegian directive	EDF Emeraude contract	NRS048-2-2007	Hydro Quebec
Purpose	Compatibility levels	Planning level	Voltage characteristics	Standard used by regulator	Voltage characteristics	Standard used by regulator	Voltage characteristics
Where it applies	International	—	Some European countries	Norway	France	South Africa	Quebec, CA
MV Very short time (3-sec)	—	2%	—	—	—	—	—
Short time (10-min)	—	2%	2% [up to 3% in some areas]	2%	2%	2% [up to 3% in some areas]	—
Other	2% [up to 3% in some areas]	—	—	—	—	—	2% (2-hr)
HV-EHV Very short time (3-sec)	Not applicable	HV =1.5% EHV=1%	Not applicable[13]	—	—	—	—
Short time (10-min)	HV =1.5% EHV = 1%	HV = 2% EHV = 1.5%		2%	2%	HV = 2% EHV = 1.5% [up to 3% in some areas]	—
Other	—	—		—	—	—	HV = 1.5% (2-hr) EHV = 1% (2-hr)

[13] The current draft of EN 50160 (March 2008) also considers HV supply characteristics.

1.3.4   Voltage fluctuations

1.3.4.1   Site indices

As mentioned in the section describing PQ disturbances, voltage fluctuations can cause light intensity fluctuations that can be perceived by our brains. This effect, popularly known as flicker, can cause significant physiological discomfort. More precisely, flicker is the impression of unsteadiness of visual sensation induced by a light stimulus whose luminance or spectral distribution properly fluctuates with time.

The UIE/IEC[14] flickermeter estimates the level of sensation of light flicker, starting from the voltage fluctuations that cause it. It is based on the model of a 230 V/60 W incandescent lamp and on a model of the human sensation system. Figure 1.12 shows a simplified block diagram of the IEC flickermeter.

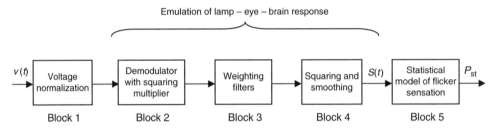

Emulation of lamp – eye – brain response

| $v(t)$ Voltage normalization | Demodulator with squaring multiplier | Weighting filters | Squaring and smoothing | $S(t)$ Statistical model of flicker sensation | P_{st} |

Block 1   Block 2   Block 3   Block 4   Block 5

Figure 1.12   Simplified block diagram of the IEC flickermeter

Block 1 performs a normalization of the input supply voltage. The aim of blocks 2, 3 and 4 is to emulate the lamp–eye–brain response. The input to block 2 of the flickermeter is the relative voltage variation that can be seen as a modulated wave superimposed on a 50 or 60 Hz wave. This modulated wave must be extracted from the carrier; it is for this reason that a demodulating block must be present in the flickermeter. The demodulation function is performed by a quadratic demodulator that reproduces the fluctuations of the squared RMS value of the voltage; this is directly related to the electrical power absorbed by the lamp, which is representative of the flicker. Block 3 is composed of two band-pass filters. The first filter eliminates the DC and double mains frequency ripple components of the demodulator output. The second filter is a weighting filter block that simulates the frequency response to sinusoidal voltage fluctuation of a coiled filament gas-filled lamp (60 W/230 V) combined with the human visual system. Block 4 emulates the response of the human brain to the fluctuations. This works by using two functions: a squaring block and a low-pass filter. At the output of block 4, the flickermeter gives a signal that is proportional to the voltage variation, weighted by the sensitivity of the lamp–eye–brain combination at different frequencies: this is the measure of the instantaneous flicker sensation, $S(t)$. The instantaneous sensation of flicker is expressed per unit of the perceptibility threshold, which is the level of flicker considered just perceptible by a significant portion of the people involved in the tests.

---

[14] UIE: International Union for Electrical Applications.

From the instantaneous flicker values, the following indices characterizing the intensity of flicker annoyance are obtained: the *short- term flicker severity* and the *long-term flicker severity*.

The short-term flicker severity (P_{st}) is measured over a period of ten minutes. This value is obtained from a statistical analysis of the instantaneous flicker value $S(t)$ by means of the following relationship:

$$P_{st} = \sqrt{0.0341P_{0.1} + 0.0525P_{1s} + 0.0657P_{3s} + 0.28P_{10s} + 0.08P_{50s}}, \qquad (1.20)$$

where the quantities $P_{0.1}$, P_1, P_3, P_{10} and P_{50} are the instantaneous flicker level values exceeded for 0.1%, 1%, 3%, 10% and 50% of the time during the observation period. The suffix 's' in the formula indicates smoothed values to be used; these are given by:

$$P_{50s} = \frac{P_{30}+P_{50}+P_{80}}{3}, \qquad P_{10s} = \frac{P_6+P_8+P_{10}+P_{13}+P_{17}}{5},$$

$$P_{3s} = \frac{P_{2.2}+P_3+P_4}{3}, \qquad P_{1s} = \frac{P_{0.7}+P_1+P_{1.5}}{3}. \qquad (1.21)$$

The term $P_{0.1}$ cannot change abruptly and no smoothing is needed for this quantity.

The long-term flicker severity (P_{lt}) is calculated from a sequence of 12 P_{st} values over a two-hour interval, according to the following relationship:

$$P_{lt} = \sqrt[3]{\frac{\sum_{i=1}^{12} P_{sti}^3}{12}}. \qquad (1.22)$$

Table 1.8 summarizes the voltage fluctuation site indices reported in various standards and guidelines [11, 12, 16, 20, 27, 38, 45, 47].

The most common reference for flicker measurement is IEC 61000-4-15–2003, and the minimum measurement period is one week [32].

Site indices adopted in some countries or suggested by working groups and not reported in Table 1.8 are outlined in the following.

CIGRE WG C4-07 recommends that the following indices be used: the 95th and 99th percentiles of P_{st} over one week and the 95th percentile of P_{lt} over one week [16].

The national authority of Argentina uses the 95th percentile of P_{st} over a period of one week [50].

In China the following indices are used: the 95th percentile of the short-term flicker severity and the maximum value of the long-term flicker severity [49].

It should be noted that experience has indicated that the indices P_{lt} and P_{st} are often correlated; the following relationship has been suggested to show this correlation [16]:

$$P_{lt95\%} = 0.84P_{st95\%}. \qquad (1.23)$$

Therefore, one of the two quantities may be seen as being redundant.

1.3.4.2   System indices

A current limitation for voltage fluctuation system indices is that, in many systems, flicker measurements are often conducted at only a few sites located near to large fluctuating loads,

Table 1.8 Voltage fluctuation site indices proposed in different standards and reference documents

Standard & Document		International			National or Regional		
		IEC 61000-3-7	IEC 61000-4-30	EN 50160	Norwegian directive	EDF Emeraude contract	NRS048-2-2007	Hydro Quebec
Purpose		Planning levels	Power quality measurement methods	Voltage characteristics	Standard used by regulator	Voltage characteristics	Standard used by regulator	Voltage characteristics
Where it applies		International	International	Some European countries	Norway	France	South Africa	Quebec, CA
Indices & assessment	Short term (10-min)	P_{st} 99th and 95th weekly percentile	P_{st} 99th weekly percentile (or X percentile as agreed)	—	P_{st} 95th weekly percentile	—	—	—
	Long term (2-h)	P_{lt} 95th weekly percentile	P_{lt} 95th weekly percentile (or X percentile as agreed)	P_{lt} 95th weekly percentile	P_{lt} Max	P_{lt} (no further specification)	P_{lt} 95th weekly percentile	P_{lt} 95th weekly percentile
	Other	—	—	—	—	—	—	—
Period for statistical assessment		One week minimum	One week minimum	One week	One week minimum	At least one week or more	At least one week	One week
Measurement method		IEC 61000-4-15	IEC 61000-4-15	IEC 61000-4-15	IEC 61000-4-15	IEC 61000-4-15 (formerly IEC 868)	IEC 61000-4-15	IEC 61000-4-15

so the measurements are unlikely to form an unbiased sample from the total population. Given this, it is obvious that a system index would not usually be recommended for general application.

1.3.4.3   Existing objectives

The definition of flicker severity somewhat dictates the objectives: under laboratory conditions and for the standard lamp (incandescent 60 W filament), a flicker severity exceeding unity will feel disturbing to the majority of individuals; a flicker between 0.7 and 1.0 is noticeable, but not disturbing for most. In Table 1.9 a comparison of voltage fluctuation objectives between different standards and reference documents is reported [11, 16, 20, 27, 33, 38, 45, 47].

In IEC 61000-3-7–2008, planning levels up to 35 kV are slightly lower than unity ($P_{st} = 0.9$ and $P_{lt} = 0.7$). The planning levels for HV and EHV systems, based on a transfer coefficient[15] equal to unity, are 0.8 for the 10-minute value and 0.6 for the 2-hour value. In particular, the 95th percentile value of P_{st} and P_{lt} should not exceed the planning level; the 99th percentile of the P_{st} value may exceed the planning level by a factor (for example, 1 to 1.5) to be specified by the system operator or owner, depending on the system and load characteristics.

Other objectives adopted in some countries or suggested by working groups and not reported in Table 1.9 are outlined below.

CIGRE WG C4-07 recommends the following values of P_{st} for planning levels: 0.9 for MV and 0.8 for HV–EHV systems.

The Australian Standard AS/NZS 61000.3.7, which specifies the limit, is based on the IEC standard of the same name, IEC 61000-3-7–1996. The compatibility levels are $P_{st} = 1$ $P_{lt} = 0.8$ planning levels fixed in the Australian standard are the same as those proposed in IEC 61000-3-7 (Table 1.9) [50].

The national authority of Argentina fixes a limit of 1.0 for the 95th percentile of P_{st} over one week [49].

In China the following objectives are fixed: the 95th percentile of $P_{st} \leq 1.0$ and the maximum value of $P_{lt} \leq 0.8$.

It is worth mentioning that many researchers report concern that a P_{st} threshold of 1.0 is too strict, since it is based on laboratory studies. In the field, this level is often exceeded significantly without known problems. The reasons for this include the fact that a daily assessment does not weight daylight hours less severely, the use of lighting technologies other than incandescent and the influence of other sources of lighting. The practical assessment of flicker, and the related indices, is an important area for further research, since the costs of meeting the P_{st} requirement of 1.0 are considered by some to be too significant for some utilities and/or their customers.

Limits for equipment emissions are given in IEC 61000-3-5–1994; the limits are expressed in terms of P_{st} and P_{lt}, even though these are limits on fluctuations in current [19]. A reference impedance, reported in the same standard, is used to obtain the voltage fluctuation from the measured current fluctuation.

---

[15] The transfer coefficient is the ratio between the flicker level at a higher voltage level and the flicker level at the terminals of the lamps.

Table 1.9 Voltage fluctuation objectives proposed in different standards and reference documents

Standard & Document	International			National or Regional			
	IEC 61000-3-7	IEC 61000-2-12	EN 50160	Norwegian directive	EDF Emeraude contract	NRS048-2-2007	Hydro Quebec
Purpose	Planning levels	Compatibility levels	Voltage characteristics	Standard used by regulator	Voltage characteristics	Standard used by regulator	Voltage characteristics
Where it applies	International	International	Some European countries	Norway	France	South Africa	Quebec, CA
MV P_{st}	$0.9^{(16)}$	$1.0^{(17)}$	—	1.2	—	—	—
P_{lt}	$0.7^{(16)}$	$0.8^{(17)}$	1.0	1.0	1.0	1.0	1.0
HV–EHV P_{st}	$0.8^{(16)}$	Not applicable	Not applicable[18]	1.0	—	Not defined	—
P_{lt}	$0.6^{(16)}$			0.8	1.0	Not defined	0.6

[16] Assuming the transfer coefficient = 1 between MV or HV systems and LV systems.

[17] No compatibility levels for flicker are defined at MV, however it refers to IEC 61000-2-2-2002 that can be transferred from LV.

[18] The current draft of EN 50160 (March 2008) also considers HV supply characteristics.

1.3.5 Mains signalling voltages

1.3.5.1 Site indices

For this type of PQ disturbance only site indices have been defined. In particular, EN 50160–2000 introduces, as a power quality index to characterize this disturbance, the 99th daily percentile of the 3-s mean of the signal voltages.

IEC 61000-4-30–2003 specifies that the aggregation algorithm used to calculate the very short, short and long-term interval values does not apply for this disturbance and defines a measurement method for signalling frequencies below 3 kHz.[19] For mains signalling frequencies above 3 kHz, IEC 61000-3-8–1997 can be consulted.

IEC 61000-4-30–2003 fixes one day as the minimum measurement period and, in Annex 6, suggests guidelines that can be used for contractual applications.

1.3.5.2 Existing objectives

EN 50160–2000 defines the limits for LV and MV distribution networks. These limits, which are functions of the signal frequency, are given in terms of percentages of the declared voltage.

For ripple control systems (signals with a frequency in the range 110–3000 Hz), some countries have officially recognized the so-called *MEISTER curve*, which gives the maximum permissible levels as a function of frequency. Where this curve is not applied, the amplitude of the signal should not exceed the compatibility levels for individual harmonic voltages for odd harmonics (non-multiples of 3) reported in IEC 61000-2-2–2002 and IEC 61000-2-12–2003.

1.3.6 Voltage dips (sags)

Voltage dips sags are PQ disturbances classified as events. The indices used to characterize this type of PQ disturbance can refer to:

- a single event (event indices);
- a site (site indices);
- a system or a part of it (system indices).

1.3.6.1 Event indices

A single voltage dip can be characterized by a pair of data (residual voltage or depth and duration) or by a single index obtained by properly handling the aforementioned pair of data (the voltage sag aggressiveness index, voltage tolerance curve-based indices, the voltage dip energy index or the voltage sag severity index).

---

[19] The voltage measurement shall be based on the 10/12 cycle RMS value interharmonic bin or the RMS of the four nearest 10/12 cycle RMS values interharmonic bin (for example, a 316,67 Hz ripple control signal in a 50 Hz power system shall be approximated by an RMS of the 310, 315, 320, 325 Hz bin available from the FFT performed on a 10-cycle time interval). The measured values are recorded during a period of time specified by the user (up to 120 s) starting from the measured value of the concerned interharmonic exceeding a pre-fixed threshold.

IEC 61000-4-30–2003 prescribes, for the voltage magnitude to be used in the residual voltage (see Section 1.2.1) calculation, the use of the RMS voltage measured over one cycle and refreshed each half cycle. In three-phase systems, the three voltages have to be considered and the residual voltage is the lowest one-cycle RMS voltage in any of the three phases.

With reference to the duration (see Section 1.2.1), it can start in one phase and finish in a different phase; in fact, in the three-phase system, the voltage dip starts when at least one of the RMS voltages drops below the dip-starting threshold, and it ends when all the three voltages have recovered above the dip-ending threshold. The value of the dip-starting threshold used to start recording the dips may affect the number of dips captured. The choice of the dip-ending threshold value affects the duration value. No specific threshold value is universally fixed, although 90% of the nominal value is typically used. The dip threshold can be a percentage of either nominal or declared voltage, or it can be a percentage of the sliding voltage reference. The user declares the reference voltage in use.

The sliding voltage reference is a voltage magnitude averaged over a specified time interval; it represents the voltage preceding a voltage dip and can be calculated using a first order filter within a one-minute time constant. The filter is given by:

$$U_{sr(n)} = 0.9967\ U_{sr(n-1)} + 0.0033 U_{(10/12)rms}, \qquad (1.24)$$

where $U_{sr(n)}$ is the present value of the sliding reference voltage, $U_{sr(n-1)}$ is the previous value of the sliding reference voltage and $U_{(10/12)rms}$ is the most recent 10/12 cycle RMS voltage. When the measurement is started, the initial value of the sliding reference voltage is set to the declared voltage.

The voltage sag aggressiveness (VSA) quantifies the severity of the sag, considering both depth and duration; in fact, shallower dips must have longer duration to be as disruptive to equipment. VSA is defined as the product *depth duration* of the dip.

Other event indices proposed in the literature can be obtained by using voltage tolerance curves. Voltage tolerance curves, also known as power acceptability curves, are plots of equipment at the maximum acceptable voltage deviation versus time duration for acceptable operation. Various voltage tolerance curves exist, but the most widely publicized are the CBEMA (Computer Business Equipment Manufacturers' Association) curve, the ITIC (Information Technology Industry Council) curve and the SEMI (Semiconductor Equipment and Materials International Group) curve.

The CBEMA curve is shown in Figure 1.13. The vertical scale shows the bus voltage amplitude in percentage of the rated value and the horizontal scale shows the duration of a disturbance event as seen at the load point. The vertical scale in Figure 1.13 is shown using percentages, although RMS values of the voltage may be used. The horizontal scale is logarithmic in seconds, but this scale may be shown linearly and/or in 'cycles.' Short disturbances correspond to the lower left corner of the CBEMA plane, while near normal operating conditions are shown close to the horizontal $V = 100\%$ line.

The more recent ITIC curve is shown in Figure 1.14. The CBEMA and ITIC curves differ in the way their regions are presented. CBEMA is a continuous curve whereas ITIC has a series of vertical and horizontal lines. The ITIC curve has an expanded acceptable region compared to the CBEMA curve.

In Figures 1.13 and 1.14, the 'acceptable' operating range is shown as the region included between the upper (overvoltage condition) limb and the lower (undervoltage condition) limb.

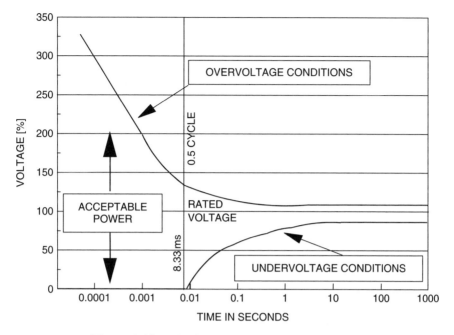

Figure 1.13   The CBEMA voltage tolerance curve

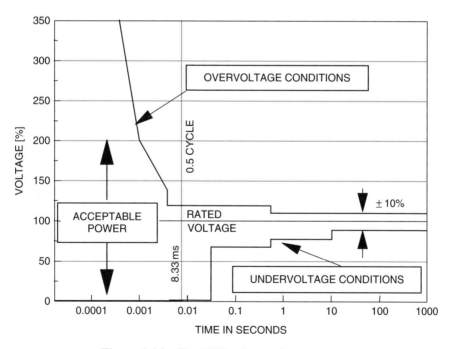

Figure 1.14   The ITIC voltage tolerance curve

It is believed that the CBEMA curve was originally developed through operating experiences using mainframe computers. However, its use has been extended to provide a measure of PQ for electric drives and solid state loads as well as a host of wide-ranging residential, commercial and industrial loads [39].

Finally, the SEMI curve is shown in Figure 1.15. Unlike the CBEMA and ITIC curves, the SEMI curve only defines dip tolerance. This curve is recognized as a superior ride-through curve compared to the ITIC curve.

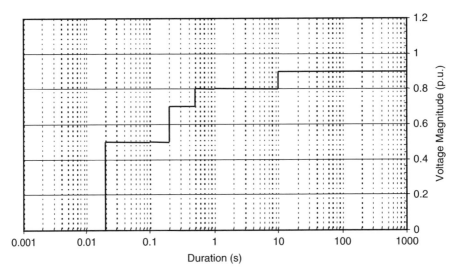

Figure 1.15   The SEMI voltage tolerance curve

An example of an event index based on a voltage tolerance curve is the power quality index (PQI), the calculation of which is based on the CBEMA curve. This was developed to cover both undervoltage and overvoltage events. In the following, we refer to the calculation of PQI in the case of a voltage dip. Each voltage dip, characterized by time duration T and residual voltage V, can be represented by a point in the voltage tolerance CBEMA plane. This point can be compared with the reference point of the CBEMA curve corresponding to the same duration T to determine the 'level' of unacceptability of the dip. It is evident that, when the point representing the dip event is in the area of unacceptability, i.e. the greater the distance from the CBEMA curve, the higher the probability that customers are experiencing problems. The power quality index (PQI) is given by:

$$\mathrm{PQI} = \left| \frac{V - 100}{V_{\mathrm{CBEMA}}(T) - 100} \right| 100, \qquad (1.25)$$

where $V_{\mathrm{CBEMA}}(T)$ is the voltage on the CBEMA curve corresponding to duration T and V is the residual voltage of the considered dip expressed as a percentage of the nominal voltage. A value of PQI higher than 100 indicates an unacceptable event for the CBEMA curve.

The IEEE P1564 draft 6 document introduces two other single-event indices: the voltage dip energy index, E_{VS} and the voltage sag severity index, S_e.

The voltage dip energy is defined as:

$$E_{VS} = \int_0^T \left[1 - V(t)^2 \right] dt, \qquad (1.26)$$

where $V(t)$ is the RMS voltage per unit versus time. The integration is taken over the duration of the event, i.e. for all values of the RMS voltage below the threshold.

The voltage sag severity S_e is defined from the residual voltage V per unit and the duration T by comparing these values with a reference curve.

From the residual voltage V and the event duration T, the voltage sag severity of the event is calculated by the following equation:

$$S_e = \frac{1 - V}{1 - V_{\text{curve}}(T)}, \qquad (1.27)$$

where $V_{\text{curve}}(T)$ is the residual voltage value of the reference curve for the same duration T. If the event is on the reference curve, the voltage sag severity is equal to 1, whereas if the event is above (below) the reference curve, the index is less (greater) than 1. For events with a residual voltage above the voltage dip threshold, the index is set to zero. In the current draft of IEEE P1564, the SEMI curve is recommended as a reference.

It is evident that the power quality index, PQI, and the voltage sag severity index, S_e, give similar information; the main difference is the voltage tolerance curve considered: PQI refers to CBEMA and S_e refers to the SEMI curve.

1.3.6.2   Site indices

Generally, site indices are calculated from single-event indices and from the frequency of dip occurrence at the considered site.

The following site indices are the most frequently used in the technical literature [9, 10, 16]:

- the RMS variation frequency index for residual voltage X at the kth site (RFI-X_k), which measures how often the residual voltage in the kth bus is below a threshold X. For example, at the kth site, to compute RFI-70$_k$ we consider voltage dips below 0.70 per unit, or 70% of the reference voltage. RFI-X_k gives information about the frequency and amplitude of dips;

- the average value of the residual dip voltage amplitude at the kth site (AVDA$_k$) or different percentiles (typically the 95th and 5th percentiles);

- the number of dips at the kth site (N_k). This index corresponds to the RFI-X_k index calculated for $X = 90\%$ (RFI-90$_k$) if dip threshold is equal to 90%;

- the mean value of the voltage sag aggressiveness index or different percentiles.

The site indices obtained using the voltage tolerance curves are:

- the SARFI-CBEMA;

- the SARFI-ITIC;

- the SARFI-SEMI;

- the PQI$_{\text{Site}}$.

The SARFI-CBEMA, SARFI-ITIC and SARFI-SEMI indices measure, at the considered site, the number of dips outside the curve. They measure the number of events, detected by points in the amplitude-duration plane, that fall into the 'not acceptable' zone of the corresponding curves.

The PQI_{Site} index is obtained by summing all the PQI values at the kth considered site:

$$\text{PQI}_{\text{Site},k} = \sum_{j=1}^{N_D} \text{PQI}_j^k, \tag{1.28}$$

where N_D is the number of dips that arise in the survey period at the considered site.

Further site indices have been derived from the single-event indices E_{VS} and S_e, in particular:

- the sag energy index, SEI_k;
- the average sag energy index, ASEI_k;
- the severity busbar index, S_k;
- the average severity busbar index, AS_k.

The sag energy index, SEI_k, sums the voltage-dip energy of all events recorded at the considered kth site during the observation period (one month or one year).

The average sag energy index, ASEI_k, is the mean value of the energy of all voltage dips that occur at the kth site during the time period considered.

The severity busbar index, S_k, sums the voltage sag severity S_e of all events recorded at the considered kth site during the observation period (one month or one year).

The average severity busbar index, AS_k, is the mean value of S_k.

Other than by site indices, the behaviour of a site is frequently characterized by magnitude-duration tables that permit a reduction in the number of indices that need to be reported and managed. The columns of the table represent ranges of voltage-dip duration; the rows represent ranges of voltage-dip residual voltage; each element of the table gives the number of events falling into a predetermined range of duration and amplitude.

The choice of the residual voltage and duration ranges for voltage-dip tables is a point of discussion. Different publications use different values. The voltage-dip table recommended by the UNIPEDE DISDIP group was shown earlier in Table 1.1.

A different voltage-dip table is reported in IEC 61000-2-8–2002. This classifies the voltage dips by residual voltage and duration in accordance with Table 1.10. Dips that involve more than one phase should be designated a single event if they overlap in time.

IEC 61000-4-11–2004 prescribes a number of duration and residual voltage values for testing equipment. These values are used to define another possible voltage-dip table (Table 1.11).

A variation on this idea is reported in NRS 048-2–2007, which defines rectangular regions in the duration/magnitude plane (Table 1.12) and gives the number of dip incidents falling into each region. The Y-type (grey) area reflects dips that are expected to occur frequently on typical HV and MV systems and against which customers should protect their plant. The X-type areas ($X1$ and $X2$) reflect 'normal' HV protection clearance times and, hence, a significant number of events are expected to occur in this area. Customers with sensitive equipment should attempt to protect against at least the $X1$ type dips, which are more

Table 1.10   Presentation of results based on IEC 61000-2-8–2002 [14]

Residual voltage [%]	Duration [s]							
	0.02-0.1	0.1-0.25	0.25-0.5	0.5-1	1-3	3-20	20-60	60-180
80-90								
70-80								
60-70								
50-60								
40-50								
30-40								
20-30								
10-20								
<10(interruptions)								

Table 1.11   Presentation of results based on IEC 61000-4-11–2004 [36]

Residual Voltage [%]	Duration [s]				
	< 0.02	0.02-0.2	0.2-0.5	0.5-5	> 5
70-80					
40-70					
10-40					
≤10					

Table 1.12   Presentation of results based on NRS 048-2–2007 [45]

Range of residual voltage U [%]	Duration		
	$20 < t \le 150$ (ms)	$150 < t \le 600$ (ms)	$0.6 < t \le 3$ (s)
$90 > U \ge 85$			
$85 > U \ge 80$	Y		Z1
$80 > U \ge 70$			
$70 > U \ge 60$	X1	S	
$60 > U \ge 40$	X2		Z2
$40 > U \ge 0$	T		

frequent. The T-type area reflects close-up faults, which are not expected to happen too regularly and which a utility should specifically address if they become excessive. S-type dips are not as common as X- and Y-type events, but they may occur where impedance protection schemes are used or where voltage recovery is delayed. Z-type dips are very uncommon on HV systems (particularly $Z2$-type events), as they generally reflect problematic protection operation. These may be more common on MV systems.

1.3.6.3   System indices

System indices are calculated from the site indices of monitored sites. There are two alternatives:

1. to define system indices considering the value of the site index not exceeded by a pre-fixed percentage (for example, 95%) of the monitored sites,

2. to obtain system indices as a weighted average of the site indices for the monitored locations. The following belong to this second category:

   - the system average voltage dip amplitude, SAVDA;

   - the system average RMS variation frequency index for retained voltage 'X', SARFI-X.

The system average voltage dip amplitude is given by the following equation:

$$ \text{SAVDA} = \frac{\sum_{1}^{M} w_k \ \text{AVDA}_k}{\sum_{1}^{M} w_k} \ , \tag{1.29} $$

where M is the total number of monitored sites (with $M \leq N$, N being the number of system sites), w_k is the weighted factor of the kth site and AVDA_k is the average value of the residual dip voltage amplitude at the kth site.

The system average RMS variation frequency index for retained voltage 'X' is given by the following equation:

$$ \text{SARFI-X} = \frac{\sum_{1}^{M} w_k \text{RFI-}X_k}{\sum_{1}^{M} w_k} \ , \tag{1.30} $$

where $\text{RFI-}X_k$ is the RMS variation frequency index for residual voltage X at the kth site.

As usual for all system indices, weighting factors are introduced to take into account the difference in importance between different sites; the weighting factors are (in most cases) taken to be equal for all sites [16].

A further way to represent the system performance in terms of voltage dip is to consider an average voltage-dip table obtained from the voltage-dip tables of the individual sites. Each element of the table representing the system performance can be obtained by averaging the values of each element over the different sites. To take into account the different monitoring periods for different sites, before the averaging operation each element of the voltage-dip table of the individual site must be divided by the corresponding time observation period of the site.

It is useful to highlight that most of the site and system indices that have been introduced to quantify the severity of voltage dips can be extended, if they are properly modified, to also quantify the severity of voltage swells. For example, when considering the indices linked to specific voltage tolerance curves, these indices have to be calculated using the curve of the overvoltage condition instead of the curve of the undervoltage condition.

1.3.6.4   Existing objectives

In spite of the very large number of site and system indices proposed for dip characterization, at present there are no specific objectives in any international standards for this kind of disturbance.

In EN 50160–2000, objectives are given, but the objectives are expressed in general and vague terms; this standard states that 'under normal operating conditions, the expected number of voltage dips in a year may be from up to a few tens to up to one thousand.'

More specific objectives are in use in some countries, e.g. South Africa and Chile. The objectives proposed by South Africa's Standard NRS 048-2–2007 are given in Tables 1.13 and 1.14 and refer to 95% and 50% of the monitored sites [45]. The limits are expressed as a function of the system voltage level and the typology of dips. The dip classification is accomplished using the pair duration/amplitude data reported in Table 1.12. Similar objectives are given in the Chilean standard DS 327/97, but with a different dip classification.

Table 1.13   Characteristic number of voltage dips (95% of sites) according to NRS 048-2–2007

| Network voltage range (nominal voltage) | Number of voltage dips per year | | | | | |
| | Dip window category | | | | | |
	X1	X2	T	S	Z1	Z2
6.6–44 kV (rural)	85	210	115	400	450	450
6.6–44 kV	20	30	110	30	20	45
44–132 kV	35	35	25	40	40	10
220–765 kV	30	30	20	20	10	5

Table 1.14   Characteristic number of voltage dips (50% of sites) according to NRS 048-2–2007

| Network voltage range (nominal voltage) | Number of voltage dips per year | | | | | |
| | Dip window category | | | | | |
	X1	X2	T	S	Z1	Z2
6.6–44 kV (rural)	13	12	10	13	11	10
6.6–44 kV	7	7	7	6	3	4
44–132 kV	13	10	5	7	4	2
220–765 kV	8	9	3	2	1	1

In France, customers connected to MV distribution networks or to the transmission network can ask for customized engagements on the maximum number of voltage dips they might suffer per year. At a transmission level of 63 kV and above, the arrangement is five voltage dips per year. Only voltage dips deeper than 30% and longer than 600 ms are counted by the operator. At MV level, this engagement is determined depending on the local conditions of the site's alimentation. A customer connected at MV level cannot have an engagement of less than five voltage dips per year. As for transmission, only voltage dips deeper than 30% and longer than 600 ms are taken into account by the operator [20].

1.3.7 Transient overvoltages

In a similar manner to voltage dips, the indices used to characterize transient overvoltages can refer to:

- a single event (event indices);
- a site (site indices);
- a system or a part of it (system indices).

1.3.7.1 Event indices

As discussed in Section 1.2, a transient overvoltage can be classified as oscillatory or impulsive. An oscillatory transient event is described by its voltage peak, predominant frequency and decay time (duration). An impulsive transient event is described by rise time, decay time and peak value.

The characterization of transient events is, in most cases, only based on peak value and duration. The peak value of transient overvoltages is the highest absolute value of the voltage waveform; the duration is the amount of time during which the voltage is above a threshold. The choice of the threshold level will affect the value of the duration. The higher the threshold level, the lower the resulting value for the duration. A suitable choice for the threshold used to calculate the duration would be the same as that used to detect the transient.

The impact of an impulsive transient event can also be characterized by its energy content, defined as $E = \int [v(t)]^2 dt$, where $v(t)$ is the transient voltage as a function of time [53]. Several instruments are able to record this value. Using the transient energy content E, it is also possible to define another parameter characterizing the transient event: the equivalent duration of the transient [35]. This value is calculated by the ratio between the energy content E and the square of the transient voltage peak: $E/(V_{\text{peak}})^2$.

As shown for voltage dips, event indices can also be defined using voltage tolerance curves. The power quality index (PQI) is an example of this type of index. In the case of a voltage transient characterized by the coordinates (T, V_{peak}), the PQI index is given by:

$$\text{PQI} = \left| \frac{V_{\text{peak}} - 100}{V_{\text{CBEMA}}(T) - 100} \right| 100, \tag{1.31}$$

where $V_{\text{CBEMA}}(T)$ is the voltage on the CBEMA curve corresponding to duration T and V_{peak} is the peak voltage of the transient. The terms $V_{\text{CBEMA}}(T)$ and V_{peak} are expressed as percentages of the nominal voltage. A value of PQI higher than 100 indicates an unacceptable event for the CBEMA curve.

1.3.7.2   Site indices

A simple site index can be obtained by measuring how often the disturbance is characterized by magnitudes (voltage peaks) greater than some percentage X. This index is analogous to the RFI-X index used for voltage-dip characterization.

Site indices can also be obtained using voltage tolerance curves. Examples of these types of index are:

- the number of events outside the tolerance curves (CBEMA and ITIC);

- the PQI$_{Site}$ index.

Each transient can be represented on the voltage tolerance curves (CBEMA or ITIC). The point representing the transient event is characterized by an abscissa value equal to the transient duration (or transient equivalent duration) and an ordinate value equal to the transient voltage peak. The pattern of points shown in Figure 1.16 is a way of presenting the full record of the transient event. The number of points outside the voltage tolerance curve can be used to quantify the performance of the considered site in terms of transients.

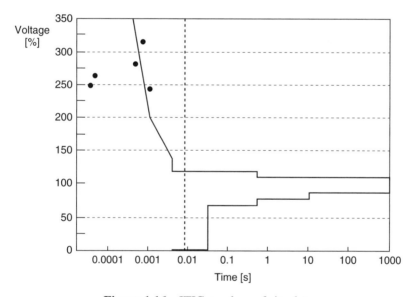

Figure 1.16   ITIC overlays of site data

The PQI$_{Site}$ index is obtained by summing all the PQIs of the considered site. (See Equation (1.28), in which the term dip is substituted by the term transient.)

Indices to characterize transients have also been proposed by the Canadian Electrical Association (CEA). The Canadian Power Quality Measurement Protocol introduces different indices: ITV (impulsive transient index), MFOT (medium-frequency transient index), LFOT (low-frequency transient index) and VRI (voltage rise index). The reader can refer to [51] for more details.

1.3.7.3   System indices

Just as with system indices that characterize voltage dips, transient system indices can be obtained by considering the value of the site index not exceeded by a fixed percentage (for example, 95%) of the monitored sites, or by a weighted average of the site indices for the monitored locations.

In [54] the authors have proposed the system average transient magnitude occurrence rate index$_{\text{Peak}}$, SATMORI$_X$, to characterize oscillatory transients in a given area of an electrical system; this system index is defined by the equation:

$$\text{SATMORI}_X = \frac{\sum N_i}{N_T},\tag{1.32}$$

where:

X is the specified peak magnitude threshold; typical values are 110%, 135%, 170% and 200%;

N_i is the number of customers experiencing transient overvoltages having magnitudes greater than X% due to measurement event i;

N_T is the number of customers served from the section of the system to be assessed.

This index provides both the utility and its customers with a measure of how often transient overvoltages exceeding the specified peak magnitude may occur.

1.3.7.4   Existing objectives

To our knowledge, suitable transient objectives are not reported in standard documents or guidelines.

1.3.8   Rapid voltage changes

Few indices are reported in the literature to characterize this type of disturbance. Frequently, a rapid voltage change is characterized by the following quantities: the 'steady-state voltage change' and the 'maximum voltage change.' The steady-state voltage change is the difference between the steady-state value reached after the change and the initial steady-state value; the maximum voltage change is the difference between the initial steady state and the lowest or highest voltage level during the event. Both quantities are typically expressed as percentages of the nominal voltage.

To date, the standards do not provide precise indications for identifying or measuring rapid voltage changes.

The existing objectives are expressed in terms of the steady-state voltage change. EN 50160–2000 suggests that, under normal operating conditions, a rapid voltage change on the LV electrical distribution system will generally not exceed 5% U_c, but a change of up to 10% U_c with a short duration might occur a few times per day. For an MV electrical distribution system, rapid voltage changes generally do not exceed 4% U_c, but short-duration changes of up to 6% U_c might occur a few times per day.

Table 1.15 shows the planning levels reported in IEC 61000-3-7–2008 for rapid voltage changes $\Delta V/V_n$ for infrequent events (expressed as percentages of the nominal value). These limits depend on the number of voltage changes in a given time period.

Table 1.15 Indicative planning level for rapid voltage changes according to IEC 61000-3-7–2008

Number of changes n	$\Delta V/V_n$ [%]	
	MV	HV/EHV
$n \leq 4$ per day	5-6	3-5
$n \leq 2$ per hour and n $>$ 4 per day	4	3
$2 < n \leq 10$ per hour	3	2.5

Table 1.16 Limits on rapid voltage changes according to the Norwegian regulation

Number of changes per day	$\Delta V/V_n$ [%]	
	0.23 kV $\leq U_n \leq 1$ kV	1 kV $< U_n$
1	10	6.0
Up to 24	5.0	4.0
More than 24	3.0	3.0

The objectives for the number of voltage changes for MV and HV customers given in the Norwegian regulations are shown in Table 1.16.

The Hydro-Quebec directive on voltage characteristic recommends that, under normal operating conditions, a rapid voltage change on the MV and LV electrical distribution systems will generally not exceed 8% of nominal voltage. In some specific environments, it may reach 10% of nominal voltage.

1.4 Conclusions

In this chapter the classifications, causes and effects of power quality disturbances have been analyzed. The indices most frequently used in standards, recommendations and guidelines for power system analysis or sizing electrical components have been shown and the objectives that should be met have been provided. Both site and system indices have been taken into account, even though objectives refer only to site indices. However, system indices are of great interest because they can be used in the new, liberalized market framework as a benchmark against which index values for different electrical systems, or for various parts of the same system, can be compared.

The main conclusion of the chapter is that an excessive number of indices have been applied in the past but, fortunately, nowadays a great deal of effort is exerted at international level to create uniform indices and procedures.

Acknowledgement

The authors wish to thank the International Electrotechnical Commission (IEC) for permission to reproduce Figure 1.11 from its International Standard IEC 61000-2-2 ed. 2 (2002). All such extracts are copyright of IEC, Geneva, Switzerland. All rights reserved. Further information on the IEC is available from www.iec.ch. IEC has no responsibility for the placement and context in which the extracts and contents are reproduced by the authors, nor is the IEC in any way responsible for the other content or accuracy herein.

References

[1] Arrillaga, J., Bollen, M. and Watson, N.R. (2000) 'Power Quality Following Deregulation', *Proceedings of the IEEE*, **88**(2), 246–261.
[2] Billinton, R. (1970) *Power System Reliability Evaluation*, New York (USA).
[3] Allan, R.N. and Billinton, R. (1988) *Reliability Assessment of Large Electric Power Systems*, Kluwer Academic Publishers.
[4] ANSI/IEEE Std C63.16–1993 (1993) *American National Standard Guide for Electrostatic Discharge Test Methodologies and Criteria for Electronic Equipment*, November.
[5] Rickem, L.W., Bridges, S.E. and Mileta, S. (1976) *EMP Radiation and Protective Techniques*, Wiley Interscience, New York.
[6] Longmire, C.L. (1978) 'On the Electromagnetic Pulse Produced by Nuclear Explosions', *IEEE Trans. on Antennas and Propagation*, **26**(1), 3–13.
[7] IEEE 1159 (1995) *Recommended Practice for Monitoring Electric Power Quality*, November.
[8] ANSI/IEEE Std C84.1 (1989) *American National Standards for Electric Power System and Equipment – Voltage Rating 60 Hz*, January.
[9] Bollen, M.H.J. (1999) *Understanding Power Quality Problems: Voltage Sags and Interruptions*, Wiley-IEEE Press, September.
[10] Bollen, M.H.J. and Yu-Hua Gu, I. (2006) *Signal Processing of Power Quality Disturbances*, IEEE PRESS/Wiley Interscience.
[11] EN 50160 (2000) *Voltage Characteristics of Electricity Supplied by Public Distribution Systems*, March.
[12] IEC 61000-4-30 (2003) *Electromagnetic Compatibility (EMC) – Part 4-30: Testing and Measurement Techniques – Power Quality Measurement Methods*, Edition 2, June.
[13] IEEE 1250 (1995) *Guide for Service to Equipment Sensitive to Momentary Voltage Disturbances*, June.
[14] IEC 61000-2-8 (2002) *Voltage Dips and Short Interruptions on Public Electric Power Supply Systems with Statistical Measurement Results*, Edition 1, November.
[15] IEC 61000-1-1 (1992) *Electromagnetic Compatibility (EMC) – Part 1: Application and Interpretation of Fundamental Definitions and Terms*, Edition 2, May.
[16] Joint Working Group Cigrè C4.07/Cired (2004)_ *Power Quality Indices and Objectives*, January, revised March.
[17] AS 2279.2 (1991) *Disturbances in Mains Supply Networks – Limitation of Harmonics Caused by Industrial Equipment*, January.
[18] IEEE 519 (1992) *Recommended Practices and Requirements for Harmonic Control in Electrical Power Systems*, May.
[19] IEC 61000-3-5 (1994) *Electromagnetic Compatibility (EMC) – Part 3-5: Limits – Limitation of Voltage Fluctuations and Flicker in Low Voltage Power Supply Systems for Equipment with Rated Current > 16 A*, Edition 1, December.
[20] EdF Emeraude Contract (1994) *Contrat Émeraude pour la Fourniture d'Electricité au Tarif Vert. Annexe 2: Qualité des Fournitures en HTA (1–50 kV) and HTB (>50 kV)*, Electricité de France.
[21] Canadian Electrical Association CEA 220 D 711 (1996) *Power Quality Measurement Protocol: CEA Guide to Performing Power Quality Surveys*, First edition, March.

[22] IEC 61000-3-8 (1997) *Electromagnetic Compatibility (EMC) – Part 3-8: Limits – Signalling on Low Voltage Electrical Installations – Emission Levels, Frequency Bands, and Electromagnetic Disturbance Levels*, Edition 1, August.

[23] IEEE 493 (1997) *Recommended Practices for the Design of Reliable Industrial and Commercial Power Systems, also known as the Gold Book*, December.

[24] IEEE 1346 (1998) *Recommended Practice for Evaluating Electric Power System Compatibility with Electronics Process Equipment*, July.

[25] IEC 61000-3-4 (1998) *Electromagnetic Compatibility (EMC) – Part 3-4: Limits – Limitation of Emission of Harmonic Currents in Low-Voltage Power Supply Systems for Equipment with Rated Current > 16 A*, Edition 1, October.

[26] Arrillaga, J., Watson, N.R. and Chen, S. (2000) *Power System Quality Assessment*, John Wiley & Sons, Ltd.

[27] Hydro-Quebec (2001) *Characteristics and Target Values of the Voltage Supplied by Hydro-Quebec Medium and Low Voltage Systems*, June, available at: www.hydroquebec.com.

[28] IEC 61000-2-2 (2002) *Electromagnetic Compatibility (EMC) – Part 2-2: Environment – Compatibility Levels for Low-Frequency Conducted Disturbances and Signalling in Public Low Voltage Power Supply Systems*, Edition 2, March.

[29] Sabin, D.D. (2002) Analysis of Harmonic Measurements Data, *IEEE Power Engineering Society Summer Meeting*, Chicago (USA), July.

[30] IEC 61000-4-7 (2002) *Electromagnetic Compatibility (EMC) – Part 4-7: Testing and Measurement Techniques – General Guide on Harmonics and Interharmonics Measurements and Instrumentation for Power Supply Systems and Equipment Connected Thereto*, Edition 2, August.

[31] Dugan, R.C., McGranaghan, M.F. and Beaty, H.W. (2002) *Electrical Power Systems Quality*, McGraw-Hill Professional, November.

[32] IEC 61000-4-15 (2003) *Electromagnetic Compatibility (EMC) – Part 4-15: Testing and Measurement Techniques – Flickermeter – Functional and Design Specifications*, Edition 1.1, February.

[33] IEC 61000-2-12 (2003) *Electromagnetic Compatibility (EMC) – Part 2-12: Environment Compatibility Levels for Low-Frequency Conducted Disturbances and Signalling in Public Medium-Voltage Power Supply Systems*, Edition 1.0, April.

[34] Zheng, T., Makram, E.B. and Girgis, A.A. (2003) 'Evaluating Power System Unbalance in the Presence of Harmonic Distortion, *IEEE Transactions on Power Delivery*, **18**(2), 393–397.

[35] Herath, C., Gosbell, V., Perera, S. and Robinson, D. (2003) A Transient Index for Reporting Power Quality Surveys, *17th International Conference on Electrical Distribution*, Barcelona (Spain), May.

[36] IEC 61000-4-11 (2004) *Electromagnetic Compatibility (EMC) – Part 4-11: Testing and Measurement Techniques – Voltage Dips, Short Interruptions, and Voltage Variations Immunity Tests*, Edition 2, March.

[37] IEEE P1564/D6 (2004) *Recommended Practice for the Establishment of Voltage Sag Indices* (Draft 6), IEEE Power Engineering Society, May.

[38] Reg. No. 1557 (2004) *Regulations Relating to the Quality of Supply in the Norwegian Power System*, November, available at: http://www.new.no.

[39] Herath, H.M.S.C., Gosbell, V.J. and Perera, S. (2005) 'Power Quality (PQ) Survey Reporting: Discrete Disturbance Limits', *IEEE Transactions on Power Delivery*, **20**(2), 851–858.

[40] Mastandrea, I., Chiumeo, R., Carpinelli, G. and De Martinis, U. (2005) *Analisi di Nuovi Indici per la Qualità del Servizio*, CESI RICERCA Rds (Ricerche di Sistema), Report A5-029554, June, available at: http://www.ricercadisistema.it.

[41] IEC 61000-3-2 (2005) *Electromagnetic Compatibility (EMC) – Part 3-2: Limits – Limits for Harmonic Current Emissions (Equipment Input Current ≤ 16 A per Phase)*, Edition 3, November.

[42] Council of European Energy Regulators (CEER), Working Group on Quality of Electricity Supply (2005) *Third Benchmarking Report on Quality of Electricity Supply*, Ref: C05-QOS-01-03, December.

[43] ERGEG (2006) *Towards Voltage Quality Regulation in Europe*. ERGEG Public Consultation Paper, Ref: E06-EQS-09-03, December.

[44] IEEE Task Force on Harmonic Modeling and Simulation (2007) 'Interharmonics: "Theory and Modeling"', *IEEE Transactions on Power Delivery*, **22**(4), 2335–2348.

[45] NRS 048-2 (2007) *Electricity Supply – Quality of Supply. Part 2: Voltage Characteristics, Compatibility Levels, Limits and Assessment Methods.*

[46] IEC 61000-3-6 (2008) *Electromagnetic Compatibility (EMC) – Part 3-6: Limits – Assessment of Emission Limits for the Connection of Distorting Installation to MV, HV and EHV Power Systems*, Edition 2, February.

[47] IEC 61000-3-7 (2008) *Electromagnetic Compatibility (EMC) – Part 3-7: Limits – Assessment of Emission Limits for the Connection of Fluctuating Installations to MV, HV and EHV Power Systems*, Edition 2, February.

[48] Bollen, M.H.J. and. Verde, P. (2008) 'A Framework for Regulation of RMS Voltage and Short-Duration Under and Overvoltages', *IEEE Transactions on Power Delivery*, **23**(4), 2015–2112.

[49] Kingham, B. (2008) National Quality of Supply Standards: Is EN 50160 the Answer?, *17th Conference of Electrical Power Supply Industry*, Macau, October.

[50] Ente Nacional Regulador de la Electricidad (enre) (no date) Controlando la Calidad del Servicio Elèctrico, available at: http://www.enre.gov.ar.

[51] Canadian Electricity Association (CEA) (no date) website: http://www.canelect.ca.

[52] IEEE 100 (2000) *The Authoritative Dictionary of IEEE Standards Terms*, December.

[53] ANSI/IEEE Std C62.41 (1991) *IEEE Recommended Practice on Surge Voltages in Low Voltage AC Power Circuits*, October.

[54] Sabin, D.D., Grebe, T.E., Brooks, D.L. and Sundaram, A. (1999) 'Rules-Based Algorithm for Detecting Transient Overvoltages due to Capacitor Switching and Statistical Analysis of Capacitor Switching in Distribution Systems', *IEEE Transmission and Distribution Conference*, New Orleans (USA), April.

[55] Gallo, D., Langella, R., Testa, A., (2002) 'On the Effects on MV/LV Component Expected Life of Slow Voltage Variations and Harmonic Distortion', 9th *International Conference on Harmonic and Quality of Power*, Rio de Janeiro (Brazil), October.

[56] Caramia, P., Carpinelli, G., Russo, A., Verde, P., (2003) 'Some Considerations on Single Site and System Probabilistic Harmonic Indices for Distribution Networks', *IEEE Power Engineering Society General Meeting*, Toronto, Ontario (Canada), July.

2

Assessing responsibilities between customer and utility

2.1 Introduction

It is well known that the degradation of power quality (PQ) can originate simultaneously at several unknown points in a power system. Therefore, it is not easy to share the responsibility for PQ degradation at the point of common coupling (PCC) between the utility company and a customer installation; this problem has been the subject of several publications in the relevant literature.

In the most general case, the indices providing such sharing information should be able:

1. to correctly identify the presence of non-ideal loads (disturbing loads) at the PCC; and

2. to properly quantify the contributions of both the customer installation and the utility to the total PQ level at the PCC.

Obviously the indices should be easy to measure and progressive, i.e. they should increase with the disturbance level introduced either by the customer or the utility.

Not all publications in the literature, however, have addressed one or both of the requirements 1 and 2 above. Some papers, for example, deal only with the problem of locating the PQ disturbance source, since they consider that when the source is identified, it is possible to assign responsibility and take suitable measures to mitigate the problem. New indices or methodologies that should be useful in identifying the so-called 'prevailing source' of disturbance are also proposed; the prevailing source is the side that contributes more to the PQ disturbance level at the PCC.

In this chapter some of the above-mentioned indices and methodologies will be presented, because they provide useful background information about the difficult, yet open, field of research dealing with responsibility assessment. Voltage unbalances and waveform distortions are considered first, followed by analyses of voltage fluctuations, voltage sags and transients. Voltage unbalances and waveform distortions are addressed together, since some

Power Quality Indices in Liberalized Markets   Pierluigi Caramia, Guido Carpinelli and Paola Verde
© 2009 John Wiley & Sons, Ltd

publications have proposed indices and methodologies that can be applied to both; in addition, we deal with the cases of a single metering section and of distributed-measurement systems.

2.2   Waveform distortions and voltage unbalances: indices based on a single metering section

Initially, for the sake of simplicity, we consider a single metering section at the PCC where a customer installation is fed by the utility supply system (Figure 2.1).

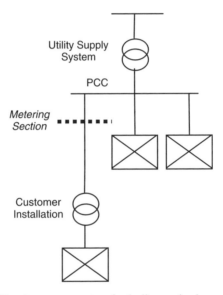

Figure 2.1   Simple power system including a single metering section

The terminals are accessible for voltage and current measurements. We assume that Fourier transform instrumentation is used to obtain voltage and current spectral components; inaccuracies in the spectral assessment, such as those due to spectral leakage, will be analyzed in depth in Chapter 3.

The presence of direct voltage and direct current is neglected; in fact, significant direct voltage and direct current terms of the distorted waveforms are rarely present in AC power systems.

Three-phase three-wire systems are considered to be the base case, since the indices of concern are defined frequently with reference to these systems; brief notes about four-wire systems will also be reported.

The indices are classified into three groups:

1. indices based on harmonic impedances [1–17];

2. indices based on powers in non-ideal conditions [18–36]; and

3. indices based on the comparison with an ideal linear load [28, 34, 37–45].

Some indices can be applied only for waveform distortions while others can be used for both waveform distortions and voltage unbalances.

2.2.1 Indices based on harmonic impedances

Generally, these indices are based on knowledge of the harmonic impedances of both the utility supply system and the customer installation seen from the PCC. They try to share the utility and customer contributions to the presence of waveform distortions.

Most of the approaches proposed in the relevant literature use the Norton model to represent both the utility and the customer in the frequency domain; a few publications include analyses based on Thevenin equivalent circuits or mixed (Norton and Thevenin) representations. In the next section some Norton-based approaches are recalled; subsequently, other representations are taken into account.

2.2.1.1 Norton-based approaches

Assuming a three-phase balanced power system, Figure 2.2 (a) shows the equivalent circuit of the system for the study of the harmonic of order h; the utility supply system and the customer installation are represented by their Norton equivalent circuits. In Figure 2.2 (a):

- $\bar{J}_{uh}$ is the equivalent harmonic current source representing the utility (background) distortion;

- $\bar{J}_{ch}$ is the equivalent harmonic current source representing the customer disturbing loads;

- $\dot{Z}_{uh}$ is the equivalent harmonic impedance of the utility system;

- $\dot{Z}_{ch}$ is the equivalent harmonic impedance of the customer installation;

- $\bar{V}_h, \bar{I}_h$ are the harmonic voltage and current measured at the PCC, respectively.

All quantities are complex and generally can also be referred to as interharmonic components.

Neglecting harmonic interactions, the principle of superposition can be applied to quantify the responsibilities of the utility and customer to the current and voltage distortions at the PCC.

By opening the equivalent harmonic current source $\bar{J}_{uh}$ representing the background distortion in Figure 2.2 (a), we obtain Figure 2.2 (b):

$$\bar{I}_{ch} = \frac{\dot{Z}_{ch}}{\dot{Z}_{uh} + \dot{Z}_{ch}} \bar{J}_{ch}, \qquad \bar{V}_{ch} = \frac{\dot{Z}_{uh}\dot{Z}_{ch}}{\dot{Z}_{uh} + \dot{Z}_{ch}} \bar{J}_{ch}. \qquad (2.1)$$

From the equivalent circuit in Figure 2.2 (a), the result is:

$$\bar{V}_h = \dot{Z}_{ch}(\bar{J}_{ch} - \bar{I}_h), \qquad (2.2)$$

and then:

$$\dot{Z}_{ch}\bar{J}_{ch} = \bar{V}_h + \dot{Z}_{ch}\bar{I}_h. \qquad (2.3)$$

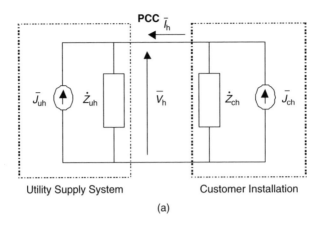

Utility Supply System   Customer Installation

(a)

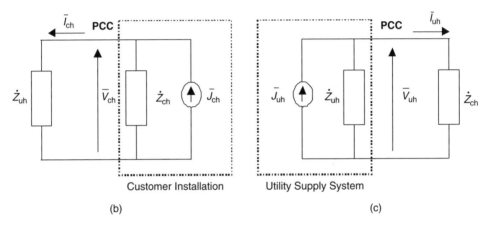

Customer Installation   Utility Supply System

(b)   (c)

Figure 2.2   Norton equivalent circuits of the system under study

By substituting Equation (2.3) into Equation (2.1), the following relationships are obtained:

$$\bar{I}_{ch} = \frac{(\overline{V}_h + \dot{Z}_{ch}\bar{I}_h)}{\dot{Z}_{uh} + \dot{Z}_{ch}}, \quad \overline{V}_{ch} = \frac{\dot{Z}_{uh}(\overline{V}_h + \dot{Z}_{ch}\bar{I}_h)}{\dot{Z}_{uh} + \dot{Z}_{ch}}, \tag{2.4}$$

which express the harmonic current $\bar{I}_{ch}$ and the harmonic voltage $\overline{V}_{ch}$ versus the harmonic voltage/current at the PCC and the equivalent impedances of both the utility supply system and the customer installation. If $Z_{ch} = |\dot{Z}_{ch}| \rightarrow \infty$, then $\overline{V}_{ch} \rightarrow \dot{Z}_{uh} \bar{I}_h$, and $\bar{I}_{ch} \rightarrow \bar{I}_h$ (symbol $|\dot{X}|$ means the magnitude of the complex quantity $\dot{X}$).

Initially, we define the harmonic voltage $\overline{V}_{ch}$ in Equation (2.4) as the *harmonic voltage emission level* and the harmonic current $\bar{I}_{ch}$ in Equation (2.4) as the *harmonic current emission level*. In practice, the harmonic voltage emission level is the harmonic voltage that would be caused by a customer installation at the PCC if the utility supply system was without background distortion; the harmonic current emission level is the harmonic current that would be injected by a customer installation at the PCC under the same condition [11]. At the end of this section we will consider a slightly different definition of the emission levels according to the recent IEC 61000-3-6–2008 [46].

The harmonic voltage and current emission levels depend on four complex quantities. Two of these ($\overline{V}_h$, $\overline{I}_h$) are directly measured at the PCC; in the next section, some techniques for calculating the remaining two ($\dot{Z}_{uh}$, $\dot{Z}_{ch}$) will be shown.

By opening the equivalent harmonic current source $\overline{J}_{ch}$ representing the customer disturbing loads in Figure 2.2 (a), we obtain Figure 2.2 (c):

$$\overline{I}_{uh} = \frac{\dot{Z}_{uh}}{\dot{Z}_{uh} + \dot{Z}_{ch}}\overline{J}_{uh}, \quad \overline{V}_{uh} = \frac{\dot{Z}_{ch}\dot{Z}_{uh}}{\dot{Z}_{uh} + \dot{Z}_{ch}}\overline{J}_{uh}. \tag{2.5}$$

From the equivalent circuit in Figure 2.2 (a), the result is:

$$\overline{V}_h = \dot{Z}_{uh}(\overline{J}_{uh} + \overline{I}_h), \tag{2.6}$$

and then:

$$\dot{Z}_{uh}\overline{J}_{uh} = \overline{V}_h - \dot{Z}_{uh}\overline{I}_h. \tag{2.7}$$

By substituting Equation (2.7) into Equation (2.5) the following relationships are obtained:

$$\overline{I}_{uh} = \frac{(\overline{V}_h - \dot{Z}_{uh}\overline{I}_h)}{\dot{Z}_{uh} + \dot{Z}_{ch}}, \quad \overline{V}_{uh} = \frac{\dot{Z}_{ch}(\overline{V}_h - \dot{Z}_{uh}\overline{I}_h)}{\dot{Z}_{uh} + \dot{Z}_{ch}}, \tag{2.8}$$

which express the harmonic current $\overline{I}_{uh}$ and the harmonic voltage $\overline{V}_{uh}$ versus the harmonic voltage/current at the PCC and the equivalent impedances of both the utility supply system and the customer installation.

We define the harmonic voltage $\overline{V}_{uh}$ in Equation (2.8) as the *harmonic voltage utility contribution* and the harmonic current $\overline{I}_{uh}$ in Equation (2.8) as the *harmonic current utility contribution*. In practice, the harmonic voltage utility contribution is the harmonic voltage that would be caused by the utility at the PCC if the customer did not have disturbing loads; the harmonic current utility contribution is the harmonic current that would be caused by the utility at the PCC under the same conditions.

The harmonic voltage and current utility contributions depend on the same four unknown complex quantities as the harmonic voltage and current emission levels.

Obviously, the measured harmonic voltage $\overline{V}_h$ and the harmonic current $\overline{I}_h$ at the PCC are given by (Figure 2.3):

$$\overline{I}_h = \overline{I}_{ch} - \overline{I}_{uh}, \quad \overline{V}_h = \overline{V}_{ch} + \overline{V}_{uh}, \tag{2.9}$$

with the harmonic voltage and current phasor contributions given by Equations (2.4) and (2.8).

Since there can be some ambiguities in quantifying the responsibilities of the customer and the utility using the phasors in Equations (2.4) and (2.8), it is useful to refer to scalar quantities. For example, referring to the customer responsibility, the harmonic voltage emission level index (HVELI) and the harmonic current emission level index (HCELI) can be introduced to quantify the customer contribution to the whole voltage and current distortionp at the PCC; these indices are defined as:

$$\text{HVELI} = |\overline{V}_{ch}| = \left|\frac{\dot{Z}_{uh}(\overline{V}_h + \dot{Z}_{ch}\overline{I}_h)}{\dot{Z}_{uh} + \dot{Z}_{ch}}\right|, \quad \text{HCELI} = |\overline{I}_{ch}| = \left|\frac{(\overline{V}_h + \dot{Z}_{ch}\overline{I}_h)}{\dot{Z}_{uh} + \dot{Z}_{ch}}\right| \tag{2.10}$$

where symbol $|\overline{X}|$ means the magnitude of the phasor $\overline{X}$.

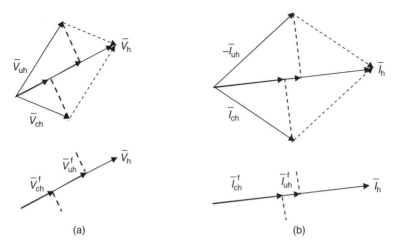

Figure 2.3 Phasor decompositions and projections: (a) harmonic voltages; (b) harmonic currents

Analogous indices could be introduced easily to quantify the contribution of the utility to the whole current and voltage distortion at the PCC (harmonic voltage utility contribution index, HVUCI, and harmonic current utility contribution index, HCUCI).

Once the utility and customer contributions are known, we can also detect the dominant (prevailing) harmonic source, i.e. the side that generates the larger contribution to the voltage or current harmonic distortions at the PCC. For example, the utility has the prevailing contribution in terms of harmonic current if HCUCI > HCELI; the customer has the prevailing contribution if HCUCI < HCELI.

An alternative proposal has been put forward to quantify the utility and the customer responsibilities using the projection phasors of the harmonic voltage or current onto the harmonic voltage $\overline{V}_h$ or current $\overline{I}_h$ at the PCC (Figure 2.3) [13]; in this way, the resulting amplitude of the harmonic voltage and current at the PCC is decomposed into two contributions, one due to the customer and the other due to the utility. When the customer and the utility contributions ($\overline{V}_{ch}^f$, $\overline{V}_{uh}^f$ and $\overline{I}_{ch}^f$, $\overline{I}_{uh}^f$) have the same arguments, they add with a positive sign to form the harmonic voltage or current at the PCC. Both sources should be discouraged. When one contribution has a π-displacement with reference to the voltage/current at the PCC, it partially compensates the other contribution and has the (positive) effect of reducing the resulting harmonic.

Example 2.1   In this example we calculate the harmonic current and voltage utility contributions, the harmonic current and voltage emission levels and the consequent indices . The results of this example have been obtained by a time domain simulation of the test system shown in Figure 2.14 (b).

Let us consider the fifth harmonic; the equivalent harmonic impedances of the customer and of the utility supply system are known in advance and have the following values:

$$\dot{Z}_{u5} = 0.219 + j10.945\ [\Omega]; \quad \dot{Z}_{c5} = 793.972 + j69.182\ [\Omega].$$

The measured harmonic voltage and current harmonics at the PCC are given by:

$$\overline{V}_5 = 75.122 - j56.754 = 94.150 \exp(-j0.647) \text{ [V]},$$
$$\overline{I}_{c5} = -0.440 - j5.255 = 5.273 \exp(-j1.654) \text{ [A]}.$$

Applying Equation (2.4), the harmonic voltage and current emission levels are:[1]

$$\overline{V}_{c5} = 58.140 - j5.809 = 58.429 \exp(-j0.099) \text{ [V]},$$
$$\overline{I}_{c5} = -0.424 - j5.320 = 5.337 \exp(-j1.650) \text{ [A]}.$$

Applying Equation (2.8), the harmonic voltage and current utility contributions are:

$$\overline{V}_{u5} = 16.981 - j50.944 = 53.700 \exp(-j1.249) \text{ [V]},$$
$$\overline{I}_{u5} = 0.016 - j0.065 = 0.067 \exp(-j1.336) \text{ [A]}.$$

Then, the result is:

$\text{HVELI} = 58.429 \text{ [V]}, \quad \text{HCELI} = 5.337 \text{ [A]}, \quad \text{HVUCI} = 53.700 \text{ [V]}, \quad \text{HCUCI} = 0.067 \text{ [A]}.$

It is interesting to note that the harmonic current utility contribution is much less than the harmonic current emission level due to the great difference between the equivalent harmonic impedances of the customer disturbing loads and of the utility supply system.

It is also interesting to note that the customer is the prevailing source, both in terms of harmonic current (HCELI > HCUCI) and in terms of harmonic voltage (HVELI > HVUCI).

The projections of the harmonic voltages and currents result in:

$$\overline{V}_{u5}^{f} = 44.259 \exp(-j0.647) \text{[V]}, \quad \overline{V}_{c5}^{f} = 49.891 \exp(-j0.647) \text{ [V]},$$

$$\overline{I}_{u5}^{f} = 0.064 \exp(j1.487) \text{[A]}, \quad \overline{I}_{c5}^{f} = 5.337 \exp(-j1.654) \text{ [A]},$$

with similar conclusions.

As previously shown, the relationships defining the indices require knowledge of the measured voltage and current waveforms at the PCC as well as the harmonic impedances of both the utility supply system and the customer installation as seen from the PCC.

In the next section some techniques are illustrated for calculating the impedances. Then, some other methods for the evaluation of the indices are shown which, in some cases, partially or totally eradicate the need for knowledge of the customer and utility harmonic impedances. A technique based on reference (or contract) impedances instead of the actual impedances is also shown in the third section below; this technique tries to avoid assigning responsibility to either the customer or the utility when the variations are located on the other side of the PCC.

a) Some methods for harmonic impedance evaluation

The calculation of the harmonic impedances has been of great interest and the object of several publications; several methods have been developed, mainly with reference to the evaluation of network harmonic impedance (online invasive and noninvasive methods). Transient-based and steady-state-based methods have been proposed and analyzed critically; transient methods are based on the injection of a transient disturbance (for example due to a

---

[1] It should be noted that some slight numerical inaccuracies can arise in all the numerical examples of this chapter due to cipher truncation.

capacitor switching) while steady-state methods are based on recording voltage and current harmonics at one or more different steady-state operating points. In the following, without any aim to be exhaustive, our interest is concentrated only on some techniques for the calculation of the harmonic impedances of both the utility supply system and the customer installation as seen from the PCC, with the aim of outlining the main problems encountered in the impedance evaluation.

A technique based on the sign of the real part of the ratio $\Delta \overline{V}_h / \Delta \overline{I}_h$

Let's consider the equivalent circuit in Figure 2.2 (a) and a measuring time interval during which the system goes from steady-state condition 1 to steady-state condition 2; applying Equations (2.2) and (2.6), the harmonic voltage $\overline{V}_h(i)$ $i=1,\ 2$ at the PCC can be calculated as:

$$\overline{V}_h(i) = \dot{Z}_{ch}(i)\left[\overline{J}_{ch}(i) - \overline{I}_h(i)\right], \quad \overline{V}_h(i) = \dot{Z}_{uh}(i)\left[\overline{J}_{uh}(i) + \overline{I}_h(i)\right]. \tag{2.11}$$

Assuming that $\dot{Z}_{uh}$, $\dot{Z}_{ch}$ and the phase angle of the fundamental voltage are constant during the time interval, we are able to analyze the conditions 1 and 2 below:

1. $\overline{J}_{uh}$ remains constant during the time interval while the equivalent harmonic current source representing the customer $\overline{J}_{ch}$ (and then $\overline{I}_h$) varies.
   Considering the second Equation of (2.11), we have:

   $$\frac{\Delta \overline{V}h}{\Delta \overline{I}_h} = \frac{\overline{V}_h(1) - \overline{V}_h(2)}{\overline{I}_h(1) - \overline{I}_h(2)} = \dot{Z}_{uh}\frac{\overline{I}_h(1) - \overline{I}_h(2)}{\overline{I}_h(1) - \overline{I}_h(2)} = \dot{Z}_{uh}. \tag{2.12}$$

2. $\overline{J}_{ch}$ remains constant during the time interval while the equivalent harmonic current source representing the utility system $\overline{J}_{uh}$ (and then $\overline{I}_h$) varies.
   Considering the first Equation of (2.11), the result is:

   $$\frac{\Delta \overline{V}h}{\Delta \overline{I}_h} = \frac{\overline{V}_h(1) - \overline{V}_h(2)}{\overline{I}_h(1) - \overline{I}_h(2)} = \dot{Z}_{ch}\frac{\overline{I}_h(2) - \overline{I}_h(1)}{\overline{I}_h(1) - \overline{I}_h(2)} = -\dot{Z}_{ch}. \tag{2.13}$$

Due to the physical constitution of impedances $\dot{Z}_{uh}$ and $\dot{Z}_{ch}$, their real parts must be positive. Then, if the real part of $\Delta \overline{V}_h / \Delta \overline{I}_h$ is positive, this ratio will be considered the utility system impedance $\dot{Z}_{uh}$; if the real part of $\Delta \overline{V}_h / \Delta \overline{I}_h$ is negative, this ratio will be considered the opposite of the customer impedance $\dot{Z}_{ch}$. Once the harmonic impedances are known, the indices of interest can be calculated. If the procedure is repeated for a long period during harmonic fluctuations, the statistics of the indices can be determined and, for example, an averaging of the successive results can be conducted.

One drawback of the above method is that, in order to get good precision, it is mandatory that simultaneous variations in the utility and customer conditions be avoided; in fact, interference effects arise and these must be minimized to obtain a good estimate of the impedances. In practice, rather short time intervals between successive measurements should be needed; in some cases, however, the voltage and current harmonics at the PCC do not change significantly within a short time interval, and the change could be less than the measurement error. In such a case, the time interval should be increased in order to ensure that the measured quantities vary more than the measurement instrumentation error. (This causes obvious contrasts in the objectives). The problem of the influence of simultaneous harmonic variations will be apparent in Example 2.2; a solution to this problem will also be analyzed.

In addition, the accuracy of the method can be affected significantly by the validity of the hypothesis concerning the constant phase angle of the fundamental voltage during the considered time interval; in fact, the voltage and current waveforms are recorded at different conditions at different times, with the consequence that in actual systems an angular difference between the fundamental voltages (the reference frame in each condition) in the two conditions generally occurs.[2] As is well known, small variations in the fundamental phase angle generate high variations in harmonic phase angles (mainly at higher orders), which can cause unacceptable inaccuracies in the evaluation of impedance.

Special attention should be paid anyway to the evaluation of voltage and current harmonics, with particular reference to their arguments, mainly when the harmonic levels are not significant.

Example 2.2   In this example the fifth harmonic impedance of the utility system $\dot{Z}_{u5} = 0.219 + j10.945$ [Ω] (see Example 2.1) is calculated by applying Equation (2.12).

Let's assume that in a measured time interval during which the system goes from steady-state condition 1 to steady-state condition 2 (due to a significant change of customer harmonic injection), the following values of the harmonic voltage and current at the PCC are measured:

$$\overline{V}_5(1) = 75.122 - j56.754 \ [V]; \quad \overline{I}_5(1) = -0.440 - j5.255 \ [A];$$
$$\overline{V}_5(2) = 47.224 - j49.732 \ [V]; \quad \overline{I}_5(2) = 0.145 - j2.695 \ [A].$$

We have:

$$\frac{\Delta\overline{V}_5}{\Delta\overline{I}_5} = \frac{\overline{V}_5(1) - \overline{V}_5(2)}{\overline{I}_5(1) - \overline{I}_5(2)} = 0.240 + j10.952 \ [\Omega].$$

The real part of the ratio $\Delta\overline{V}_5/\Delta\overline{I}_5$ is positive, and so the ratio is considered the utility supply system impedance $\dot{Z}_{u5}$.

It should be noted that the considered technique gives a result near the true value but not coincident to it. This is due to the hypothesis that $\overline{J}_{u5}$ remains constant during the time interval when the equivalent harmonic current source representing the customer $\overline{J}_{c5}$ (and then $\overline{I}_5$) varies. Obviously, when the customer condition changes, the electrical state of the system also changes and the equivalent harmonic current source representing the background distortion can change in turn.

---

[2] Generally, in each condition the reference frame for the harmonic voltages and currents is their own fundamental voltage; if the phase angle of the fundamental voltage does not remain constant during two different conditions, it follows that the angular difference between the fundamental voltages of the two conditions should be taken into account to assess correctly the reference for all voltage and current harmonics. Some authors suggest that the Thevenin fundamental voltage be considered the common reference frame and that this voltage should be assumed to be constant during the considered interval. (The phase shift between the fundamental voltage at the PCC and the Thevenin fundamental voltage is calculated with proper analytical expressions).

In fact, the time-domain simulation of the test system demonstrates the following variation in the equivalent harmonic current source representing the utility system:

$$\overline{J}_{u5}(1) = -4.606 - j1.709[\text{A}], \quad \overline{J}_{u5}(2) = -4.601 - j1.709 \ [\text{A}].$$

Taking into account the above slight variation in $\overline{J}_{u5}$, we have (see Equation (2.11)):

$$\dot{Z}_{u5} = \frac{\Delta\overline{V}_5}{\Delta\overline{I}_5 + \Delta\overline{J}_{u5}} = 0.219 + j10.948 \ [\Omega],$$

which is a result practically coincident with the true value.

A technique based on the hypothesis that $|\dot{Z}_{uh}| << |\dot{Z}_{ch}|$: the step method

As previously demonstrated, it is desirable to avoid simultaneous variations in the utility and customer conditions in order to get good precision in the impedance evaluation. However, it is obvious that, in actual power systems, simultaneous variations may occur, making it difficult to discriminate between them in the impedance evaluation. Generally, it is difficult to determine the origin of fluctuations with certainty, so a method has been proposed that helps to better determine the conditions for the impedance evaluation [5].

The method relies mainly on the assumption that the magnitude of the customer impedance is frequently much greater than that of the utility system, in practical cases at nonresonance frequencies. When $|\dot{Z}_{uh}| << |\dot{Z}_{ch}|$, a variation in the customer equivalent harmonic current source $\overline{J}_{ch}$ causes a variation in the harmonic current at the PCC that is much greater than that caused by the same variation in the utility equivalent harmonic current source $\overline{J}_{uh}$ (see Equations (2.1) and (2.5)); obviously, this effect becomes even greater if the considered customer equivalent harmonic current source variation is greater than that of the utility. In practice, the greater (lower) the harmonic current variations at the PCC are, the higher is the probability that the variations are caused by the customer (utility) and they can be used to calculate $\dot{Z}_{uh}(\dot{Z}_{ch})$. This is the principle used to determine the side from which the changes in currents and voltages mainly originate.

Figure 2.4 helps in understanding this principle. Let's consider M harmonic current and voltage variations at the PCC. The horizontal axis gives the sum $\Sigma\Delta I_{5i}$ ($i = 1, \ldots, M$) of the 5th harmonic current variations at the PCC arranged from smaller to larger values, while the vertical axis gives the corresponding sum $\Sigma\Delta V_{5i}$ ($i = 1, \ldots, M$) of the 5th harmonic voltage variations at the PCC. A point X can be seen in Figure 2.4, and this point shares the curve in two zones with different slopes. The left zone has a much greater slope than the right zone.

Considering the harmonic impedance values, we can conclude that the slope on the right side may be related to the equivalent harmonic impedance of the supply system $\dot{Z}_{uh}$ while the slope on the left side may be related to the equivalent harmonic impedance of the customer loads $\dot{Z}_{ch}$.

In particular, from Equation (2.11) and considering harmonic current and voltage variations, the result is:

$$\Delta\overline{V}_h = \dot{Z}_{ch}\left(\Delta\overline{J}_{ch} - \Delta\overline{I}_h\right), \quad \Delta\overline{V}_h = \dot{Z}_{uh}\left(\Delta\overline{J}_{uh} + \Delta\overline{I}_h\right). \tag{2.14}$$

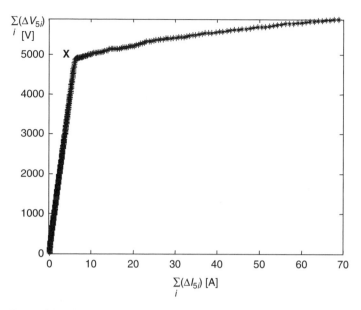

Figure 2.4 Sum of the 5th harmonic voltage variations versus the sum of harmonic current variations

By introducing two complex interference coefficients $\overline{\sigma}_{ch}$ and $\overline{\sigma}_{uh}$ in Equation (2.14), we obtain:

$$\frac{\Delta \overline{V}_h}{\Delta \overline{I}_h} = \dot{Z}_{ch}\left(\frac{\Delta \overline{J}_{ch}}{\Delta \overline{I}_h} - 1\right) = \dot{Z}_{ch}(\overline{\sigma}_{ch} - 1) \ , \quad \frac{\Delta \overline{V}_h}{\Delta \overline{I}_h} = \dot{Z}_{uh}\left(\frac{\Delta \overline{J}_{uh}}{\Delta \overline{I}_h} + 1\right) = \dot{Z}_{uh}(\overline{\sigma}_{uh} + 1).$$

$$(2.15)$$

A threshold value of the harmonic current variation amplitude $\Delta I_{h,cv}$ is derived in relation to the maximum change at X in Figure 2.4; mean values of the equivalent harmonic impedances $\dot{Z}_{uh}$ and $\dot{Z}_{ch}$ can be then obtained as:

$$\frac{1}{M_1}\sum_{i=1}^{M_1}\frac{\Delta \overline{V}_h(i)}{\Delta \overline{I}_h(i)} = \dot{Z}_{ch}\left(\frac{1}{M_1}\sum_{i=1}^{M_1}\overline{\sigma}_{ch}(i) - 1\right) \cong -\dot{Z}_{ch}$$

under the conditions that:

$$|\Delta \overline{I}_h(i)| < \Delta I_{h,cv} \quad \text{and} \quad \mathrm{Re}\left\{\frac{\Delta \overline{V}_h(i)}{\Delta \overline{I}_h(i)}\right\} < 0 \qquad (2.16)$$

and:

$$\frac{1}{M_2}\sum_{i=1}^{M_2}\frac{\Delta \overline{V}_h(i)}{\Delta \overline{I}_h(i)} = \dot{Z}_{uh}\left(\frac{1}{M_2}\sum_{i=1}^{M_2}\overline{\sigma}_{uh}(i) + 1\right) \cong \dot{Z}_{uh}$$

under the conditions that:

$$|\Delta \overline{I}_h(i)| > \Delta I_{h,cv} \quad \text{and} \quad \text{Re}\left\{\frac{\Delta \overline{V}_h(i)}{\Delta \overline{I}_h(i)}\right\} > 0. \qquad (2.17)$$

Eventually, in Equations (2.16) and (2.17) the interference coefficients are considered to be fluctuating randomly, so that their mean values are assumed to be approximately equal to zero, if M_1 and M_2 are great enough. A better smoothing can be obtained by performing the average on a recursive basis.

A technique based on the Correlation

This technique is based on the assumption that the variations $\Delta \overline{V}_h(i)$ and $\Delta \overline{I}_h(i)$ $(i = 1, 2, \ldots)$ of the voltage and current harmonics measured at the PCC are components of two stochastic vectors $\Delta \mathbf{V_h}$ and $\Delta \mathbf{I_h}$. The technique can be used to obtain an estimation of the mean value of the utility and customer harmonic impedances using the cross- and auto-correlation functions between the stochastic vectors.

An estimation of the cross-correlation function $\hat{R}_{\Delta \mathbf{X} \Delta \mathbf{Y}}$ between two stochastic vectors $\Delta \mathbf{X}$ and $\Delta \mathbf{Y}$ can be calculated using the following unbiased estimator:

$$\hat{R}_{\Delta \mathbf{X} \Delta \mathbf{Y}}(m) = \frac{1}{N-m} \sum_{i=1}^{N-m} \Delta X(i) \Delta Y^*(i+m), \qquad (2.18)$$

where the symbol * represents the complex conjugate, m represents the delay variable and N the variations to be taken into account.

Equation (2.18) can be used to estimate the auto-correlation functions $\hat{R}_{\Delta \mathbf{X} \Delta \mathbf{X}}$ and $\hat{R}_{\Delta \mathbf{Y} \Delta \mathbf{Y}}$, if it is applied using only the vector $\Delta \mathbf{X}$ or $\Delta \mathbf{Y}$, respectively. In the next part of this section, the vectors $\Delta \mathbf{X}$ and $\Delta \mathbf{Y}$ in Equation (2.18) are assumed to be the two stochastic vectors $\Delta \mathbf{V_h}$ and $\Delta \mathbf{I_h}$.

Now, let us remember that the harmonic voltage at the PCC, $\overline{V}_h$, is a function of the three variables $\dot{Z}_{uh}$, $\overline{J}_{uh}$, and $\overline{I}_h$ (see Equation (2.6)). Then, the harmonic voltage variation $\Delta \overline{V}_h$ can be approximated using the following first order term of the Taylor series:

$$\Delta \overline{V}_h \cong \frac{\partial \overline{V}_h}{\partial \overline{I}_h} \Delta \overline{I}_h + \frac{\partial \overline{V}_h}{\partial \dot{Z}_{uh}} \Delta \dot{Z}_{uh} + \frac{\partial \overline{V}_h}{\partial \overline{J}_{uh}} \Delta \overline{J}_{uh} = \dot{Z}_{uh} \Delta \overline{I}_h + (\overline{I}_h + \overline{J}_{uh}) \Delta \dot{Z}_{uh} + \dot{Z}_{uh} \Delta \overline{J}_{uh}. \quad (2.19)$$

Using Equation (2.19) and multiplying for the complex conjugate of the current harmonic variation $\Delta \overline{I}_h^*(i+m)$, the result is:

$$\Delta \overline{V}_h(i)\Delta \overline{I}_h^*(i+m) = \dot{Z}_{uh}(i)\Delta \overline{I}_h(i)\Delta \overline{I}_h^*(i+m) + \overline{I}_h(i)\Delta \dot{Z}_{uh}(i)\Delta \overline{I}_h^*(i+m)+$$
$$+\overline{J}_{uh}(i)\Delta \dot{Z}_{uh}(i)\Delta \overline{I}_h^*(i+m) + \dot{Z}_{uh}(i)\Delta \overline{J}_{uh}(i)\Delta \overline{I}_h^*(i+m). \qquad (2.20)$$

By substituting Equation (2.20) into Equation (2.18) (with the assumptions that $\Delta \mathbf{X} = \Delta \mathbf{V_h}$ and $\Delta \mathbf{Y}^* = \Delta \mathbf{I_h^*}$) and assuming that the values of $\dot{Z}_{uh}$ are statistically independent of the variations $\Delta\bar{I}_h$, $\Delta\bar{J}_{uh}$ and that the values of $\bar{J}_{uh}$ and $\bar{I}_h$ are statistically independent of the variations $\Delta\dot{Z}_{uh}$, $\Delta\bar{I}_h$, we have:

$$\hat{\bar{R}}_{\Delta\bar{V}_h\Delta\bar{I}_h(m)} = \frac{1}{N-m}\sum_{i=1}^{N-m}\dot{Z}_{uh}(i)\left[\frac{1}{N-m}\sum_{i=1}^{N-m}\Delta\bar{I}_h(i)\Delta\bar{I}_h^*(i+m)\right]$$

$$+\frac{1}{N-m}\sum_{i=1}^{N-m}\bar{I}_h(i)\left[\frac{1}{N-m}\sum_{i=1}^{N-m}\Delta\dot{Z}_{uh}(i)\Delta\bar{I}_h^*(i+m)\right]$$

$$+\frac{1}{N-m}\sum_{i=1}^{N-m}\bar{J}_{uh}(i)\left[\frac{1}{N-m}\sum_{i=1}^{N-m}\Delta\dot{Z}_{uh}(i)\Delta\bar{I}_h^*(i+m)\right]$$

$$+\frac{1}{N-m}\sum_{i=1}^{N-m}\dot{Z}_{uh}(i)\left[\frac{1}{N-m}\sum_{i=1}^{N-m}\Delta\bar{J}_{uh}(i)\Delta\bar{I}_h^*(i+m)\right]. \tag{2.21}$$

Using the definitions of the mean value and of the cross- and auto-correlation functions, Equation (2.21) can be expressed as:

$$\hat{\bar{R}}_{\Delta\bar{V}_h\Delta\bar{I}_h}(m) = E[\dot{Z}_{uh}]\hat{\bar{R}}_{\Delta\bar{I}_h\Delta\bar{I}_h}(m) + E[\bar{I}_h]\hat{\bar{R}}_{\Delta\dot{Z}_{uh}\Delta\bar{I}_h}(m)+$$
$$E[\bar{J}_{uh}]\hat{\bar{R}}_{\Delta\dot{Z}_{uh}\Delta\bar{I}_h}(m) + E[\dot{Z}_{uh}]\hat{\bar{R}}_{\Delta\bar{J}_{uh}\Delta\bar{I}_h}(m) \tag{2.22}$$

where the symbol E[X] represents the mean value of X.

On the basis of the considerations in the previous sections, when the utility harmonic current source is varying far less than the customer source, a weak correlation can be assumed between the utility harmonic current and the harmonic current at the PCC, and the variations $\Delta\bar{J}_{uh}$ and $\Delta\bar{I}_h$ can be considered such that $\hat{\bar{R}}_{\Delta\bar{J}_{uh}\Delta\bar{I}_h}(m) \cong 0$. If we also assume that the variations $\Delta\dot{Z}_{uh}$ and $\Delta\bar{I}_h$ are such that $\hat{\bar{R}}_{\Delta\dot{Z}_{uh}\Delta\bar{I}_h}(m) \cong 0$, Equation (2.22) becomes:

$$\hat{\bar{R}}_{\Delta\bar{V}_h\Delta\bar{I}_h}(m) = E[\dot{Z}_{uh}]\hat{\bar{R}}_{\Delta\bar{I}_h\Delta\bar{I}_h}(m), \tag{2.23}$$

and it is possible to estimate the mean value of the utility harmonic impedance using the following equation:

$$E[\dot{Z}_{uh}] = \frac{\hat{\bar{R}}_{\Delta\bar{V}_h\Delta\bar{I}_h}(m)}{\hat{\bar{R}}_{\Delta\bar{I}_h\Delta\bar{I}_h}(m)}. \tag{2.24}$$

Starting from Equation (2.2) instead of Equation (2.6), similar considerations can be applied for the estimation of customer harmonic impedance. In fact, when the customer harmonic current source is not present or is varying far less than the utility source, a weak correlation between it and the harmonic current at the PCC can be assumed, and the values of the variations $\Delta \overline{\mathbf{J}}_{ch}$ and $\Delta \overline{\mathbf{I}}_h$ can be considered such that $\hat{\overline{R}}_{\Delta \overline{J}_{ch} \Delta \overline{I}_h}(m) \cong 0$. If we also assume that the variations $\Delta \dot{\mathbf{Z}}_{ch}$ and $\Delta \overline{\mathbf{I}}_h$ are such that $\hat{\overline{R}}_{\Delta \dot{Z}_{ch} \Delta \overline{I}_h}(m) \cong 0$, it is possible to estimate the mean value of the customer harmonic impedance using the following relationship:

$$E[\dot{\mathbf{Z}}_{ch}] = -\frac{\hat{\overline{R}}_{\Delta \overline{V}_h \Delta \overline{I}_h}(m)}{\hat{\overline{R}}_{\Delta \overline{I}_h \Delta \overline{I}_h}(m)}. \tag{2.25}$$

In order to assess the utility and customer impedances using Equations (2.24) and (2.25), it should be noted that the available measurements of the variations $\Delta \overline{I}_h(i)$ and $\Delta \overline{V}_h(i)$ have to be separated into the group for which $\mathrm{Re}\left\{ \frac{\Delta \overline{V}_h(i)}{\Delta \overline{I}_h(i)} \right\} > 0$ and the group for which $\mathrm{Re}\left\{ \frac{\Delta \overline{V}_h(i)}{\Delta \overline{I}_h(i)} \right\} < 0$, respectively.

One advantage of this statistical method is that the variations are measured for long periods, thus allowing a statistical characterization of harmonic emission level indices.

An iterative technique
The iterative technique consists of assigning a tentative value, for example to the equivalent harmonic impedance of the utility supply system $\dot{Z}_{uh}$, and then calculating the equivalent harmonic current source $\overline{J}_{uh}$ (or vice versa) with the relationship $\overline{V}_h = \dot{Z}_{uh}(\overline{J}_{uh} + \overline{I}_h)$, for known voltage and current harmonics at the PCC. To achieve convergence, a very good initial value is needed. A good initial value for $\dot{Z}_{uh}$ would be that obtained by applying the method based on the sign of the real part of the ratio $\Delta \overline{V}_h / \Delta \overline{I}_h$.

This technique allows us to follow the values of the equivalent harmonic impedance $\dot{Z}_{uh}$ (and current $\overline{J}_{uh}$) versus time.

The same authors that proposed the above iterative technique have also proposed a nonlinear regression technique [14]. This technique requires a number of measurements during an interval in which the values of the harmonic impedances $\dot{Z}_{ch}$, $\dot{Z}_{uh}$ and currents $\overline{J}_{ch}$ and $\overline{J}_{uh}$ have no significant variations. The technique allows us to find the values of these quantities that minimize estimation error.

b) Some other methods for the evaluation of indices
Other methods for the evaluation of indices exist; in some cases, these methods either partially or totally eradicate the need for knowledge of the utility and customer harmonic impedances.

First let us consider the case of switching at the customer installation (Figure 2.5) and assume that both $\dot{Z}_{uh}$ and $\overline{J}_{uh}$ are unchanged after S_1 is open.

When S_1 is open (Figure 2.5), the voltage harmonic at the PCC is given by:

$$\overline{V}_{h0} = \dot{Z}_{uh} \overline{J}_{uh}, \tag{2.26}$$

while when S_1 is closed (Figure 2.2 (a)), the well-known result is:

$$\overline{V}_h = \dot{Z}_{uh}(\overline{J}_{uh} + \overline{I}_h) = \dot{Z}_{ch}(\overline{J}_{ch} - \overline{I}_h). \tag{2.27}$$

With trivial manipulations of Equations (2.26) and (2.27), the customer equivalent harmonic current source is given by:

$$\overline{J}_{ch} = \frac{\dot{Z}_{uh} + \dot{Z}_{ch}}{\dot{Z}_{uh} \dot{Z}_{ch}} \overline{V}_h - \frac{\overline{V}_{h0}}{\dot{Z}_{uh}}; \tag{2.28}$$

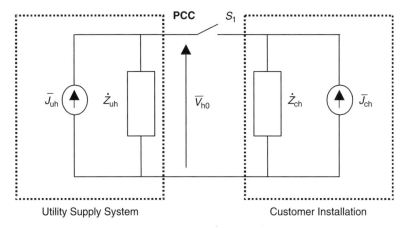

Figure 2.5   Switching at the customer installation

thus, bearing in mind Equation (2.1), the harmonic voltage and current emission levels become:

$$\overline{V}_{ch} = \frac{\dot{Z}_{uh}\dot{Z}_{ch}}{\dot{Z}_{uh}+\dot{Z}_{ch}}\,\overline{J}_{ch} = \frac{\dot{Z}_{uh}\dot{Z}_{ch}}{\dot{Z}_{uh}+\dot{Z}_{ch}}\left(\frac{\dot{Z}_{uh}+\dot{Z}_{ch}}{\dot{Z}_{uh}\dot{Z}_{ch}}\overline{V}_{h}-\frac{\overline{V}_{h0}}{\dot{Z}_{uh}}\right) = \overline{V}_{h} - \frac{\dot{Z}_{ch}}{\dot{Z}_{uh}+\dot{Z}_{ch}}\overline{V}_{h0}$$

$$\overline{I}_{ch} = \frac{\dot{Z}_{ch}}{\dot{Z}_{uh}+\dot{Z}_{ch}}\,\overline{J}_{ch} = \frac{\dot{Z}_{ch}}{\dot{Z}_{uh}+\dot{Z}_{ch}}\left(\frac{\dot{Z}_{uh}+\dot{Z}_{ch}}{\dot{Z}_{uh}\dot{Z}_{ch}}\overline{V}_{h}-\frac{\overline{V}_{h0}}{\dot{Z}_{uh}}\right) = \frac{\overline{V}_{h}}{\dot{Z}_{uh}} - \frac{\dot{Z}_{ch}}{\dot{Z}_{uh}\left(\dot{Z}_{uh}+\dot{Z}_{ch}\right)}\overline{V}_{h0}.$$

$$(2.29)$$

If $|\dot{Z}_{uh}| << |\dot{Z}_{ch}|$, it follows that:

$$\overline{V}_{ch} \approx \overline{V}_{h} - \overline{V}_{h0}, \quad \overline{I}_{ch} \approx \frac{\overline{V}_{h}-\overline{V}_{h0}}{\dot{Z}_{uh}} \ . \tag{2.30}$$

Starting from Equation (2.30), the indices of interest can be calculated. The application of this method requires, once again, measurements under two sets of conditions at different times with the problems already demonstrated for the frame reference. In the literature it has been proposed that we determine the phase difference between the harmonic voltages in Equation (2.30) by using a measurement of the current $\overline{I}_{h}$ or by directly processing the difference between the instantaneous values of the voltages measured before and after the opening of S_1 [11].

In its most general formulation, this method requires accurate knowledge of the harmonic impedances of both the utility supply system and the customer loads. However, the simplified case, when $|\dot{Z}_{uh}| << |\dot{Z}_{ch}|$, does not require any knowledge of impedance for the evaluation of $\overline{V}_{ch}$ and only knowledge of the utility harmonic impedance is required for the evaluation of $\overline{I}_{ch}$.

In all cases, the method provides results that correspond only to the switching test operation, with consequent problems for providing a statistic of harmonic level indices.

Example 2.3   In this example the harmonic voltage emission and the harmonic current emission levels are calculated using Equations (2.29) and (2.30) and by considering once again the test system of Example 2.1.

Let us assume that in a measured time interval during which the customer installation is switched from on (S_1 is closed) to off (S_1 is open), the following values of the harmonic voltages at the PCC have been measured:

$$\overline{V}_5 = 75.122 - j56.754[\text{V}], \quad \overline{V}_{50} = 16.940 - j50.858[\text{V}].$$

Applying Equation (2.29) and assuming the values for the harmonic impedances of Example 2.1, it follows that:

$$\overline{V}_{c5} = 57.575 - j6.042[\text{V}], \quad \overline{I}_{c5} = -0.445 - j5.269\,[\text{A}].$$

Then:

$$\text{HVELI} = 57.891[\text{V}], \quad \text{HCELI}{=}5.288\,[\text{A}].$$

Applying Equation (2.30), we get:

$$\overline{V}_{c5} = 58.857 - j6.207[\text{V}], \quad \overline{I}_{c5} = -0.463 - j5.204\,[\text{A}].$$

Then:

$$\text{HVELI} = 57.194[V], \quad \text{HCELI} = 5.225[\text{A}].$$

It should be noted that Equation (2.30) gives a result close to the true values (see Example 2.1: HVELI = 58.429 [V], HCELI = 5.337 [A]). On the other hand, in the considered case, the hypothesis that $|\dot{Z}_{u5}| << |\dot{Z}_{c5}|$ is clearly verified.

Now we consider a method that allows direct calculation of all indices, completely eradicating the need for knowledge of the utility and customer impedances [11]. This method requires knowledge of only the phasors (magnitude and argument) relative to the same conditions, so it also avoids the reference frame uncertainties resulting from measurements at different times.

To explain the method, let us consider the case when a customer shunt impedance $\dot{Z}_{sh}$ that can be switched on (S_2 closed) and off (S_2 open) is available and is known as exactly as possible (Figure 2.6).

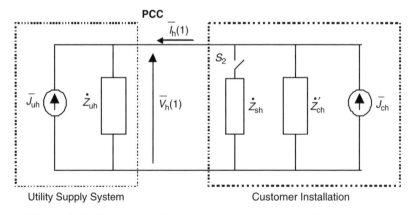

Figure 2.6   Switching of a known shunt impedance of the customer

When S_2 is open, the harmonic voltage at the PCC is given by (Figure 2.6):

$$\overline{V}_h(1) = \dot{Z}_{uh}\left[\overline{I}_h(1) + \overline{J}_{uh}\right], \quad \overline{V}_h(1) = \dot{Z}'_{ch}\left[\overline{J}_{ch} - \overline{I}_h(1)\right]. \tag{2.31}$$

When S_2 is closed, and by assuming the harmonic impedance and harmonic source values are unchanged, we get:

$$\overline{V}_h(2) = \dot{Z}_{uh}\left[\overline{I}_h(2) + \overline{J}_{uh}\right], \quad \overline{V}_h(2) = \frac{\dot{Z}_{sh}\dot{Z}'_{ch}}{\dot{Z}_{sh} + \dot{Z}'_{ch}}\left[\overline{J}_{ch} - \overline{I}_h(2)\right]. \tag{2.32}$$

By eliminating the two harmonic impedances $\dot{Z}'_{ch}$ and $\dot{Z}_{uh}$ from Equations (2.31) and (2.32), the following expressions can be obtained for the equivalent harmonic current source $\overline{J}_{ch}$ and for the equivalent impedance $\dot{Z}_{toth}$:

$$\overline{J}_{ch} = \frac{\overline{V}_h(2)\overline{I}_h(1) - \overline{V}_h(1)\overline{I}_h(2) - \overline{V}_h(2)\overline{V}_h(1)/\dot{Z}_{sh}}{\overline{V}_h(2) - \overline{V}_h(1)}$$

$$\dot{Z}_{toth} = \frac{1}{\dfrac{1}{\dot{Z}_{sh}} + \dfrac{1}{\dot{Z}'_{ch}} + \dfrac{1}{\dot{Z}_{uh}}} = \dot{Z}_{sh}\left[1 - \frac{\overline{V}_h(2)}{\overline{V}_h(1)}\right]. \tag{2.33}$$

Eventually, the harmonic voltage emission level becomes:

$$\overline{V}_{ch} = \dot{Z}_{toth}\overline{J}_{ch} = \overline{V}_h(2) + \dot{Z}_{sh}\overline{I}_h(2) - \dot{Z}_{sh}\overline{I}_h(1)\frac{\overline{V}_h(2)}{\overline{V}_h(1)}; \tag{2.34}$$

it is expressed only as a function of the shunt impedance $\dot{Z}_{sh}$ and of the voltage and current harmonics measured at the previously defined conditions. Then, as previously demonstrated, no knowledge of the utility and customer harmonic impedances is required.

Once again, the main difficulty of this method is that the harmonic voltage and current at the PCC are not measured at the same time, i.e. the two conditions are different. However, if we assume the harmonic voltage phasor $\overline{V}_h(1)$ is the phase-reference, such that:

$$\dot{Z}_{sh} = Z_{sh}\exp{(j\psi)}, \quad \overline{V}_h(1) = V_h(1)\exp{(j0)}, \quad \overline{I}_h(1) = I_h(1)\exp{(j\alpha_1)},$$

$$\overline{V}_h(2) = V_h(2)\exp{(j\theta_0)}, \quad \overline{I}_h(2) = I_h(2)\exp{[j(\alpha_2 + \theta_0)]}, \tag{2.35}$$

we can evaluate the harmonic voltage emission level index, HVELI, as:

$$\mathrm{HVELI} = |\overline{V}_{ch}| = \left|V_h(2) + Z_{sh}I_h(2)\exp{[j(\alpha_2 + \psi)]} - \frac{Z_{sh}I_h(1)V_h(2)\exp{[j(\alpha_1 + \psi)]}}{V_h(1)}\right|. \tag{2.36}$$

Equation (2.36) allows a more simple calculation of HVELI, because it only requires that the phasor magnitudes and angular differences (α_1 and α_2) relative to phasors of the same conditions be known.

Moreover, if the current flowing in the shunt impedance $\dot{Z}_{sh}$ is measured too, Equation (2.36) can be further simplified to obtain a relationship in which the harmonic voltage emission level does not depend on any impedance values, and knowledge of the shunt impedance $\dot{Z}_{sh}$ is not necessary.

Obviously, the application of this method requires the presence of an auxiliary shunt element to be switched on and off and, once again, this method provides results that correspond only to the switching test operation, with consequent problems related to statistical analysis. Moreover, as is well known, the current harmonics depend on the state of the power system and the harmonics change when the state changes due to a switching operation.

Example 2.4 In this example the harmonic voltage emission level is calculated using Equations (2.34) and (2.36), and considering once again the test system of Example 2.1 with $\dot{Z}_{ch} = \dot{Z}'_{ch}$. The shunt impedance $\dot{Z}_{sh}$ that is added and switched on is a capacitor with $C = 0.5\ \mu\text{F}$.

Let us assume that the following values of the harmonic voltage and current at the PCC were measured before and after the switching:

$$\overline{V}_5(1) = 75.122 - j56.754\ [\text{V}]; \quad \overline{I}_5(1) = -0.440 - j5.255\ [\text{A}];$$
$$\overline{V}_5(2) = 73.888 - j59.293 [\text{V}]; \quad \overline{I}_5(2) = -0.642 - j5.289\ [\text{A}].$$

Applying Equation (2.34) it follows that:

$$\overline{V}_{c5} = 52.913 - j2.178\ [\text{V}]$$
$$\text{HVELI} = |\overline{V}_{c5}| = 52.958\ [\text{V}].$$

Applying Equation (2.36), and taking into account the new phasor references, the result is the same:

$$\text{HVELI} = |\overline{V}_{c5}| = 52.958\ [\text{V}].$$

It is interesting to note that the HVELI values are far from the actual values; the true value, obtained from the analysis of the whole system simulation, is, in fact, given by:

$$\text{HVELI} = 58.847\ [\text{V}].$$

The difference between the two values is due to the hypothesis that in applying Equations (2.34) and (2.36), $\overline{J}_{u5}$ and $\overline{J}_{c5}$ are considered to remain constant. Obviously, when the capacitor is switched, the electrical state of the system also changes and, in turn, the equivalent harmonic currents change.

A further simplified method calculates the harmonic voltage emission level $\overline{V}_{ch}$ and the corresponding HVELI directly, in the presence of a dominant disturbing load [11]. This method

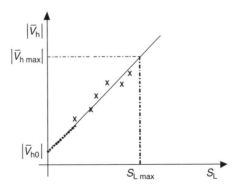

Figure 2.7 Qualitative correlation between harmonic voltage and apparent load power

assumes that a linear correlation exists between the amplitude of the harmonic voltage at the PCC $|\overline{V}_h|$ and the apparent power of the load S_L. HVELI is calculated as (Figure 2.7):

$$\text{HVELI} = |\overline{V}_{h\,\max}| - |\overline{V}_{h0}|,\qquad(2.37)$$

which can also be obtained by performing a linear regression.

Simple and double correlation between harmonics at the PCC can also be considered [11].

c) A method based on reference (or contract) impedances

The main objective of this method is to clearly separate the customer and utility responsibilities, providing attribution of the harmonic variations located on either side to the other side. To better understand this objective, let us consider the equivalent circuits in Figure 2.2 and assume that the equivalent harmonic impedance of the supply system varies, e.g. it decreases. Assuming that $\dot{Z}_{ch}$ and $\overline{J}_{ch}$ are constant (i.e. nothing changes in the customer installation), it follows that the customer harmonic current contribution at the PCC varies and, specifically, it increases (see Equation (2.1)) so that the customer appears to be indirectly responsible for a harmonic variation located on the utility side. Obviously, analogous considerations for the utility can occur in cases in which there is variation in the customer impedance.

To overcome the above problem and to reach the stated objective, the key idea of the method relies on using reference (contract) impedances in the Norton equivalent circuit of Figure 2.2 instead of the actual utility and customer impedances; the impedance changes (with respect to the reference values) are converted into equivalent current source changes that can be evaluated on the basis of the voltage and current harmonics measurements at the PCC. In particular, as will be shown in the following discussion, to be sure of essentially attributing to the customer only the customer's own responsibility, the utility impedance change is converted by changing only the equivalent harmonic current source of the utility itself. Likewise, to be sure of essentially attributing to the utility only the utility's own responsibility, the customer impedance change is converted by changing only the equivalent harmonic current source of the customer itself.

So, in the circuit of Figure 2.2, let us assume a change in only the utility supply system impedance. This change (with respect to the reference value) is converted to an equivalent harmonic current source variation and the new utility harmonic current source (Figure 2.8) is given by:

$$\overline{J}_{uh-\text{new}} = \frac{\overline{V}_h}{\dot{Z}_{uh-\text{ref}}} - \overline{I}_h,\qquad(2.38)$$

where $\dot{Z}_{uh-\text{ref}}$ is the chosen utility reference impedance.

The same procedure can be applied when a customer impedance variation occurs; the result is:

$$\overline{J}_{ch-\text{new}} = \frac{\overline{V}_h}{\dot{Z}_{ch-\text{ref}}} + \overline{I}_h,\qquad(2.39)$$

where $\dot{Z}_{ch-\text{ref}}$ is the chosen customer reference impedance.

The Norton harmonic equivalent circuit, including the reference impedances, is shown in Figure 2.9 with the harmonic sources given by Equations (2.38) and (2.39). The analysis of this circuit clearly demonstrates that both customer and utility are essentially responsible only for their own harmonic variations, because all the customer variations (in terms of both harmonic current and impedance) are converted in changes of the customer equivalent harmonic currents and the same occurs for the utility variations.

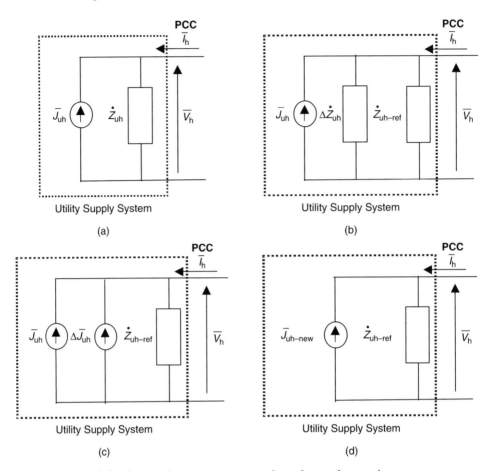

Figure 2.8   Successive steps to convert impedance changes into current

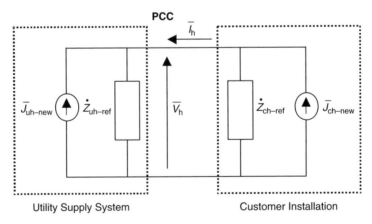

Figure 2.9   Norton equivalent circuit with reference impedances

A similar approach can easily be applied to quantify the customer and utility harmonic contributions to measured voltages.

Once the reference impedances had been chosen and the voltage and current harmonics at the PCC measured, the authors who first proposed the method calculated the new harmonic current sources and then applied Equations (2.1) and (2.5) to determine all the indices of interest [13].

The main disadvantage of this method is that it relies on the requirement to choose reference impedances. In fact, the choice of the reference impedances directly influences the consequent attribution of responsibility. The authors who first proposed this method [13] thought that the verification of compliance to any harmonic standards and recommendations, including IEEE 519, would be conducted on the basis of certain reference impedances. Many utilities provide information to customers about a reference value of the equivalent impedance of the supply system; the same applies to the impedance of customers, since the harmonic limit compliance check also needs this impedance. So, if the interconnection contract between the utility and the customer is based on these reference harmonic impedances, a legally defensible method to bill the customer should be based on these impedances.

Recently, a resistance calculated directly from the measurements at the PCC has been proposed in the literature as the customer reference impedance; in addition, for the utility reference impedance, the sum of the short-circuit network impedance and the impedance of the last transformer before the PCC has been used [17].

Numerical applications of this method have evidenced its robustness in achieving the objective of maintaining the same responsibility for the customer in the presence of supply system changes and for the utility in the presence of customer changes. The method also seems to handle resonance conditions quite well.

Example 2.5 In this example the harmonic current emission levels due to the utility and due to the customer are calculated, considering once again the simple test system of Example 2.1. For the sake of simplicity, the reference impedances of the customer and the utility are assumed to be equal to $\dot{Z}_{u5-ref} = 0.219 + j10.945 \ [\Omega]$ and $\dot{Z}_{c5-ref} = 793.972 + j69.182 \ [\Omega]$, respectively.

We assume a variation in the customer impedance with respect to the values of Example 2.1 such that $\dot{Z}_{c5} = \frac{\dot{Z}_{c5-ref}}{4}$.

The following values of the harmonic voltage and current at the PCC were measured:

$$\overline{V}_5 = 69.955 - j61.575 \ [V], \quad \overline{I}_5 = -0.852 - j4.956 \ [A].$$

The phasors $\overline{I}_{c5}$ and $\overline{I}_{u5}$ and the harmonic current emission levels due to the utility and the customer, calculated by applying Equations (2.4) and (2.8), are:

$$\overline{I}_{c5} = -0.809 - j5.219 \ [A], \quad \overline{I}_{u5} = 0.043 - j0.263 \ [A],$$

$$HCELI = |\overline{I}_{c5}| = 5.282 \ [A]; \quad HCUCI = |\overline{I}_{u5}| = 0.267 \ [A].$$

Now, applying Equations (2.38) and (2.39), the harmonic current sources are:

$$\overline{J}_{c5-new} = 5.099 \exp(-j1.723) \ [A]$$

$$\overline{J}_{u5-new} = 4.894 \exp(-j \ 2.820) \ [A]$$

Finally, using Equations (2.1) and (2.5) and taking into account the phasors $\bar{J}_{c5-\text{new}}$ and $\bar{J}_{u5-\text{new}}$, we get:

$$\bar{I}_{c5} = -0.839 - j5.022 \text{ [A]},$$
$$\bar{I}_{u5} = 0.013 - j0.065 \text{ [A]},$$
$$\text{HCELI} = |\bar{I}_{c5}| = 5.091 \text{ [A]}, \quad \text{HCUCI} = |\bar{I}_{u5}| = 0.067 \text{ [A]}.$$

A comparison between the above results and the results obtained in Example 2.1 leads to the following considerations:

- using the present impedances, HCUCI increases ($0.267 > 0.067$), so that the utility appears indirectly responsible for the variation located on the customer's side;

- using the reference impedances, HCUCI remains constant, which suggests the practical invariability of the utility's responsibility.

2.2.1.2   Thevenin-based and mixed approaches

Other approaches use Thevenin equivalent circuits to represent either the utility supply system and the customer installation or just one of them; obviously, these representations are equivalent to the Norton representations.

Figure 2.10 shows other possible representations to those shown in Figure 2.2. In Figure 2.10 (a) the Thevenin equivalent circuit is used to represent the utility side, while in Figure 2.10 (b) it is used for both sides.

Bearing in mind the well-known relationships that link Thevenin and Norton circuit parameters ($\bar{V}_{ch}^{T} = \dot{Z}_{ch}\bar{J}_{ch}, \bar{V}_{uh}^{T} = \dot{Z}_{uh}\bar{J}_{uh}$), we can easily obtain the utility and customer contributions for the new system representation. For example, by applying the superposition method to the harmonic current in Figure 2.10 (b), the following expressions for the harmonic current emission and utility levels $\bar{I}_{ch}, \bar{I}_{uh}$ can be obtained:

$$\bar{I}_{ch} = \frac{\bar{V}_{ch}^{T}}{\dot{Z}_{uh} + \dot{Z}_{ch}} = \frac{\bar{V}_{h} + \dot{Z}_{ch}\bar{I}_{h}}{\dot{Z}_{uh} + \dot{Z}_{ch}}, \quad \bar{I}_{uh} = \frac{\bar{V}_{uh}^{T}}{\dot{Z}_{uh} + \dot{Z}_{ch}} = \frac{\bar{V}_{h} - \dot{Z}_{uh}\bar{I}_{h}}{\dot{Z}_{uh} + \dot{Z}_{ch}}, \quad (2.40)$$

and the condition $|\bar{V}_{ch}^{T}| < |\bar{V}_{uh}^{T}|$ means that $|\bar{I}_{ch}| < |\bar{I}_{uh}|$, so we can say that the utility side generates the larger contribution to the harmonic current distortions at the PCC (harmonic source detection problem). Similar considerations can be effected in the other possible case.

Among other possible techniques, an interpolation method based on the injection of interharmonic current components [10] and a so-called increment-based approach [7] have been applied to calculate all the circuit elements and, subsequently, all the quantities and indices that depend on them (e.g. $\bar{V}_{ch}, \bar{I}_{ch}$, HCELI and HVELI).

With reference to the interpolation method, let $\bar{J}_{h+}$ and $\bar{J}_{h-}$ represent the two interharmonic components close to the harmonic of order h and injected by an auxiliary active device at the PCC. Assuming negligible interharmonic components at these interharmonic orders from the utility and customer sides, the following interharmonic voltage components can be measured at the PCC (Figure 2.11 (a)), where only the equivalent circuit for interharmonic $h+$ is shown):

$$\bar{V}_{h+} = \dot{Z}_{ch+}\bar{I}_{ch+}, \quad \bar{V}_{h-} = \dot{Z}_{ch-}\bar{I}_{ch-}. \quad (2.41)$$

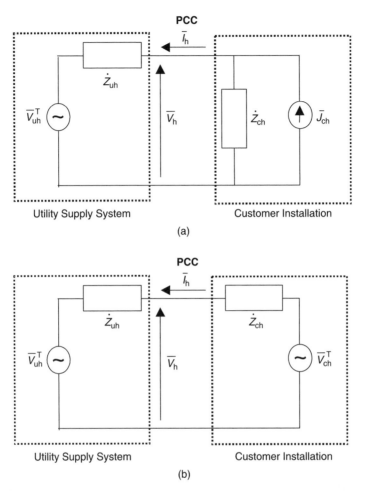

Figure 2.10   Thevenin equivalent circuits (a) for the utility supply system; (b) for the utility supply system and the customer installation

The equivalent harmonic impedance of the customer loads $\dot{Z}_{ch}$ is calculated, using a linear interpolation, as:

$$\dot{Z}_{ch} = \frac{\dot{Z}_{ch+} + \dot{Z}_{ch-}}{2} = \frac{\dfrac{\overline{V}_{h+}}{\overline{I}_{ch+}} + \dfrac{\overline{V}_{h-}}{\overline{I}_{ch-}}}{2}. \tag{2.42}$$

The accuracy of the impedance calculation can be improved by increasing the number of interharmonics injected around the harmonic of order h.

Once the equivalent harmonic impedance $\dot{Z}_{ch}$ is known, the equivalent harmonic current source of the customer can be calculated as $\overline{J}_{ch} = \overline{I}_h + \frac{\overline{V}_h}{\dot{Z}_{ch}}$ (Figure 2.10 (a)).

A similar procedure can be applied to calculate the utility parameters $\dot{Z}_{uh}$ and $\overline{V}_{uh}^{T}$.

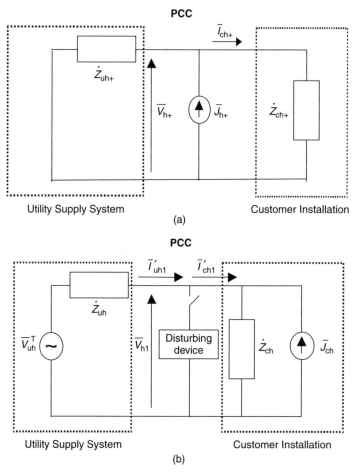

Figure 2.11 Equivalent circuits (a) for interharmonic impedance calculation; (b) for the increment-based approach application

The main disadvantages of this method are that:

- fairly powerful signal generators and suitable transformers are needed;

- spectral leakage problems must be avoided (due, for example, to an incorrect choice of time window length, which can cause the presence of false interharmonics).

The increment-based approach (Figure 2.11 (b)) requires the measurement (before and after the switching of a disturbing device, such as a capacitor or a harmonic-generating device) of the harmonic voltages at the PCC $\overline{V}_{hi}$ ($i = 1, 2$) and of the harmonic currents $\overline{I}'_{uhi}$ and $\overline{I}'_{chi}$ ($i = 1, 2$). With trivial calculations, we obtain:

$$\dot{Z}_{uh} = \frac{\overline{V}_{h1} - \overline{V}_{h2}}{\overline{I}'_{uh2} - \overline{I}'_{uh1}}, \quad \dot{Z}_{ch} = \frac{\overline{V}_{h1} - \overline{V}_{h2}}{\overline{I}'_{ch1} - \overline{I}'_{ch2}}. \tag{2.43}$$

There is also a method that uses the whole Thevenin representation of Figure 2.10 (b) and this is called the *critical impedance method*. The method is applied to determine whether the utility or the customer has more contributions to harmonic distortions at the PCC. This method is based on the calculation of a critical impedance value $\dot{Z}_{hcr}$ that, when compared with the actual total impedance $(\dot{Z}_{uh} + \dot{Z}_{ch})$, allows the determination of the main harmonic source [15].

Before concluding the discussion of harmonic impedance-based methods, we want to emphasize that, from a theoretical point of view, the impedance calculation as well as the sharing of responsibilities can also be assessed by simulations of the power system under study with the help of computer programs.

Also, we want to emphasize that the recent IEC 61000-3-6–2008 [46] introduced a new definition of harmonic emission levels, which is slightly different from the definition considered at the beginning of this section. Specifically, with reference to the harmonic voltages, the emission level was defined as the magnitude of the phasor equal to the difference between the measured voltage $\overline{V}_h$ and the voltage $\overline{V}_{uh}^T$ in Figure 2.10 (a) (in practice, the difference between the post- and pre-connection harmonic voltages). When this difference causes an increase in the harmonic distortion, it is requested to be less than the emission limits. The two definitions practically coincide in the hypothesis $|\dot{Z}_{uh}| \ll |\dot{Z}_{ch}|$ (see Equation 2.30).[3]

2.2.2   Indices based on powers in non-ideal conditions

These indices are based on the evaluation of the powers in the presence of waveform distortions and/or unbalances.

Starting with the Fryze and Budeanu theories,[4] the definition of powers in non-ideal conditions has been discussed widely in the literature, but to date there is no universally accepted definition for all powers.

Confining our interest to the indices for sharing responsibility, active power-based indices are the most common in the literature and they look mainly at active power losses as the PQ detrimental effect. A reactive power-based index has also been proposed for three-phase, balanced systems, with the goal of overcoming the problems associated with the active power-based indices. Other indices based on nonactive power (e.g. Fryze's reactive power) have been proposed recently.

---

[3] We note that both IEC 61000-3-6–2008 and IEC 61000-3-7–2008 introduce three points of interest. The point of evaluation (POE) is defined as the point where the emission levels of a customer are to be assessed against the emission limits. The point of common coupling (PCC) is defined as the point which is electrically closest to the customer and to which other customers are, or may be, connected. The point of connection (POC) is defined as the point where the customer is, or can be, connected. For more details, see [46, 47].

[4] The Fryze theory separates the time domain current waveform in the 'active' current (characterized by the same waveform as the voltage) and the 'nonactive' or 'reactive' current (the remaining part). The Budeanu theory assumes that a power system, in the frequency domain, can be subdivided into elementary sinusoidal circuits, one for each frequency included in the nonsinusoidal voltage and current spectra. Then, each circuit is characterized by RMS values of voltage and current, active and reactive powers. The Fryze and Budeanu theories have been discussed extensively in the relevant literature, and several other important approaches have been proposed and different definitions of powers formulated (e.g. Czarnecki, Emanuel, Sharon, Kusters and Moore) [22, 48].

The following indices are considered in this section:

- the supply and loading quality index;
- the utility and customer total harmonic distortion factors;
- the harmonic phase index and the harmonic global index;
- the critical impedance index and the critical admittance index.

The first four are active, power-based indices. The supply and loading quality index was one of the first indices that appeared in the literature dealing with active power direction; thus, this index will be analyzed in more detail below to demonstrate some of the problems associated with the use of active power in the definition of an index. The critical impedance and the critical admittance indices are reactive, power-based indices.

2.2.2.1   The supply and loading quality index

This index takes into account the contemporaneous presence of waveform distortions and unbalances at the PCC. In spite of this, and in order to better demonstrate the behaviour of this index, the waveform distortions and unbalances are first considered separately and then they are considered together. Three-phase, three-wire systems are taken into account.

Initially, let us consider a balanced system in which the voltage and current have nonsinusoidal waveforms. The supply and loading quality index is defined as:

$$\xi_{\text{slq}} = \frac{P}{P_1} = \frac{P_1 + P_{\text{H}}}{P_1} = \frac{P_1 + \sum_{h \neq 1} P_h}{P_1} = \frac{P_1 + 3\sum_{h \neq 1} V_h I_h \cos(\alpha_h - \beta_h)}{P_1}, \qquad (2.44)$$

where P is the (total) active power, P_1 is the fundamental active power, P_h is the hth-order harmonic active power, P_{H} is the (total) harmonic active power, V_h and I_h are the RMS values of voltage and current harmonics and α_h and β_h are their phase angles.

The supply and loading quality index ξ_{slq} is equal to 1 if we consider the case of a linear load in an ideal supply system ($P_h = 0 \; \forall h$, and then $P_{\text{H}} = 0$).

If the section x-x feeds a linear load and the utility has background distortion (Figure 2.12 (a)), the harmonic active powers are all positive and delivered to the load ($P_h > 0 \; \forall h$, and then $P_{\text{H}} > 0$), so the supply and loading quality index ξ_{slq} is greater than 1. Clearly, the index increases as the harmonic active power delivered to the load increases.

If the section x-x feeds a static converter and the utility is without background distortion (Figure 2.12 (b)), all the harmonic active powers are negative and reflected backwards in the supply system ($P_h < 0 \; \forall h$, and then $P_{\text{H}} < 0$), so the supply and loading quality index ξ_{slq} is less than 1. In this condition, as is well known, the fundamental active power delivered to the static converter is in part transferred to its DC load while the remaining power is reflected backwards in the supply system as harmonic active power and is dissipated at the supply resistance; the fundamental active power delivered to the static converter is higher than its DC load active power (Figure 2.13, where the static converter is considered free of losses). Clearly, the index decreases as the harmonic active power that is dissipated at the supply resistance increases.

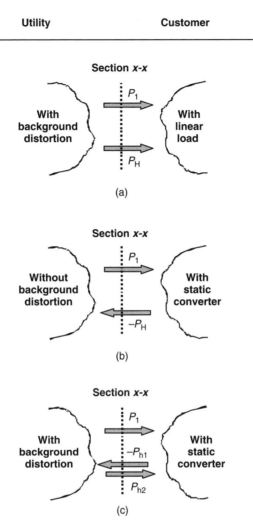

(a)

(b)

(c)

Figure 2.12   Active powers in electrical systems with waveform distortions [5] (see Plate 1)

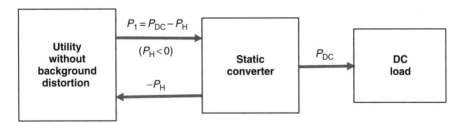

Figure 2.13   Diagram of active power balance flow

---

[5] In Figure 2.12 the arrow directions indicate the positive values of powers.

If the section x-x feeds a static converter and the supply system has background distortion due to the presence of converters at the same bus or at one or more different buses in the power system (Figure 2.12 (c)), some harmonic active powers P_h (for example P_{h1}) can be negative (reflected backwards) and some others (for example P_{h2}) can be positive (delivered to the static converter), depending on the sign of the cosines in Equation (2.44); obviously, the sign depends on the phase shift between the harmonic current and voltage. It follows that the resulting harmonic active power P_H can be positive or negative; theoretically, it can also be equal to zero if the positive and negative components of the power are equal to each other. Because of its dependence on the value and sign of P_H, the supply and loading quality index ξ_{slq} can be greater than 1, less than 1 or equal to 1. The same case could occur if the customer were to include both a static converter and a linear load.

Example 2.6   In this example the harmonic active powers and the supply and loading quality index values are calculated considering the simple test systems of Figure 2.14, simulated in the time domain. A 20 kV medium voltage supply system ($X_{u1} = 2.189\ \Omega$, $R_{u1} = 0.219\ \Omega$) feeds one (Figure 2.14 (a)) or two (Figure 2.14 (b)) 1.0 MW, 6-pulse AC/DC line-commutated converters, the firing angles of which are $\alpha_1 = 15°$ and $\alpha_2 = 30°$, respectively. Each AC/DC converter is connected to the medium voltage bus with a 1.25 MVA transformer ($V_{cc,\%} = 4.5\%$) and the system frequency is 50 Hz. Only waveform distortions are considered, given that the system is balanced.

If the system in Figure 2.14 (a) is considered initially, a time-domain simulation furnishes the fundamental and the hth-order harmonic active powers given in Table 2.1,[6] where the

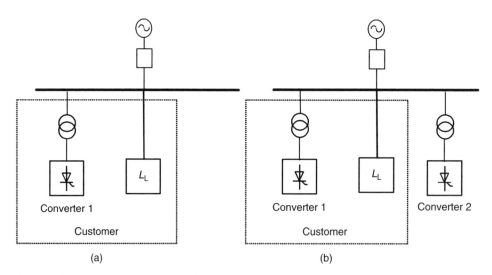

(a)   (b)

Figure 2.14   Simple test system (a) with one static converter and one linear load; (b) with two static converters and one linear load

---

[6] It should be noted that several ciphers have been considered for the indices without rounding in the last one in order to make it possible to show the behaviour of the indices in the various cases.

Table 2.1 Active powers and supply and loading quality indices of the test system in Figure 2.14(a)

	Converter 1	Linear load	Customer
P_1 [kW]	887796.419	496408.346	1384204.765
P_5 [W]	−31.701	12.880	−18.821
P_7 [W]	−20.895	11.969	−8.926
P_{11} [W]	−14.486	11.126	−3.360
P_{13} [W]	−12.479	10.260	−2.219
P_{17} [W]	−9.887	8.777	−1.110
P_{19} [W]	−8.660	7.864	−0.796
P_H [W]	−120.557	84.222	−36.335
ξ_{slq}	0.99986	1.00017	0.99997

static converter and the linear load powers are shown separately and where only harmonic orders $h = 5, 7, 11, 13, 17$ and 19 are reported. The harmonic active power P_H and the supply and loading quality index ξ_{slq} given by Equation (2.44) are also reported, considering the contributions due to the characteristic harmonic orders up to $h = 40$.

From the analysis of the data in Table 2.1, it clearly appears that, as is predictable theoretically, the harmonic active powers of the static converter are all negative ($\xi_{slq} < 1$) while the linear load powers are all positive ($\xi_{slq} > 1$). All the harmonic active powers have very low values compared to the active power at fundamental frequency, so that the supply and loading quality index values are close to 1. A more critical situation happens if we consider the customer (both the linear load and the converter together), due to the compensating effects of the active powers; in this case, ξ_{slq} is slightly less than 1.

Finally, it can be verified easily that the sum of the harmonic active powers of converter 1 and the linear load (third column in Table 2.1) coincides with the harmonic active power dissipated at the supply resistance, since the supply system is without background distortion. For example, if we consider the 5th harmonic order ($I_5 = 5.352$ A $R_{u5} = 0.219\ \Omega$), the result is:

5th harmonic active power dissipated at supply resistance $= 3R_{u5}I_5^2 = 3(0.219)(5.352)^2$

$$= 18.819,$$

which is practically the value of $(-P_5)$ in Table 2.1.

Consider now the simple test system of Figure 2.14 (b), in which the static converter 2 generates a distortion for the first converter. This simulates the utility distortion. Table 2.2 shows the same quantities reported in Table 2.1.

From the analysis of the results reported in Table 2.2, it clearly appears that the harmonic active powers of both converters are much larger than the powers of converter 1 in Table 2.1. This is due to the fact that the converter harmonic active powers are directed towards passive impedances of the network (linear load and supply system), and also transferred from one converter to the other. In practice, converter 1 supplies converter 2 with harmonic active powers or vice versa. For example, converter 1 supplies converter 2 with 746.772 W at the 5th harmonic while converter 2 supplies converter 1 with 145.758 W at the 19th harmonic.

Table 2.2 Active powers and supply and loading quality index of the test system in Figure 2.14 (b)

	Converter 1	Linear load	Customer	Converter 2
P_1 [W]	883894.497	493615.967	1377510.464	718153.872
P_5 [W]	−828.590	33.242	−795.348	746.772
P_7 [W]	−588.199	20.993	−567.206	551.551
P_{11} [W]	−205.335	3.865	−201.470	200.302
P_{13} [W]	−43.498	0.246	−43.252	43.198
P_{17} [W]	132.357	5.038	137.395	−138.051
P_{19} [W]	145.758	11.015	156.773	−157.888
P_H [W]	−1400.341	170.220	−1230.120	1158.072
ξ_{slq}	0.99841	1.00034	0.99910	1.00161

Due to the aforementioned effect, the harmonic active power P_H of converter 2 is positive and its supply and loading quality index ξ_{slq} is greater than 1, so that the condition $\xi_{slq} < 1$ cannot be used to distinguish a converter from a linear load. It is interesting to note that the harmonic active power of converter 1 is still negative (in spite of the positive sign of some harmonic active powers) and that $\xi_{slq} < 1$; however, converter 1 is characterized by a ξ_{slq} value lower than the value in the absence of converter 2 ($0.99841 < 0.99986$), so it appears to be a more perturbing load due to the increased value of harmonic active power linked to the presence of converter 2.

Now, let us consider the case of an unbalanced system in which the voltage and current at the x-x terminals have sinusoidal waveforms. The supply and loading quality index can be defined as:

$$\xi_{slq} = \frac{P_{+1} + P_{-1}}{P_{+1}}, \tag{2.45}$$

where P_{+1} is the fundamental positive-sequence component active power and P_{-1} the fundamental negative-sequence component active power. The fundamental negative-sequence power P_{-1} in Equation (2.45) can have positive, negative or zero values.

Similar to the case of harmonic active power, if the section x-x feeds a balanced load in a supply system with background unbalances, the fundamental negative-sequence power P_{-1} is positive and delivered to the load. If the section x-x feeds an unbalanced load and the supply system is without background unbalances, the fundamental negative-sequence power P_{-1} is negative and reflected backwards in the supply system where it is dissipated on its resistance; this is a case similar to the case in Figure 2.12 (b) when a supply system without background distortion feeds a static converter. If the section x-x feeds an unbalanced load in a supply system with background unbalances, the fundamental negative-sequence power P_{-1} can theoretically be positive, negative or zero. Because of its dependence on the value and sign of the fundamental negative-sequence power P_{-1}, the supply and loading quality index ξ_{slq} is less than, greater than or equal to 1.

If we consider the most general case of a system with unbalances and distortions, the following equation holds:

$$\xi_{slq} = \frac{P_{+1} + P_{-1} + \Delta P}{P_{+1}},$$
(2.46)

where ΔP includes the remaining active powers delivered to the load or reflected backwards in the supply system.

Example 2.7   In this example the supply and loading quality index values are calculated considering the presence of both waveform distortions and unbalances.

Consider the simple test system of Figure 2.14 (a), in which both a converter and a linear balanced load are present. The supply system has background unbalances, with an unbalance factor K_d equal to 1.8%. Table 2.3 shows the value of supply and loading quality index ξ_{slq} obtained by applying Equation (2.46)

Table 2.3   Supply and loading quality index values in the presence of background unbalances

	Converter 1	Linear load	Customer
ξ_{slq}	1.00002	1.00051	1.00019

From the analysis of the results in Table 2.3, it appears that the value of the supply and loading quality index ξ_{slq} for the converter is greater than 1; this is due to the overcompensating effect between the fundamental negative-sequence power and the harmonic active power (151.558 W and –120.704 W, respectively). So, once again the condition $\xi_{slq} < 1$ cannot be used to distinguish the converter presence.

In addition, it should be noted that the value of the index ξ_{slq} for the customer is greater than 1 due to a significant value of the fundamental negative-sequence power $P_{-1} = 312.695$ W while the harmonic active power is $P_H = -36.375$ W.

From the analysis of the supply and loading quality index ξ_{slq} definitions and from the analysis of the numerical examples, the following considerations arise.

First, the indices based on harmonic impedances considered in Section 2.2.1 are based on the principle of superposition and the customer and utility contributions are determined *separately*, i.e. the contributions are quantified with the values of two different indices for current and two different indices for voltage. Here, only one index, ξ_{slq}, is used and no superposition is applied; the result is a limit in separating customer and utility responsibilities.

Theoretically, from the condition $\xi_{slq} = 1$, one cannot conclude that the quality is perfect as far as distortion and unbalance are concerned, because of the possible mutual cancellation of active powers (either at harmonic frequencies or associated with the presence of unbalances). For the same reason, one can fail in locating a source of disturbance; for example, the condition $\xi_{slq} > 1$ can occur both in the presence of a linear balanced load fed by a supply system with background distortions (Example 2.6) and in the presence of a static converter fed by a supply system with background unbalance (Example 2.7).

On the other hand, large harmonic and/or negative sequence currents can produce small associated active powers if they are shifted by about $\pi/2$ with respect to the corresponding voltage components, with the consequence that the values of the index are close to unity; these cases do not represent the amount of injected disturbance adequately.

It is also worth noting that the values of index ξ_{slq} vary with time, even if the perturbing load characteristics are unchanged. For example, if we consider the case of an unbalanced load with fixed electrical parameters fed by a supply system with varying background unbalances, the negative-sequence active power interchanges between the utility and the customer vary in turn and the ξ_{slq} values also vary, with the consequence that the unbalanced load appears to have different perturbing characteristics depending on the utility unbalances.

Actually, the same authors that proposed the index ξ_{slq} emphasized that it can be used to help locate the *dominant (prevailing)* source of disturbance [26]. Their intention was that this index should allow the determination of which side (utility or customer) contributes more to the presence of the disturbance at the PCC, subject to the constraint that measurements are taken only at the PCC. They say that, in terms of active powers, when the utility side's disturbing effects prevail over the customer side's disturbing effects, the result is that $\xi_{slq} > 1$, whereas when the customer's effects prevail over the utility's effects, the result is that $\xi_{slq} < 1$. This criterion for identifying the prevailing source is different from that used when the superposition principle is applied (impedance-based indices). In that case, we decide on the basis of the values assumed by the contributions of the utility and customer to the whole current and voltage distortion at the PCC.

We note that publications in the relevant literature have shown that the conclusions to be drawn from applying the two criteria (addressed above) for locating the prevailing source of the disturbance can be different. For example, if we refer to the simple case of only one harmonic, it can be demonstrated that, if the condition $P_h < 0$ holds, it may be that:

- $\xi_{slq} < 1$, in which case the customer side should be the prevailing source on the basis of the active power-based criterion;

- $I_{uh} > I_{ch}$ so that the utility side should be the prevailing source on the basis of the superposition-based criterion.

In fact, if we neglect the resistances, the harmonic active power flowing between the utility and the customer depends on the difference between the phase angles of the two harmonic sources. The direction of this power is related to the phase angles. On the other hand, on the basis of Equations (2.1) and (2.5), and neglecting the resistances once again, the result is that the condition required for $I_{uh} > I_{ch}$ to hold is:

$$X_{uh}J_{uh} > X_{ch}J_{ch}; \tag{2.47}$$

so it does not depend on the phase angle difference between the two harmonic sources. It follows that, even if the condition (2.47) holds and $I_{uh} > I_{ch}$, there can be some values of the phase angle difference between the two harmonic sources for which the result is $P_h < 0$. For more details on this subject see [31].

It can also be demonstrated that, even if the condition $P_h < 0$ holds in a static converter, the harmonic voltage V_h (Figure 2.2) at the PCC (and also in other system busbars) does not necessarily increase, so the presence of a static converter could, in some cases, improve and does not worsen the harmonic voltage at the PCC. For more details on this subject see [21].

2.2.2.2   The utility and customer total harmonic distortion factors

Let us consider initially a balanced system in which the voltage and current at the x-x terminals have nonsinusoidal waveforms.

Let N_A denote the set of harmonic orders h for which harmonic active power is not negative ($P_h \geq 0$) and N_B denote the set of harmonic orders for which harmonic active power is negative ($P_h < 0$) [18]. In practice, it is assumed that N_A includes all harmonic orders for which the supply system is the source of energy and the customer appears as a passive load; although the customer may contain harmonic sources for the harmonics of this set, the customer is assumed to be replaced by a passive element. Meanwhile, N_B includes all harmonic orders for which the customer is the source of energy and the supply system is replaced by a passive element. Sets N_A and N_B are separable.

The following expressions for voltage $v(t)$ and current $i(t)$ at the section x-x can be written:

$$v(t) = v_1 + \sum_{h \in N_A} v_h + \sum_{h \in N_B} v_h = \sqrt{2} V_1 \, \sin(\omega t + \alpha_1) + \sqrt{2} \sum_{h \in N_A} V_h \, \sin(h\omega t + \alpha_h)$$

$$+ \sqrt{2} \sum_{h \in N_B} V_h \, \sin(h\omega t + \alpha_h)$$

$$i(t) = i_1 + \sum_{h \in N_A} i_h + \sum_{h \in N_B} i_h = \sqrt{2} I_1 \, \sin(\omega t + \beta_1) + \sqrt{2} \sum_{h \in N_A} I_h \, \sin(h\omega t + \beta_h)$$

$$+ \sqrt{2} \sum_{h \in N_B} I_h \, \sin(h\omega t + \beta_h). \tag{2.48}$$

In Equation (2.48), V_1 and I_1 are the RMS values of the fundamental components of the voltage and current, and α_1 and β_1 are their phase angles.

The utility total harmonic distortion factors for current and voltage (THDI_U, THDV_U) and the customer total harmonic distortion factors for current and voltage (THDI_C, THDV_C) are the usual total harmonic distortion factors (see Equation 1.5 in Chapter 1) obtained by considering *only* the harmonics of set N_A for THDI_U and THDV_U and *only* the harmonics of set N_B for THDI_C and THDV_C. The result is:

$$\left.\begin{array}{l} \text{THDI}_U = \sqrt{\dfrac{\displaystyle\sum_{h \in N_A} I_h^2}{I_1^2}} \\[4ex] \text{THDV}_U = \sqrt{\dfrac{\displaystyle\sum_{h \in N_A} V_h^2}{V_1^2}} \end{array}\right\} \quad \text{Utility Total Harmonic Distortion Factors} \qquad (2.49)$$

$$\left.\begin{array}{l} \text{THDI}_C = \sqrt{\dfrac{\displaystyle\sum_{h \in N_B} I_h^2}{I_1^2}} \\[4ex] \text{THDV}_C = \sqrt{\dfrac{\displaystyle\sum_{h \in N_B} V_h^2}{V_1^2}} \end{array}\right\} \quad \text{Customer Total Harmonic Distortion Factors}$$

If the section x-x feeds a linear load and the utility has background distortion (Figure 2.12 (a)), the harmonic active powers are all positive and delivered to the load ($P_h > 0 \; \forall \; h$) and the utility total harmonic distortion factors coincide with the usual total harmonic distortion factors, while the customer total harmonic distortion factors are equal to zero. Clearly, the factors increase as the corresponding voltage or current harmonics increase.

On the other hand, if the section x-x feeds a static converter and the supply system is without background distortions (Figure 2.12 (b)), the harmonic active powers are all negative and delivered to the supply system ($P_h < 0 \; \forall \; h$); in this case, the customer total harmonic distortion factors coincide with the usual total harmonic distortion factors while the utility total harmonic distortion factors are equal to zero. Once again, the factors increase as the corresponding voltage and current harmonics increase.

If the section x-x feeds a static converter and the supply system has background distortions (Figure 2.12 (c)), all the indices given by Equation (2.49) can assume values different from zero.

Example 2.8   Consider initially the simple test system of Figure 2.14 (a). The time-domain simulation of the system used for Example 2.6 furnishes the utility total harmonic distortion factors for current and voltage (THDI$_U$, THDV$_U$) and the customer total harmonic distortion factors for current and voltage (THDI$_C$, THDV$_C$), which are given in Table 2.4. We note that the word 'customer' is present in the index name; to avoid confusion, in the third columns of Tables 2.4 and 2.5 we use 'converter 1 and linear load' instead of the usual 'customer'.

Table 2.4   Utility total harmonic distortion factors and customer total harmonic distortion factors for current and voltage of the test system in Figure 2.14 (a)

	Converter 1	Linear load	Converter 1 and linear load
THDV$_U$	0.00000	0.01302	0.00000
THDI$_U$	0.00000	0.01195	0.00000
THDV$_C$	0.01302	0.00000	0.01302
THDI$_C$	0.27486	0.00000	0.17395

Bearing in mind the harmonic active power values reported in Table 2.1 (Example 2.6) and from the analysis of the data in Table 2.4, it clearly appears that, as is predictable theoretically, the utility total harmonic distortion factors are equal to zero for the static converter ($P_h < 0 \; \forall \; h$) while a value different from zero characterizes the linear load ($P_h > 0 \; \forall \; h$); an opposite behaviour characterizes the customer total harmonic distortion factors. If we consider both the linear load and converter 1 together, they can be considered globally as a perturbing equivalent load ($P_h < 0 \; \forall \; h$), but they are characterized by a lower customer total harmonic distortion factor for current due to a reduced value for harmonic currents and a greater value of the fundamental component.

Table 2.5 Utility total harmonic distortion factors and customer total harmonic distortion factors for current and voltage of the test system in Figure 2.14 (b)

	Converter 1	Linear load	Converter 1 and linear load	Converter 2
$THDV_U$	0.01128	0.01856	0.01128	0.01474
$THDI_U$	0.07432	0.01704	0.04915	0.27387
$THDV_C$	0.01474	0.00000	0.01474	0.01128
$THDI_C$	0.26471	0.00000	0.16523	0.08989

Now, consider the simple test system of Figure 2.14 (b). Table 2.5 shows the same quantities reported in Table 2.4.

Bearing in mind the harmonic active power values in Table 2.2 (Example 2.6) and from the analysis of the data in Table 2.5, it clearly appears that both converters are characterized by utility total harmonic distortion factors and customer total harmonic distortion factors different from zero. This is due to the fact that in both cases, some harmonic active powers are positive and some are negative. Once again, a value different from zero characterizes correctly the utility total harmonic distortion factors of the linear load ($P_h > 0 \ \forall$ h).

Now let us consider the case of an unbalanced nonsinusoidal system. The new utility and customer total harmonic distortion factors are defined as:

$$THDI_{U^+} = \sqrt{\dfrac{3\lambda_U I_{-1}^2 + \displaystyle\sum_{p=a,b,c}\sum_{h\in N_A} I_{ph}^2}{3I_{+1}^2}}$$

$$THDV_{U^+} = \sqrt{\dfrac{3\lambda_U V_{-1}^2 + \displaystyle\sum_{p=a,b,c}\sum_{h\in N_A} V_{ph}^2}{3V_{+1}^2}}$$

Utility Total Harmonic Distortion Factors

$$THDI_{C^+} = \sqrt{\dfrac{3\lambda_C I_{-1}^2 + \displaystyle\sum_{p=a,b,c}\sum_{h\in N_B} I_{ph}^2}{3I_{+1}^2}}$$

$$THDV_{C^+} = \sqrt{\dfrac{3\lambda_C V_{-1}^2 + \displaystyle\sum_{p=a,b,c}\sum_{h\in N_B} V_{ph}^2}{3V_{+1}^2}}$$

Customer Total Harmonic Distortion Factors

$$(2.50)$$

where V_{+1} (I_{+1}) is the fundamental positive sequence of the voltage (current), V_{-1} (I_{-1}) is the fundamental negative sequence of the voltage (current), and p is the phase code ($p = a, b, c$); now, the sets N_A and N_B include the harmonic orders for which the three-phase harmonic active power is not negative (N_A) and for which the three-phase harmonic active power is negative (N_B). In addition:

- $\lambda_U = 1$ and $\lambda_C = 0$ if the negative-sequence fundamental active power is not negative (delivered to the load);

- $\lambda_U = 0$ and $\lambda_C = 1$ if the negative-sequence fundamental active power is negative (delivered back to the supply).

The following considerations arise.

First, we outline once again a difference with the harmonic impedance-based indices, which, as previously demonstrated, use the superposition principle. The indices based on the separable sets N_A and N_B share all harmonics in two separate groups: the first group is linked to non-negative active powers, for which the utility bears total responsibility, and the second group is linked to negative active powers, for which the customer bears total responsibility. Thus, strictly speaking, each harmonic is attributed exclusively either to the utility or to the customer. In practice, the utility total harmonic distortion factors for current and voltage and the customer total harmonic distortion factors for current and voltage try to separate customer and utility contributions, although in a different way from the harmonic impedance-based indices.

The presence of two separate indices for the utility and the customer, which include harmonic current and voltages derived separately from active powers with different signs, avoids the possibility of mutual cancellation (either at harmonic frequencies or associated with the presence of unbalances). Moreover, they seem to be more robust than ξ_{slq} in identifying a source of disturbance, obviously in terms of active power effects. In fact, a static converter fed by a supply system with background distortion is characterized by customer total harmonic distortion factors different from zero, so it is identified as a disturbing load, under the condition that at least one of the harmonic active powers P_h is negative. In other words, if, for example, all harmonic active powers are positive except the 13th order and $P_H > 0$, ξ_{slq} is greater than 1 and the load is identified as a nondisturbing load, but the customer total harmonic distortion factors are different from zero due to the sign of the 13th order harmonic active power.

In addition, the values of the utility and customer total harmonic distortion factors vary with time, in a similar manner to the case analyzed for the supply and loading quality index.

Finally, it should be noted that the measurements of all the harmonic active power-based indices require the analysis of the signs of the harmonic and fundamental negative-sequence active powers, which depend on the phase shift between the corresponding voltage and current components; obviously, particular problems can arise when the phase shift gets close to 90°. To overcome these potential problems, measurement uncertainty has to be taken into account and estimated correctly. The overall uncertainty results from the combination of all the elements of the measurement chain. It is appropriate to make use of a nondecisional range around zero (uncertainty range) for the active powers, the amplitudes of which will be related to the accuracy of the instruments employed; the powers in this range are considered neither positive nor negative and more accurate identification methods should be applied.

For more details on the subject of this index see [18, 26, 28, 33, 34, 35].

2.2.2.3   The harmonic phase index and the harmonic global index

Initially, let us consider a balanced three-wire system, in which the voltage and current at the x-x terminals have nonsinusoidal waveforms.

The harmonic phase index makes use of the sets N_A and N_B that were introduced previously for the utility and customer harmonic distortion factors. The harmonic phase index is defined as:[7]

$$\xi_{HPI} = \frac{\|\mathbf{I}_C\|^2}{\|\mathbf{I}_U\|^2} = \left(\frac{\sqrt{\sum_{h \in N_B} I_h^2}}{\sqrt{I_1^2 + \sum_{h \in N_A} I_h^2}} \right)^2 = \frac{(THDI_C)^2}{1 + (THDI_U)^2}, \tag{2.51}$$

where $\mathbf{I}_C$ and $\mathbf{I}_U$ are two vectors, the components of which include the current harmonic phasors associated with the sets N_B and N_A, respectively, and the symbol $\|\ \|$ indicates the square root of the sum of the squared magnitudes of the vector elements. Vector $\mathbf{I}_U$ also includes the fundamental component.

If the section x-x feeds a linear load and the utility has background distortions (Figure 2.12 (a)), the harmonic active powers are all positive ($P_h > 0 \ \forall h$) and the customer total harmonic distortion factor for the current is equal to zero, as is the harmonic phase index.

If the section x-x feeds a static converter and the supply system is without background distortions (Figure 2.12 (b)), the harmonic active powers are all negative ($P_h < 0 \ \forall h$). The utility total harmonic distortion factors for the current are equal to zero and the harmonic phase index coincides with the square of the usual total harmonic distortion factor and, clearly, increases with it.

If the section x-x feeds a static converter and the supply system has background distortions (Figure 2.12 (c)), the harmonic phase index is generally greater than zero. The greater the index, the greater the responsibility of the customer may be; conversely, the lower the index, the greater the responsibility of the utility may be.

Example 2.9   Consider the simple test system of Figure 2.14 (a). The time-domain simulation of the system used for Example 2.6 furnishes the harmonic phase index given in Table 2.6.

Table 2.6   Harmonic phase index of the test system in Figure 2.14 (a)

	Converter 1	Linear load	Customer
ξ_{HPI}	0.07555	0.00000	0.03026

From the analysis of the data in Table 2.6 it clearly appears that, as is predictable theoretically, the harmonic phase index of the linear load is equal to zero while it coincides with the square of the usual total harmonic distortion factor for converter 1 (see Table 2.4). If

---

[7] This index is sometimes presented in the literature without the squares.

we consider the customer (both linear load and converter 1 together), it is characterized by a lower harmonic phase index than converter 1 due to a lower value of the index THDI_C ($0.17395 < 0.27486$) and the value $\text{THDI}_U = 0$ for both.

Now consider the simple test system of Figure 2.14 (b). Table 2.7 shows the same quantities reported in Table 2.6.

Table 2.7 Harmonic phase index of the test system in Figure 2.14 (b)

	Converter 1	Linear load	Customer	Converter 2
ξ_{HPI}	0.06968	0.00000	0.02723	0.00751

From the analysis of the results reported in Table 2.7, it clearly appears that both converters are characterized by harmonic phase index values different from zero (see Table 2.5). Converter 1 is characterized by an index value greater than converter 2 due to a lower value of the utility total harmonic distortion factor of current and a greater value of the customer total harmonic distortion factor of current (see Table 2.5).

Now let us consider the case of an unbalanced system in which voltage and current at the x-x terminals have nonsinusoidal waveforms.

The harmonic phase index is defined as:

$$\xi_{\text{HPI}} = \frac{\lambda_C I_{-1}^2 + \|\mathbf{I}_C\|^2}{\|\mathbf{I}_U\|^2},$$
(2.52)

where I_{-1} is the fundamental negative-sequence component of the phase currents and λ_C is the parameter introduced in the previous section. The sets N_A and N_B are extended to the phase powers P_{ph} ($p = a, b, c$; $h = 1, 2, \ldots, H_{\max}$), so that $\mathbf{I}_C$ and $\mathbf{I}_U$ are two vectors that properly include the fundamental and harmonic phase current phasors $\bar{I}_{ph}$ ($p = a, b, c$; $h = 1, 2, \ldots, H_{\max}$).

In the case of four-wire systems, an imaginary fourth conductor is considered corresponding to the zero-sequence. On the basis of the sign of a properly calculated homopolar power for each harmonic, the corresponding homopolar current is included in the vector $\mathbf{I}_C$ or $\mathbf{I}_U$.

This index makes use of the two separable sets N_A and N_B; however, limits arise in separating customer and utility responsibilities because all information about their behaviour is included in only one index. Moreover, it refers only to currents.

In addition, this index once again requires analysis of the sign of the active power. As previously discussed, the sign may be estimated incorrectly and the corresponding current component assigned to the wrong set. Clearly, a double-error effect on the index value arises, because both the numerator and the denominator of its relationship are affected by errors and these errors combine to amplify the error ratio. If, for example, the 5th harmonic active power is estimated (P_5 is estimated to be < 0 instead of > 0), the corresponding harmonic current is included as an $\mathbf{I}_C$ vector component instead of an $\mathbf{I}_U$ vector component, resulting in a double-error effect in the index calculation.

It should be noted that the harmonic phase index values are based on harmonic current values instead of harmonic active powers. So, the amount of the injected disturbance might be better represented.

In the case of four-wire systems, great attention is required when dealing with homopolar powers.

The harmonic global index is conceptually very similar to the harmonic phase index and similar considerations can be effected; the difference is that it uses the Park transformation. For the sake of brevity, we consider the case of an unbalanced system.

If the Park transformation is applied to phase voltages and currents, the well-known d-q Park components are obtained and, subsequently, the voltage and current Park vectors and their complex Fourier series can be applied:

$$u_d(t) + ju_q(t) = \sum_{k = -\infty}^{k = +\infty} \overline{U}k e^{jk\omega t}, \quad i_d(t) + ji_q(t) = \sum_{k = -\infty}^{k = +\infty} \overline{I}k e^{jk\omega t}, \tag{2.53}$$

where the harmonic components of the Park vector for $k > 0$ are related to the positive symmetrical components, while for $k < 0$ they are related to the negative symmetrical components.

The following active powers associated with the voltage and current components in Equation (2.53) can be calculated as shown:

$$P_k = \mathrm{Re}\{\overline{U}k\overline{I}_k^*\} \quad \text{for } k \geq 0, \quad P_k = \mathrm{Re}\{\overline{U}_k^*\overline{I}k\} \quad \text{for } k < 0, \tag{2.54}$$

and then two vectors $\mathbf{I}_{CP}$ and $\mathbf{I}_{UP}$ can be introduced, the components of which are the current components associated with positive ($\mathbf{I}_{UP}$) and negative ($\mathbf{I}_{CP}$) active powers in Equation (2.54), respectively. Finally, the harmonic global index is given by:

$$\xi_{HGI} = \frac{\|\mathbf{I}_{CP}\|^2}{\|\mathbf{I}_{UP}\|^2}. \tag{2.55}$$

In the case of four-wire systems, the homopolar components are also included in the procedure to calculate Equation (2.55).

For more details about these indices see [27].

Indices based on Park vectors, but which use the active powers directly, have also been proposed [20].

2.2.2.4   The critical impedance and admittance indices

The critical impedance index (CI index) has been proposed to determine whether the utility or customer is mainly responsible for the harmonic current distortions measured at the PCC ($I_{ch} >$ or $< I_{uh}$). Equivalently, the critical admittance index (CA index) has been proposed to determine whether the utility or customer is mainly responsible for the harmonic voltage distortions measured at the PCC ($V_{ch} >$ or $< V_{uh}$). These indices are based on the evaluation of harmonic reactive power.

In order to calculate the CI and CA indices, the utility impedance $\dot{Z}_{uh}$ and a range of the customer impedance $\dot{Z}_{ch}$ are assumed to be approximately known; then, in the following, an approximate range $[\dot{Z}_{h\ min}, \dot{Z}_{h\ max}]$ of the total impedance ($\dot{Z}_h = \dot{Z}_{ch} + \dot{Z}_{uh}$) is assumed to be known.

First let us consider the CI index and the Thevenin equivalent circuit for the utility and the customer installation reported in Figure 2.10 (b). Initially, it is assumed, for the sake of simplicity, that the impedances are purely reactive impedances ($\dot{Z}_{uh} = jX_{uh}$ and $\dot{Z}_{ch} = jX_{ch}$, with $X_{uh} + X_{ch} > 0$). This assumption provides the following result for the utility harmonic reactive power (Figure 2.10 (b)):

$$Q_{uh} = Im(\overline{V}_{uh}^{T}\overline{I}_{h}^{*}) = \frac{V_{uh}^{T}V_{ch}^{T}\cos(\psi_h) - (V_{uh}^{T})^2}{(X_{uh} + X_{ch})}, \tag{2.56}$$

where ψ_h is the phase angle difference between customer and utility voltage sources.

From the analysis of Equation (2.56) it appears that:

1. if $Q_{uh} > 0$, the relation $V_{ch}^{T} > V_{uh}^{T}$ is verified. Thus, bearing in mind Equation (2.40), we can conclude that the main contribution to the harmonic current distortion at the PCC is due to the customer ($I_{ch} > I_{uh}$);

2. if $Q_{uh} < 0$, it is not possible to decide the main contribution to the harmonic distortion at the PCC, because the utility 'generates' reactive power that may not reach the customer side since the utility and customer reactances 'absorb' reactive power. A further step of investigation is needed.

The key idea for solving the problem in case 2 is to find how far the reactive power can 'travel' along the combined utility and customer impedances $\dot{Z}_h \cong jX_h = j(X_{uh} + X_{ch})$, assuming this to be a 'line reactance' that is uniformly distributed between the harmonic voltage sources $\overline{V}_{uh}^{T}$ and $\overline{V}_{ch}^{T}$. The authors who proposed this method in [32] defined X_h^0 as the reactance from the utility voltage source $\overline{V}_{uh}^{T}$ to the lowest voltage point along the 'line'; then, they assumed that:

- if X_h^0 is located closer to the customer side $(X_h^0 > \frac{X_h}{2})$, the utility source is expected to have the larger magnitude $(V_{uh}^{T} > V_{ch}^{T})$, since the utility source can 'push' its reactive power output beyond the halfway point of the 'line reactance'; and

- if X_h^0 is located closer to the utility side $(X_h^0 < \frac{X_h}{2})$, the customer source is expected to have the larger magnitude $(V_{uh}^{T} < V_{ch}^{T})$.

It can be demonstrated that the reactance X_h^0 is given by the following relationship:

$$X_h^0 = \frac{(V_{uh}^{T})^2 - V_{uh}^{T}V_{ch}^{T}\cos(\psi_h)}{(V_{uh}^{T})^2 + (V_{ch}^{T})^2 - 2V_{uh}^{T}V_{ch}^{T}\cos(\psi_h)}X_h. \tag{2.57}$$

Bearing in mind Equation (2.56) and using the Thevenin equivalent circuit of Figure 2.10 (b) with $\dot{Z}_{ch} + \dot{Z}_{uh} = jX_h$, Equation (2.57) can be rewritten as follows:

$$X_h^0 = -\frac{Q_{uh}}{I_h^2}. \tag{2.58}$$

Starting from Equation (2.58), the CI index is defined as follows:

$$CI = -2X_h^0 = 2\frac{Q_{uh}}{I_h^2} = 2\frac{V_{uh}^{T}\sin(\theta_h)}{I_h} = 2\frac{|\overline{V}_h - I_h\dot{Z}_{uh}|\sin(\theta_h)}{I_h}, \tag{2.59}$$

where θ_h is the phase angle by which $\overline{V}_{uh}^{T}$ leads $\overline{I}_h$.

Then, in order to determine whether the utility or customer is mainly responsible for the harmonic current distortion, and assuming that $[X_{h\ min}, X_{h\ max}]$ is the known approximate range of X_h, the following procedure is proposed:

1. if $CI > 0$ ($Q_{uh} > 0$), the customer side is the dominant harmonic current source at the PCC.

2. If $CI < 0$ ($Q_{uh} < 0$), the following three cases can result:

   (a) if $|CI| > X_{h\ max}$, then $X_h^0 > \frac{X_{h\ max}}{2}$, and the utility side is the dominant harmonic current source at the PCC;

   (b) if $|CI| < X_{h\ min}$, then $X_h^0 < \frac{X_{h\ min}}{2}$, and the customer side is the dominant harmonic current source at the PCC; and

   (c) if $X_{h\ min} < |CI| < X_{h\ max}$, it is not possible to conclude which entity is the dominant harmonic current source at the PCC.

The above procedure can be generalized by taking into account the presence of the non-negligible resistances of the impedances $\dot{Z}_{uh}$ and $\dot{Z}_{ch}$. Then, Equation (2.56) can be extended, taking into account the total impedance $\dot{Z}h$ and its angle β. In fact, using a rotation transformation matrix based on angle β, the transformed utility harmonic reactive power can be calculated. Once the transformed utility harmonic reactive power is known, and assuming that the approximate range of the total impedance $\dot{Z}h$ is also known, the CI index value can be obtained and the above procedure can be applied by substituting the impedances for the reactances [32].

On the same conceptual ground, the critical admittance (CA) index was proposed to determine whether the utility or the customer is mainly responsible for the harmonic voltage distortions at the PCC ($V_{ch} >$ or $< V_{uh}$). The CA index uses the Norton equivalent circuits reported in Figure 2.2 (a).

In this case, the key requirement for solving the aforementioned problem 2 is to find how far the reactive power can 'travel' along the combined utility and customer admittance, assuming that it is distributed between the harmonic current sources $\overline{J}_{uh}$ and $\overline{J}_{ch}$ as a 'line susceptance' B_h. The authors who proposed this method in [32] introduced B_h^0 as the susceptance from the utility current source $\overline{J}_{uh}$ to the lowest harmonic current point along the 'line', and then they assumed that, if $B_h^0 > \frac{B_h}{2}$, the utility side is the main contributor to the harmonic voltage distortion at the PCC, and that, if $B_h^0 < \frac{B_h}{2}$, the customer side is the main contributor to the harmonic voltage distortion at the PCC.

Analogously to the CI index, the CA index was defined as a function of the harmonic reactive power and a similar procedure was proposed in order to determine whether the utility or customer is mainly responsible for the harmonic voltage distortion.

The main problems with the application of the CI and CA indices are the accuracy of the knowledge of the utility impedance $\dot{Z}_{uh}$ and the accuracy of the knowledge of the range of the impedances $\dot{Z}_{ch}$; however, in the literature, it has been shown that the method can tolerate significant impedance errors. We experienced some problems when the utility side source $\overline{J}_{uh}$ (or $\overline{V}_{uh}$) and the customer side source $\overline{J}_{ch}$ (or $\overline{V}_{ch}$) had similar values.

Example 2.10   In this example we consider once again the test system of Figure 2.14 (b) and we calculate the CI index by applying the procedure outlined above. Let us consider the 5th harmonic and, initially and for the sake of simplicity, assume that we have good knowledge of the actual values of the equivalent harmonic impedances of the utility and the customer. We assume that the impedances are purely reactive impedances:

$$\dot{Z}_{u5} = j10.945 \ [\Omega]; \quad \dot{Z}_{c5} = j69.182 \ [\Omega].$$

Then, the combined utility and customer impedance value is:

$$\dot{Z}5 = j80.127 \ [\Omega].$$

Table 2.8 shows that the value of the CI index for the customer (converter 1 and linear load together) is positive. Then, it can be concluded that the customer is the dominant harmonic current source contributor at the PCC. (See Example 2.1.)

Table 2.8   Value of the CI index

	Customer
CI [Ω]	4.480

Now let us assume an approximate value for the utility impedance ($\dot{Z}_{u5} \approx j15.5 \ [\Omega]$) and a range of customer impedance values ($\dot{Z}_{c5} \in [j10, j150] \ [\Omega]$), so that the combined utility and customer impedance value is included in the following interval: $\dot{Z}_5 \in [jX_{5min}, jX_{5max}] = [j25.5, j165.5] \ [\Omega]$. Table 2.9 shows the new value of the CI index for the customer (converter 1 and linear load together). This value is negative and its absolute value is lower than the minimum reactance value X_{5min}; so, once again, it can be concluded that the customer is the dominant harmonic current source contributor at the PCC.

Table 2.9   Value of the CI index

	Customer
CI [Ω]	−4.629

Note that a further single-point proposal based on harmonic powers different from the active power was published recently in the literature [36]. This proposal has the aim of identifying the harmonic source by comparing some nonactive powers, i.e. the fundamental reactive power, Fryze's reactive power and the quadrature reactive power proposed by Sharon.

2.2.3   Indices based on comparison with an ideal linear load

These indices were designed to quantify the extent to which a nonideal load deviates from being an ideal load. The indices considered in this section are:

- current-based indices; and
- the load nonlinearity indicator.

They differ from each other in the definitions of the ideal load to be considered as reference and in the electrical quantities to be quantified.

2.2.3.1   Current-based indices

The first current-based approach proposed in the literature was based on the concept of conforming and nonconforming currents [37–39, 41]; initially, it was applied only for attributing harmonic responsibility.

The conforming current is the portion of the current $i(t)$ that retains the same level of distortion as the voltage $v(t)$ at the PCC. The conforming current can be described mathematically as:

$$i_C(t) = \sqrt{2} \sum_{h=1}^{H_{max}} \frac{I_1}{V_1} V_h \sin[h\omega t + \alpha_h + h(\beta_1 - \alpha_1)], \qquad (2.60)$$

where V_h is the RMS value of the voltage harmonic, α_h is its phase angle, V_1 and I_1 are the RMS values of the fundamental components of the voltage and current, and α_1 and β_1 are their phase angles.

From the analysis of Equation (2.60), it clearly appears that $i_C(t)$ accounts for all the active and reactive powers at the fundamental frequency. (The fundamental frequency component of the conforming current coincides with that of the current $i(t)$ at the PCC). At harmonic frequencies, it has spectral component amplitudes in the same proportion as their counterparts in the voltage.

The nonconforming current is the remaining part of the current at the PCC:

$$i_{NC}(t) = i(t) - i_C(t). \qquad (2.61)$$

It is clear that both conforming and nonconforming currents can contain harmonics.

Starting from the aforementioned separation of current, it is stated that the nonconforming current is attributable to the customer and that the distortion in the conforming current is attributable to the utility. With such a sharing of responsibility, a pure resistance is assumed to be the ideal load for the customer; in fact, it draws harmonics in the same proportion as in the voltage since it has the same impedance at all frequencies. Thus, nonconforming current is generated by nonlinear loads, time-varying loads and linear loads which have 'equivalent impedance' that depends on the frequency.

Example 2.11   In this example, considering the test system of Figure 2.14 (a), we calculate conforming and nonconforming currents by applying Equations (2.60) and (2.61). The time-domain simulation of the system furnishes the conforming and nonconforming harmonic currents (RMS values) given in Table 2.10 (only harmonic orders $h = 5, 7, 11, 13$ and fundamental are shown).

From the analysis of the data in Table 2.10 it clearly appears that the nonconforming current harmonics of the converter are significantly greater than those of the linear load. However, we should note the presence of nonconforming current harmonics for the linear load, due to the R-L linear load model adopted in the simulation, and of conforming current harmonics for the converter due to the presence of voltage harmonics at the PCC (see Equation (2.60)).

The criterion used for sharing conforming and nonconforming currents (and consequently attributing responsibilities), and the fact that this sharing involves only current (for all

Table 2.10   Conforming and nonconforming current harmonics of the test system in Figure 2.14 (a)

h	Converter 1 Conforming current [A]	Converter 1 Nonconforming current [A]	Linear load Conforming current [A]	Linear load Nonconforming current [A]
1	27.109	0.000	15.687	0.000
5	0.138	5.499	0.079	0.127
7	0.133	3.795	0.077	0.145
11	0.128	2.215	0.074	0.111
13	0.123	1.734	0.071	0.064

current-based indices), have generated intensive discussion; the influence of the supply system has also been investigated.

Current sharing discourages customers with linear loads which have 'nonconforming' currents. For example, although one may expect only a conforming current when a resistive-inductive load (linear load) is subjected to background distortions, to the contrary it is characterized by a nonconforming current because its impedance varies with frequency. This effect is more critical for linear loads characterized by a significant variation in their impedance versus frequency.[8]

Another problem is that the current of a static converter in a supply system without background distortion has conforming current spectral components, in spite of the fact that the utility is free of its own distortions, due to voltage drops in the supply system.

In the most general case, it should be noted that the voltage waveform at the PCC influences the current absorbed by customer loads and, at the same time, it is influenced by these currents so that, for example, the conforming current harmonics of the customer (attributed to the utility responsibility) are influenced also by the customer load characteristics.

A modified conforming criterion based on a virtual current waveform which has the same harmonic amplitudes of current but the arguments of the voltage was proposed recently. In this way, the conforming current completely conforms to the voltage at the PCC [44].

The original aforementioned separation current criterion for attributing harmonics was also proposed for attributing unbalances.

With reference to unbalances, let $\overline{V}p, \overline{V}n, \overline{V}z, \overline{I}p, \overline{I}_n, \overline{I}z$ be the symmetrical components of the three-phase voltages and currents at the PCC.

The conforming current is the portion of the total current $i(t)$ at the PCC that retains the same level of unbalance as the voltage while at the same time accounts for all the positive-sequence current. It follows that the conforming current has:

- the same positive-sequence $\overline{I}_{Cp}$ as the total current:

$$\overline{I}_{Cp} = \overline{I}_p \tag{2.62}$$

---

[8] In the discussion section of the paper in which this method was proposed [37], a penalty was proposed for capacitive loads (considered to be loads that amplify distortions) and a credit was proposed for inductive loads (considered to be loads that attenuate distortions).

- the negative and zero sequences $\bar{I}_{Cn}, \bar{I}_{Cz}$ in the same proportion as their counterparts in the voltage:

$$\frac{\bar{I}_{Cn}}{\bar{I}_{Cp}} = \frac{\bar{V}_n}{\bar{V}_p}, \quad \frac{\bar{I}_{Cz}}{\bar{I}_{Cp}} = \frac{\bar{V}_z}{\bar{V}_p}. \tag{2.63}$$

Then, the following relationships define the conforming current in the presence of unbalances:

$$\bar{I}_{Cp} = \bar{I}_p, \quad \bar{I}_{Cn} = \bar{I}_p \left(\frac{\bar{V}_n}{\bar{V}_p} \right), \quad \bar{I}_{Cz} = \bar{I}_p \left(\frac{\bar{V}_z}{\bar{V}_p} \right). \tag{2.64}$$

The nonconforming current is the balance of the current. In terms of symmetrical components, it is given by the following relationship:

$$\bar{I}_{NCp} = 0, \quad \bar{I}_{NCn} = \bar{I}_n - \bar{I}_{Cn}, \quad \bar{I}_{NCz} = \bar{I}z - \bar{I}_{Cz}. \tag{2.65}$$

Starting from the symmetrical components in Equations (2.64) and (2.65), the three-phase conforming and nonconforming currents can both be obtained easily with the well-known transformation relationships.

We note that no specific index has been clearly introduced to synthetically quantify customer and utility responsibilities, but the conforming and nonconforming currents are considered as a whole and as a starting point for attributing responsibilities.

The second current-based approach (index) tries to evaluate separately the customer contribution to waveform distortions and unbalances at the PCC [42, 43, 45]. It assumes that an ideal load is an 'equivalent linear and balanced load.' This 'equivalent linear and balanced load' is comprised of three identical R-L series circuits with the resistance R and inductance L values assumed to be constant with frequency; their values are estimated on the basis of the voltage and current measured at the PCC.

If only waveform distortion evaluation is dealt with, the R-L parameters of the ideal load are estimated on phase a with the following relationships:

$$R_a = \frac{V_{a1}}{I_{a1}} \cos(|\alpha_{a1} - \beta_{a1}|), \quad L_a = \frac{V_{a1}}{2\pi f_1 I_{a1}} \sin(|\alpha_{a1} - \beta_{a1}|), \tag{2.66}$$

where V_{a1} and I_{a1} are the RMS values of voltage and current fundamental components at phase a, α_{a1} and β_{a1} are the phase angles of the same quantities and f_1 is the fundamental frequency. The same method, of course, can be applied to phases b and c.

Once the ideal load parameters are known, an ideal linear current is calculated as:

$$i_{aL}(t) = \sum_{h=1}^{H_{max}} \frac{\sqrt{2} V_{ah}}{Z_{ah}} \sin[h\omega t + (\alpha_{ah} - \varphi_{ah})] \tag{2.67}$$

where:

$$\dot{Z}_{ah} = Z_{ah} \exp(j\varphi_{ah}) = R_a + jh\omega L_a.$$

Then, a nonlinear current for phase a can be calculated as the difference between the actual and ideal linear currents:

$$i_{aNL}(t) = i_a(t) - i_{aL}(t). \tag{2.68}$$

Finally, the following phase a nonlinear current index is introduced to determine the level of nonlinearity of a disturbing load:

$$\text{NLCI}_a = \frac{I_{a\text{NL}}}{I_a} \, 100, \qquad (2.69)$$

where:

$$I_{a\text{NL}} = \sqrt{\sum_{h=1}^{H_{max}} I^2_{a\text{NL},h}}, \quad I_a = \sqrt{\sum_{h=1}^{H_{max}} I^2_{a,h}}$$

and where $I_{a,h}$ is the RMS value of the hth harmonic of the actual current at phase a and $I_{a\text{NL},h}$ is the same quantity of the nonlinear current.[9]

In an electrical system with unbalances, values of three different indices such as Equation (2.69) can be calculated, with the nonlinear currents for phases b and c $i_{b\text{NL}}(t)$, $i_{c\text{NL}}(t)$ obtained by applying the same procedure applied for phase a; it is also possible to evaluate the waveform distortion on a three-phase basis by combining the three different index values calculated for each phase into a unique index defined as the ratio between the norm of the vector of the spectral components (RMS) of nonlinear currents $i_{p\text{NL}}(t)$ $(p = a, b, c)$ and the norm of the vector of the spectral components (RMS) of the phase currents $i_p(t)$ $(p = a, b, c)$.

Example 2.12   In this example, considering the test system of Figure 2.14 (a), we calculate the phase a nonlinear current index by applying Equation (2.69). The time-domain simulation of the system furnishes the NLCI_a values given in Table 2.11.

Table 2.11   Phase a nonlinear current index values of the test system in Figure 2.14 (a)

	Converter 1	Linear load	Customer
$\text{NLCI}_a[\%]$	26.859	1.148	17.479

From the analysis of the data in Table 2.11 it clearly appears that, as is predictable theoretically, the phase a nonlinear current index value for converter 1 is much greater than that for the linear load. If we consider the customer (linear load and converter 1 together), it can be globally considered as a perturbing equivalent load, but characterized by a lower *phase a nonlinear current index* value due to the partial compensating effect of the linear load. The linear load shows an index value different from zero due to an R-L parallel circuit adopted in the simulation instead of the R-L series circuit characterizing the ideal load.

It should be noted that the index values are obviously strongly affected by the ideal load model choice. For example, in the case of a linear load in the presence of background distortion, the index values are zero (customer not penalized) only if the ideal load model approaches the actual load model perfectly. If the actual linear load shows a different

---

[9] To take into account the effect of the nonlinear current on the voltage waveform, a weighted index defined by the ratio of $I_{a\text{NL}}$ and the short circuit current (RMS value) was also proposed.

model behaviour (for example, a resistive-capacitive load), the index values are different from zero.

Some experiments published in the relevant literature have shown that the index values have low sensitivity to the network voltage conditions. The choice of linear load model and the sensitivity of the converter harmonic currents to the supply voltage distortions can influence the results.

If only unbalance evaluation is dealt with, the three identical R-L series parameters of the ideal load are estimated on the fundamental voltage and current of the phase characterized by the minimum current RMS value. Assuming that this is phase a, the following balanced linear currents are calculated for phases b and c:

$$i_{pBL}(t) = \sum_{h=1}^{H_{max}} \frac{\sqrt{2}V_{ph}}{Z_{ah}} \sin\left[h\omega t + \left(\alpha_{ph} - \varphi_{ah}\right)\right] \quad p = b, c; \tag{2.70}$$

the supply system is the only cause of their unbalances and distortions.

Then, an unbalanced nonlinear current can be calculated as the difference between the actual and balanced linear currents:

$$i_{pUNL}(t) = i_p(t) - i_{pBL}(t) \quad p = b, c. \tag{2.71}$$

The currents given by Equation (2.71) should take into account the extent to which phases b and c differ from the ideal reference conditions in terms of distortion and unbalances. Then, if it is assumed that the contribution of waveform distortion on phases b and c can be eliminated by subtracting the nonlinear currents for phases b and c $i_{pNL}(t)$ ($p = b$, c) in Equation (2.71), the following currents can be obtained, which should take into account only the unbalances:

$$i_{pU}(t) = i_{pUNL}(t) - i_{pNL}(t) \quad p = b, c. \tag{2.72}$$

Finally, the following unbalance current index is introduced:

$$UCI = \frac{\sqrt{\sum_{p=b,c} I_{pU}^2}}{\sqrt{\sum_{p=a,b,c} I_{pNL}^2}}. \tag{2.73}$$

where:

$$I_{pU} = \sqrt{\sum_{h=1}^{H_{max}} I_{pU,h}^2}, \quad I_{pNL} = \sqrt{\sum_{h=1}^{H_{max}} I_{pUNL,h}^2},$$

and $I_{pU,h}$ and $I_{pNL,h}$ are the RMS values of the hth harmonic of currents $i_{pU}(t)$ and $i_{pNL}(t)$.

Some experiments published in the literature have shown that the index value is zero or nearly zero in the presence of a balanced load and almost constant in the face of supply system variations. Once again, problems can arise due to the choice of the load model.

2.2.3.2 The load nonlinearity indicator

Initially, let us consider a balanced system in which the voltage and current at the x-x terminals have nonsinusoidal waveforms.

The load nonlinearity indicator, LNI, is defined as [40]:

$$\text{LNI} = \sqrt{\frac{1}{N^*} \sum_{n=1}^{N^*} \left[\frac{v(n) - v_c(n)}{V_p} \right]^2},$$

(2.74)

where N^* is the total number of samples in one fundamental cycle, n is the sample code, v is the measured voltage, V_p is the peak value of v and v_c is a reference voltage obtained by applying the following procedure:

1. Calculate the fundamental components of v and i using Fourier transforms.

2. Calculate the fundamental impedance $\dot{Z}_1 = Z_1 \exp(j\varphi_1) = R + j\omega L$ of an ideal load using the fundamental voltage and current components obtained in step 1.

3. Calculate the reference voltage v_c using the current i measured at the PCC and the resistance R and inductance L obtained in step 2; the reference voltage corresponds to the voltage that the actual current would generate if flowing into the ideal load impedance.

A normalization to the voltage v was also proposed.

Example 2.13   In this example, considering the test system of Figure 2.14 (a), we calculate the load nonlinearity index applying Equation (2.74). The time-domain simulation of the system furnishes the load nonlinearity indicator given in Table 2.12.

Table 2.12   Load nonlinearity indicator values of the test system in Figure 2.14 (a)

	Converter 1	Linear load	Customer
LNI	0.197	0.003	0.126

From the analysis of the data in Table 2.12 it clearly appears that, as predictable theoretically, the load nonlinearity indicator value for the converter is much greater than that for the linear load. If we consider the customer (linear load and converter 1 together), it can be globally considered a perturbing equivalent load, but characterized by a lower load nonlinearity indicator value due to the partial compensating effect of the linear load.

The load nonlinearity indicator is calculated for different loading conditions and the average is used to represent the load.

In an electrical system with unbalances, three different index values for the three phases can be calculated – with the three indicators obtained by applying the procedure outlined previously – then the load nonlinearity on a three-phase basis can be computed by applying the following relationship:

$$\text{LNI}_{3\varphi} = \frac{\displaystyle\sum_{p=a,b,c} I_p \text{LNI}_p}{\displaystyle\sum_{p=a,b,c} I_p},$$

(2.75)

where I_p is the average RMS value of the phase current.

It should be noted that this index refers to voltage rather than current at the PCC; low values should be assumed in the case of linear loads and high values should be assumed in the case of static converters, with the above-mentioned problems linked to the ideal load model chosen. Once again, no index has been proposed to quantify the utility responsibility explicitly.

2.3 Waveform distortions and voltage unbalances: indices based on distributed measurement systems

These indices are based on simultaneous multi-point measurements. Two indices are analyzed below:

1. the global index [49, 50];

2. the cost of deleterious effects index [51, 52].

2.3.1 The global index

This index takes into account both waveform distortions and unbalances.

Let us consider N lines (with loads) connected to the same bus and fed by a utility supply system (Figure 2.15); the supply system can be either with or without background distortion and/or unbalance. The lines and supply system terminals are accessible for measurements, i.e. several metering sections MS_j are placed on every line and on the utility's supply MS_u.

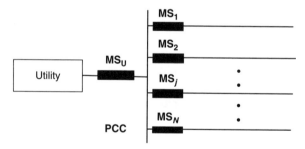

Figure 2.15 Multiple lines connected to the same bus: the case of several metering sections

The supply and loading quality index ξ_{slq} and the harmonic global index ξ_{HGI} introduced in Section 2.2.2 present some limits in terms of meeting all the requirements needed to be an exhaustive index for sharing responsibilities between the utility and the customer. The basic idea of the new global index is to use a proper combination of indices (with the addition of an extra index) and to perform the measurements on all lines connected to the PCC (both those feeding the loads and the supply line); the aim of this idea is to compensate for the different causes that can result in failure of the indices.

The global index is defined for each line j leaving the PCC as follows:

$$\nu_j = \frac{1}{k_{1,j} + k_{2,j} + k_{3,j}} \left(k_{1,j} \frac{\xi_{\text{slq,U}}}{\xi_{\text{slq},j}} + k_{2,j} \frac{\xi_{\text{HGI},j}}{\xi_{\text{HGI,U}}} + k_{3,j} \frac{\eta_j^+}{\eta_U^+} \right), \tag{2.76}$$

where subscript U refers to the utility supply line and $j = 1, \ldots, N$.

The index η^+ is given by:

$$\eta^+ = \frac{\text{GTHD}_{I^+}}{\text{GTHD}_{U^+}}, \tag{2.77}$$

with:

$$\text{GTHD}_{I^+} = \sqrt{\frac{I_\Sigma^2}{I_{\Sigma+1}^2} - 1}, \quad \text{GTHD}_{U^+} = \sqrt{\frac{U_\Sigma^2}{U_{\Sigma+1}^2} - 1},$$

where I_Σ, U_Σ are the collective RMS values of the current and voltage, respectively, and $I_{\Sigma+1}, U_{\Sigma+1}$ are the collective RMS values of the fundamental, positive-sequence components of the current and voltage, respectively.[10] The index given in Equation (2.77) reflects the tendency of a load to amplify the voltage into the current distortion because of the presence of nonlinearity or resonance in the load itself.

The idea of conducting measurements on both the supply side and the lines feeding the loads derives from some theoretical and experimental evidence that allowed the authors who proposed the index to say that one may expect the ratios in Equation (2.76) to increase if the disturbance derives from the customer and decrease if the disturbance derives from the utility [49].

With reference to the quantities $k_{1,j}$, $k_{2,j}$ and $k_{3,j}$ in Equation (2.76), their values are determined on the basis of the following considerations.

From a general point of view, the simplest way to assign $k_{1,j}$, $k_{2,j}$ and $k_{3,j}$ values is to consider the global index ν_j as a simple average of the three ratios in Equation (2.76) ($k_{1,j} = k_{2,j} = k_{3,j} = 1$). However, experience in the application of Equation (2.76) in several cases has demonstrated that the three terms can have values that are significantly different from each other, resulting in their influencing the global index value nonuniformly; as will be shown, this is due mainly to the different influences of the operating conditions on the value of each term.

First, it has been noted that the first term in Equation (2.76) should not assume values significantly far from 1; in this case, wide variations are unlikely and there is no need to apply any particular criterion in choosing the value of $k_{1,j}$. Contrastingly, the second term in Equation (2.76) is very sensitive to load operating conditions (due to the presence of the fundamental current component in the expression ξ_{HGI}); thus, high variations can be expected. This means that, under some conditions, the contribution of this term is significantly higher than the contributions of other terms, while under other conditions, the opposite may be true. This consideration created the incentive to tune

---

[10] In a p-conductor system, the collective RMS value of the vector $\mathbf{X} = (X_1, \ldots \ldots, X_p)^T$ of the RMS values of the conductor quantities (currents or voltages) is given by $X_\Sigma = \sqrt{\sum_{j=1}^{p} X_j^2}$ [53].

the $k_{2,j}$ value to the condition to be tested, i.e. the index mean value and the fraction of rated load absorbed.

With regard to the third ratio in Equation (2.76), it is worth noting that it is also sensitive to the load operating conditions, but it seems less sensitive than the second ratio. The $k_{3,j}$ value is tuned only to the fraction of rated load absorbed.

Finally, the expressions most usually adopted for the $k_{1,j}$, $k_{2,j}$ and $k_{3,j}$ quantities of line j are:

$$k_{1,j} = 1$$

$$k_{2,j} = k'_{2,j} k''_{2,j} = \frac{N}{\displaystyle\sum_{j=1}^{N} \frac{\xi_{\mathrm{HGI},j}}{\xi_{\mathrm{HGI},U}}} \frac{I_{\Sigma_j}/I_{\Sigma_{nj}}}{I_{\Sigma_U}/I_{\Sigma_{nU}}} \tag{2.78}$$

$$k_{3,j} = k''_{2,j}$$

where the subscript n refers to the rated value.

The authors who proposed the index forecast that:

- when the supply system voltages are sinusoidal and balanced and the loads connected to the lines are linear, balanced, time-invariant and operating at their full rated power, the global index ν_j should be equal to 1, for all j values;

- when the load connected to the jth line is a disturbing load, then the global index ν_j should be greater than 1; if more lines have ν_j greater than 1, the line with the greatest values of ν_j should be the most disturbing load. On the other hand, when the global index ν_j is less than or equal to 1, the load connected to the jth line should be disturbed.

Example 2.14   In this example, considering the test system of Figure 2.14 (b), we calculate the global index values by applying Equation (2.76). Table 2.13 shows the values of the global index.

Table 2.13   Global index values of the test system in Figure 2.14 (b)

	Converter 1	Linear load	Customer	Converter 2
ν_j	1.86969	0.39614	1.06643	1.07521

From the analysis of the results presented in Table 2.13, it clearly appears that both converters are characterized by ν_j values greater than 1 while the linear load is characterized by a value less than 1, thus clearly identifying the converters' presence. The customer (linear load and converter 1 together) can be globally considered a perturbing equivalent load.

Several tests of the global index published in the relevant literature [33, 49, 50, 54, 55] have shown its effectiveness in identifying the source of disturbance even though some critical situations can occur.

2.3.2 The cost of deleterious effects index

An approach that uses instrumentation to take synchronous and simultaneous measurements of harmonics at several buses is based on the so-called 'toll road' model.

In this approach an index ξ_{cde} quantifies the customer responsibility with reference to the harmonic pollution. It is a function of the stress the customer's load causes on network components; in particular, the cost of the deleterious effect is considered the reference index.

Let us consider a generic component C in a power system; let $\bar{I}_{Ch}$ be the total harmonic current flowing in the component. If the component of impedance $\dot{Z}'_{pqh}$ is connected between nodes p and q, the harmonic current phasor $\bar{I}_{Ch} = \bar{I}_{pqh}$ that flows from p to q is given by:

$$\bar{I}_{Ch} = \bar{I}_{pqh} = \frac{\overline{V}_{ph} - \overline{V}_{qh}}{\dot{Z}'_{pqh}} = \frac{(\dot{Z}_{p1h} - \dot{Z}_{q1h})\bar{I}_{1h} + (\dot{Z}_{p2h} - \dot{Z}_{q2h})\bar{I}_{2h} + \ldots\ldots + (\dot{Z}_{pMh} - \dot{Z}_{qMh})\bar{I}_{Mh}}{\dot{Z}'_{pqh}}$$

$$= \dot{\alpha}_{C1h}\bar{I}_{1h} + \dot{\alpha}_{C2h}\bar{I}_{2h} + \ldots\ldots + \dot{\alpha}_{CMh}\bar{I}_{Mh}, \tag{2.79}$$

where $\dot{Z}_{ijh}$ is the $(i\text{-}j)$ term of the bus impedance matrix linking the harmonic bus currents to the harmonic phase voltages. The term $\dot{\alpha}_{Cih}\bar{I}_{ih}$ in Equation (2.79) represents the contribution of the ith static converter to the total harmonic current $\bar{I}_{Ch}$.

The basic idea of this approach is that the nonlinear loads present in the power system are shared by two groups: a P-group, including the nonlinear loads that cause an increasing value of $\bar{I}_{Ch}$, and an N-group, causing a decreasing value. As shown in Figure 2.16, the positive group is the set of phasors that are in the same half plane as $\bar{I}_{Ch}$ and the negative group is in the other half plane.

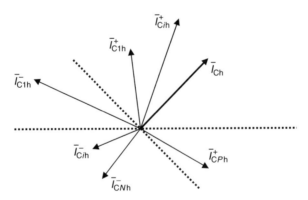

Figure 2.16 Qualitative graphical representation of P-group and N-group phasors

The RMS value of the total current $\bar{I}_{Ch}$ flowing in the component C is given by (Figure 2.16):

$$I_{Ch} = I_{Ph} - I_{Nh} = \sum_{i=1}^{P} I_{pih} - \sum_{i=1}^{N} I_{nih} = \sum_{i=1}^{P} I_{Cih}^{+}\cos(\theta_{Cih}^{+} - \theta_{Ch}) + \sum_{i=1}^{N} I_{Cih}^{-}\cos(\theta_{Cih}^{-} - \theta_{Ch}),$$

$$\tag{2.80}$$

where:

$$\bar{I}_{Cih}^{+} = I_{Cih}^{+}\exp\left(j\theta_{Cih}^{+}\right)$$

$$\bar{I}_{Cih}^{-} = I_{Cih}^{-}\exp\left(j\theta_{Cih}^{-}\right)$$

$$\bar{I}_{Ch} = I_{Ch}\exp\left(j\theta_{Ch}\right).$$

Remembering Equation (2.79) and Figure 2.16, Equation (2.80) produces this result:

$$\bar{I}_{Cih}^{+} = \dot{\alpha}_{Cih}^{+}\bar{I}_{ih}^{+}$$
$$\bar{I}_{Cih}^{-} = \dot{\alpha}_{Cih}^{-}\bar{I}_{ih}^{-}, \tag{2.81}$$

where $\bar{I}_{ih}^{+}$ is the harmonic current phasor injected by the ith nonlinear load with $i \in P$ and $\bar{I}_{ih}^{-}$ is the harmonic current phasor injected by the ith nonlinear load with $i \in N$.

In the 'toll road' model, it is assumed that the cost of the deleterious effects (stress) caused by the harmonic current $\bar{I}_{Ch}$ in component C is a function of the square of the harmonic RMS value; more precisely, the cost of the deleterious effects due to harmonic currents is expressed as a function F of the sum of the squared weighted harmonics:

$$\xi_{cde} = F\left(\sum_{h \neq 1} w_h^2 I_{Ch}^2\right), \tag{2.82}$$

where w_h is a weighting factor that is dependent on both the harmonic order and the component's characteristics. However, as is well known, this is not a general rule, but Equation (2.82) can be considered to provide reasonable approximations for several problems associated with harmonics [52].

Then, for only the harmonic of order h, Equation (2.80) is squared to obtain:

$$I_{Ch}^2 = \left(\sum_{i=1}^{P} I_{pih} - \sum_{i=1}^{N} I_{nih}\right)^2 = \left[\sum_{i=1}^{P} I_{Cih}^{+}\cos\left(\theta_{Cih}^{+} - \theta_{Ch}\right) + \sum_{i=1}^{N} I_{Cih}^{-}\cos\left(\theta_{Cih}^{-} - \theta_{Ch}\right)\right]^2$$

$$= \left[\sum_{i=1}^{P} \alpha_{Cih}^{+}I_{ih}^{+}\cos\left(\theta_{Cih}^{+} - \theta_{Ch}\right) + \sum_{i=1}^{N} \alpha_{Cih}^{-}I_{ih}^{-}\cos\left(\theta_{Cih}^{-} - \theta_{Ch}\right)\right]^2. \tag{2.83}$$

The analysis of Equation (2.83) shows that the ith nonlinear load contributes to the overall detrimental effect due to the harmonic of order h with distinct types of contributions, i.e. individual contributions and joint contributions. The individual contributions are $\left[\alpha_{Cih}^{+}I_{ih}^{+}\cos\left(\theta_{Cih}^{+} - \theta_{Ch}\right)\right]^2$ (for a P-group nonlinear load) and $\left[\alpha_{Cih}^{-}I_{ih}^{-}\cos\left(\theta_{Cih}^{-} - \theta_{Ch}\right)\right]^2$ (for an N-group nonlinear load); this kind of contribution can be allocated easily to the corresponding nonlinear load. The several other joint contributions in Equation (2.83) involve two different nonlinear loads (a couple of P-group loads or a couple of N-group loads or, finally, one P-group load and one N-group load); it is apparent that the allocation of these joint contributions between the associated nonlinear loads is less immediate. The author who proposed the 'toll road' model has analyzed different options (the joint term is divided evenly between the loads, proportionally with the harmonic current contribution and proportionally with the harmonic current squared) and has concluded that the current-squared criterion yields the most equitable division of the joint contribution [52]. So, for example, the joint contribution $2I_{pih}I_{pjh}$ ($j \neq i$) in the first part of Equation (2.83) can be shared between loads i and j of the P group in the following manner:

$$2I_{pih}I_{pjh}\frac{I_{pih}^2}{I_{pih}^2 + I_{pjh}^2} \text{ for load } i, \quad 2I_{pih}I_{pjh}\frac{I_{pjh}^2}{I_{pih}^2 + I_{pjh}^2} \text{ for load } j. \tag{2.84}$$

Using this sharing criterion, trivial manipulations of Equation (2.83) produce:

$$I_{Ch}^2 = A_h I_{Ph}^2 + B_h I_{Nh}^2, \tag{2.85}$$

with:

$$A_h = B_h = 1 - 2\frac{I_{Ph}I_{Nh}}{I_{Ph}^2 + I_{Nh}^2} \tag{2.86}$$

and:

$$I_{Ph}^2 = \left(\sum_{i=1}^{P} I_{pih}\right)^2 = \left(\sum_{i=1}^{P} J_{pih}^2\right) = \sum_{i=1}^{P}\left(I_{pih}^2 + 2I_{pih}\sum_{j=1,j\neq i}^{P}\frac{I_{pih}^2}{I_{pih}^2 + I_{pjh}^2}I_{pjh}\right)$$

$$I_{Nh}^2 = \left(\sum_{i=1}^{N} I_{nih}\right)^2 = \left(\sum_{i=1}^{N} J_{nih}^2\right) = \sum_{i=1}^{N}\left(I_{nih}^2 + 2I_{nih}\sum_{j=1,j\neq i}^{N}\frac{I_{nih}^2}{I_{nih}^2 + I_{njh}^2}I_{njh}\right). \tag{2.87}$$

Eventually, the contribution of the ith nonlinear load of the positive (negative) group to the overall detrimental effect due to harmonic of order h is given by $A_h J_{pih}^2 (B_h J_{nih}^2)$.

This procedure has been applied numerically in the relevant literature to share the cost of the additional losses caused by harmonics in the transformer feeding an electrical distribution system [55]. It has also been applied to share the cost of mitigation equipment installed by the utility (in particular a passive filter) among the users that cause the harmonic pollution [51]. In this case it is assumed that the rating of the passive filter depends on the filter harmonic currents squared and that the filter cost is divided according to the current contribution of each individual nonlinear load; this contribution is obtained following the procedure described previously.

The application of the 'toll road' method requires knowledge of the harmonic impedance matrix terms; such knowledge can be either a priori or obtained by measurements for some elements. Moreover, it requires the measurement or estimation of harmonic current phasors at different points in the electrical system under study. A harmonic state estimation can be performed in order to estimate the harmonic quantities needed for the application of the method; a multi-point measurement system that is synchronized with a unique common trigger signal (e.g. a GPS signal) should be employed. If synchronized measurements are not available or are considered significantly too expensive, single-point independent measurements can be substituted; the author who proposed the 'toll road' method showed that in these cases it is possible to divide the cost of harmonic pollution among the nonlinear loads in a fair manner, using single-point measurements based on the following indices [51]:

1. the harmonic apparent power : $S_H = 3V_H I_H = 3\left(\sum_{h\neq 1} V_h^2\right)^{1/2}\left(\sum_{h\neq 1} I_h^2\right)^{1/2}$;

2. the harmonic active power P_H;

3. the total harmonic current squared I_H^2; and

4. the nonfundamental apparent power $S_N = 3\sqrt{(V_1 I_H)^2 + (V_H I_1)^2 + (V_H I_H)^2}$ squared.

Numerical applications have shown that the most adequate index seems to be the non-fundamental apparent power squared index [51].

In the literature, an extension of the 'toll road' method to unbalanced power systems has also been proposed [54, 55]. In addition, other approaches based on a sharing criterion such as the one reported in Figure 2.16 have been published [56, 57], for example to share responsibilities among customers for the harmonic voltages at a given point of a network.

2.4   Voltage fluctuations

Except for a few cases, the index taken into account in the literature to assess responsibility for voltage fluctuations is the classical short-term flicker severity index P_{st} (see Section 1.3.4 in Chapter 1); more precisely, a percentile value obtained from the cumulative probability function (for example, the 99th percentile) is used.

The main problem considered is the calculation of how much an individual fluctuating customer contributes to the overall flicker level and how much the utility itself contributes to it, assuming that the flicker emission level from a fluctuating load is the flicker level which would be produced in the power system if no other fluctuating loads were present. The flicker emission level for the fluctuating load at the ith bus is synthetically quantified using the short-term flicker severity index which we will denote P_{sti}. The problem of identifying and locating the flicker source has also been analyzed in depth.

Several approaches have been proposed in the literature either for sharing responsibilities or for identifying and locating the flicker source [47, 58–72]; they, refer mainly to arc furnaces as the flicker source. In the next sections, the following will be presented:

- an approach based on the correlation between flicker and load power;
- an approach based on Gaussian probability functions;
- summation law-based approaches;
- voltage-based approaches;
- voltage and current-based approaches;
- power-based approaches;
- a simplified approach.

2.4.1   An approach based on the correlation between flicker level and load power

Let us consider some arc furnaces working in the vicinity. Both the flicker level and the power of each load are simultaneously measured.

The method is based on determining the time intervals when each arc furnace is working alone; these situations can be obtained by analyzing the continuous measurements on each arc furnace. This might be a reliable method to assess flicker level, but it requires time-consuming measurements in order to assess the flicker levels for each arc furnace. In the literature, an experience was reported in which several weeks were necessary to obtain a reliable assessment of the flicker emission level for a given arc furnace [58].

A long measurement interval can also increase the risk of network topology changes with consequent short-circuit level variations at the PCC. If the short-circuit power levels corresponding to each flicker value were known (and reliable), the results could be processed and related to the same reference short-circuit power levels (see Section 2.4.4).

2.4.2   An approach based on Gaussian probability functions

The method is based on the assumption that the short-term flicker severity indices at the PCC with and without the considered fluctuating load ($P_{sti,1}$ and $P_{sti,2}$, respectively) are Gaussian independent random variables [65]. It is also assumed that:

$$P_{sti} = P_{sti,1} - P_{sti,2}. \tag{2.88}$$

From all the assumptions, the result is that P_{sti} is also a Gaussian random variable whose mean μ_i and standard deviation σ_i are given by:

$$\mu_i = \mu_1 - \mu_2$$
$$\sigma_i = \sqrt{\sigma_1^2 + \sigma_2^2}, \tag{2.89}$$

where μ_1, μ_2 and σ_1, σ_2 are the means and standard deviations of the $P_{sti,1}$ and $P_{sti,2}$ Gaussian probability density functions (pdfs), respectively.

This approach is not without its problems either. The first problem is linked to the approximations that can be derived from the assumption made in Equation (2.88), as will be shown in Section 2.4.3. In addition, the hypothesis of Gaussian pdfs may not be valid, mainly when dominating fluctuating loads are present.

Another problem is that we are interested in percentiles derived from P_{sti} probability functions. If the actual distribution does not fit a Gaussian distribution perfectly, the emission level assessments obtained with this method may be very different from the actual values due to the unavoidable inaccuracies that can exist in the 95th and 99th percentile regions of the distribution functions.

In addition, we have to consider the influence of changes in the operating conditions of the utility system with and without the considered fluctuating load, and that this method can require a significant number of P_{st} measurements in order to obtain reliable results.

2.4.3   Summation law-based approaches

IEC 61000-3-7–2008 [47] introduces the following relationship to obtain the short-term flicker severity index values caused by several fluctuating loads:

$$P_{st} = \sqrt[m]{\sum_j P_{stj}^m}, \tag{2.90}$$

where P_{stj} is the contribution of the jth fluctuating load. The same relationship is suggested for P_{lt} (see Section 1.3.4 in Chapter 1).

The coefficient m in Equation (2.90) can assume the following values:

- $m = 1$, when there is a very high occurrence of coincident voltage changes;
- $m = 2$, when coincident fluctuations are probable, for example coincident melts on arc furnaces;
- $m = 3$, when the risk of coincident voltage changes occurring is small (the majority of studies combining uncorrelated disturbances fall into this category);
- $m = 4$, when simultaneous fluctuations are very unlikely.

It should be noted that experimental results on arc furnaces published in the literature have shown that the summation law that best fits measurement results also depends on the percentile chosen for limiting purposes. In particular, the cubic law generally gives conservative predictions. As an example, in the case of two arc furnaces, it has been shown that for percentiles $\geq$ the 95th percentile, there is practically no summation and the flicker level is caused almost entirely by the most disturbing fluctuating load.

The summation law given in Equation (2.90) has also been suggested to evaluate the short-term flicker severity index values of a fluctuating load in a utility with an existing background disturbance level due to the presence of other disturbing loads [60, 65]. It has been assumed that $m = 3$, so that the P_{sti} value is calculated as follows:

$$P_{sti} = \sqrt[3]{P_{sti,1}^3 - P_{sti,2}^3},\tag{2.91}$$

with $P_{sti,1}$ and $P_{sti,2}$ being the flicker levels at the PCC with and without the considered fluctuating load.

Once again, a problem in the application of the summation law given in Equation (2.91) is due to the above-mentioned fact that the measurements at the PCC with and without the considered fluctuating load are not performed simultaneously, so the results are influenced by changes in the operating conditions of the network and/or other fluctuating loads. Moreover, attention should be paid to the use of the coefficient $m = 3$ for all disturbing loads.

2.4.4   Voltage-based approaches

These methods require the measurement of the voltage waveforms in two different busbars.

The original method was based on the consideration that the flicker is linked to voltage fluctuations, so that it is possible to apply the superposition principle to the voltage fluctuation sources.

The circuit configuration shown in Figure 2.17 is considered the reference, where the impedance between busbars i and j is assumed to be known; typically, it is assumed to represent the impedance (practically, the reactance) of the HV/MV transformer supplying the customer installation, typically an arc furnace.

The method consists of evaluating the flicker related to the difference in voltage fluctuations in the two busbars i and j, because this is considered to be an 'image' of the flicker emission level related to the known chosen reference impedance.[11]

---

[11] In practice, the flicker level so obtained has to be considered the level due to the voltage changes caused by the customer loads on the reference impedance.

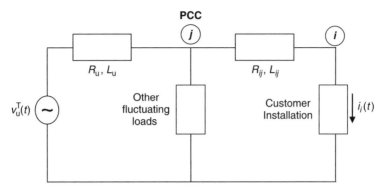

Figure 2.17   Circuit configuration for voltage-based approaches

The flicker emission level so obtained, $P_{st,ref}$, is related to a chosen reference impedance (e.g., the reactance of the HV/MV transformer) and then to a reference short-circuit power; so $P_{st,ref}$ has to be transposed to the actual or contractual short-circuit power at the PCC in order to obtain the necessary final value of the flicker emission level.

The following relation can be used to obtain the flicker emission level at the PCC, if resistances are neglected:

$$P_{st}(HV) = \frac{X_u}{X_{ij}} P_{st,ref}, \qquad (2.92)$$

where X_u is the short-circuit reactance of the HV supply system and X_{ij} is the HV/MV transformer reactance [66].

Actually, experimental results have shown that in some cases the reactance that should be used in Equation (2.92) is lower (and the corresponding short-circuit power greater), due to the voltage dependence of the loads in the supply system. To overcome this problem, a proposal was made to substitute the reactance X_u in Equation (2.92) with an empirical 'virtual network reactance' $X_u^* < X_u$, so that Equation (2.92) becomes:

$$P_{st}(HV) = \frac{X_u^*}{X_{ij}} P_{st,ref}. \qquad (2.93)$$

The ratio X_u^*/X_{ij} in Equation (2.93) can be obtained by plotting the measured $P_{st}(HV)$ values against the $P_{st,ref}$ values obtained with the proposed method; a plot such as that shown in Figure 2.18 is obtained, in which the straight line approximates the lowest levels of the HV flicker level, which most probably correspond to situations involving no other fluctuating loads at the PCC other than the one under study; the slope of this line should indicate the desired ratio [66].

It should be noted that additional measurements on actual plants are needed to verify that Equation (2.93) is conclusive. Difficult cases, tested experimentally, generated problems in identifying the best ratio to take into account; for example, in some cases a greater reactance was determined experimentally (i.e. a lower short-circuit power). In addition, active power variations, mainly when the reactive ones are well compensated with proper devices, can

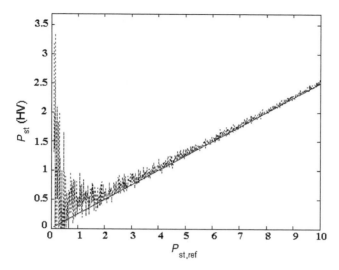

Figure 2.18 Qualitative values of the flicker level P_{st}(HV) versus the $P_{st,ref}$ values

influence the flicker emission level, with the consequence that the above method, which concentrates on fluctuations across the reactance of the transformer, must be improved.

Once Equation (2.93) has been obtained, a further step may be necessary in order to relate the flicker level to the contractual short-circuit power, i.e. the value that is used as the contractual reference in the flicker emission assessment. One suggestion for effecting the transposition has been to multiply Equation (2.93) by the ratio between the short-circuit powers corresponding to the actual operating conditions and the contractual condition.

Another approach (the voltage-drop approach), based on a known reference impedance and the simultaneous measurement of the voltage waveforms $v_i(t)$ and $v_j(t)$ at busbars i and j, respectively, allows the calculation of the flicker emission level P_{sti} on the basis of the following procedure [67]:

1. the voltage waveforms $v_i(t)$ and $v_j(t)$ are reported in per-unit values using the RMS value over ten minutes as the reference value;

2. the voltage fluctuations $\Delta v_{ij}(t) = v_i(t) - v_j(t)$ on the reference impedance (once again, usually the impedance of the transformer feeding the customer, which is almost purely reactive) are calculated in the time domain;

3. the voltage fluctuations $\Delta v_{ij}(t)$ are subtracted from an ideal sinusoidal voltage source with amplitude equal to 1 p.u. and with the same argument as the measured voltage at bus j;

4. the resulting waveform in step 3 is assumed to be the input voltage for the flickermeter to calculate the P_{sti} value.

The calculated P_{sti} value is the emission level of the customer related to the chosen impedance (in practice, an inductance). Then, the transposition procedure is applied to obtain the final flicker emission level with reference to the contractual short-circuit level.

The procedure involves some approximations. The most significant ones are in neglecting the phase shift between the source and the PCC voltage (in fact, in step 3 the source sinusoidal voltage has the argument of the voltage at bus j) and in assuming as reference impedance the transformer impedance, which is almost purely reactive; this second assumption results in only reactive power variations being visible even though active variations exist and cause voltage fluctuations on the supply system resistance, thus influencing the emission level at the PCC.

2.4.5   Voltage and current-based approaches

Several methods have been proposed based on the simultaneous measurement of both voltage and current waveforms. Generally, they do not require a known impedance.

The first method requires the measurement of the voltage waveforms $v_j(t)$ at the PCC and of the load current $i_i(t)$ at bus i. This method calculates the flicker emission voltage at the PCC $v_e(t)$, i.e. the voltage waveform at the PCC if the only fluctuating load in the power system is the fluctuating load at bus i. Once known, this voltage can be used as the input to a flickmeter to obtain P_{sti} values.

To obtain the voltage $v_e(t)$ the following procedure is suggested [67]:

1. The measured load current $i_i(t)$ at bus i and voltage $v_j(t)$ at the PCC are used to obtain a reference voltage source $v^*(t)$ that takes into account the measured voltage at bus j and the voltage drop caused on the supply system resistance R_u and inductance L_u due to the current at bus i:

$$v^*(t) = v_j(t) + R_u i_i(t) + L_u \frac{di_i(t)}{dt}. \tag{2.94}$$

2. The following sinusoidal voltage source, free of disturbances and having the same argument as the fundamental of the voltage $v^*(t)$ in step 1, is calculated:

$$v_{uf}^T(t) = \frac{\sqrt{2}}{\sqrt{3}} U_f \sin(\omega t + \alpha). \tag{2.95}$$

The voltage in Equation (2.95) is an ideal voltage source, with U_f the nominal or reference voltage.

3. Finally, the voltage $v_e(t)$ is obtained using the following equation (Figure 2.19):

$$v_e(t) = v_{uf}^T(t) - R_u i_i(t) - L_u \frac{di_i(t)}{dt}. \tag{2.96}$$

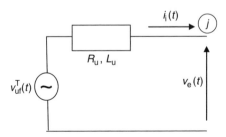

Figure 2.19   Circuit configuration for the voltage and current-based approach

It should be noted that, thanks to Equations (2.94) and (2.96), better consideration of the actual phase shift between voltage and load current is made (both reactive and active load power demands are taken into account); in fact, the sinusoidal voltage source has the argument of the fundamental of the voltage $v^*(t)$ instead of the argument of the voltage at bus j.

This approach is suggested when active power fluctuations are not negligible (compared to the reactive ones) and/or when the supply system resistance is not negligible.[12]

Example 2.15   This example is taken from [67] and has the aim of comparing the flicker emission levels obtained by applying the voltage-drop (VD) approach presented in Section 2.4.4 and the first voltage and current (VI) approach presented above in this section. The considered disturbing load is a 140 MVA DC arc furnace compensated with a 60 MVar Static Var System.

Table 2.14 shows the 99th percentile of P_{st} as calculated by both methods. The VI method shows a greater value of the index.

Table 2.14 $P_{st,99}$ values for the VD and VI approaches

	VD Approach	VI Approach
$P_{st,99}$	0.86	1.16

It should be noted that the difference in the index values is due to the fact that, in the VD method, the reference impedance is almost purely a reactance, so the presence of active power variations is hidden; in fact, if only the reactance is used in the VI approach, similar results can be obtained for both methods.

A second approach that requires the measurement of the voltage waveform $v_j(t)$ at the PCC and of the load current $i_i(t)$ at bus i has the aim of assessing both the flicker level P_{sti} and the background flicker coming from the utility [64]. This approach assumes that the customer is responsible for fluctuations correlated with its current, with the utility impedance being the correlation coefficient. The 'background' voltage is calculated as the difference between the actual measured voltage at the PCC (including fluctuations due to the utility and due to the customer) and the voltage variations that the customer load causes on the supply system impedance.

Examples have shown that the proposed technique is useful for separating the flicker caused by the customer and the flicker present in the background; attention must be paid to situations in which, even though the customer is responsible for virtually all the flicker measured at a site, occasional background events occur which, if not identified properly, could significantly influence the flicker attributed to the customer.

---

[12] In the literature it is evidenced that neglecting the resistance R can cause significant errors when the active current variations are important and/or when the network impedance angle is $< 85°$. In fact, if we remember the well-known voltage-drop relationship $\Delta V = RI \cos\phi + XI \sin\phi$ (ϕ is the current argument) and we neglect the resistance R, we have an underestimation or an overestimation of the voltage drop.

A further voltage-current approach (the V-I slope method) was proposed recently in the literature, with the only aim being to detect the flicker source [70]. In fact, as previously mentioned, one can only take suitable measures to mitigate the problem of sharing the responsibility when the flicker source is well identified.

Let us consider the simple situations depicted in Figures 2.20 (a) and (b). In Figure 2.20 (a) a source of flicker is supposed to be upstream from the measurement section x-x while in Figure 2.20 (b) the source is downstream from the measurement section.

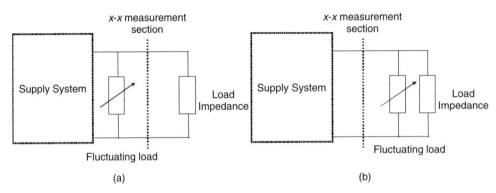

Figure 2.20 The V-I slope method (a) upstream flicker source; (b) downstream flicker source

In the case of Figure 2.20 (a), the current measured at the x-x section decreases as the measured voltage decreases, since the voltage supplies the load impedance. Contrastingly, in the case of Figure 2.20 (b) when the measured current increases, the voltage decreases.

The above-mentioned considerations can be expressed graphically (Figure 2.21), where only the case of a flicker source upstream from section x-x is shown for the sake of simplicity.

Once the voltage and current have been measured and the data have been plotted as in Figure 2.21, a positive slope is indicative of a flicker source that is upstream from the

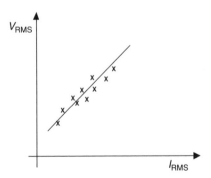

Figure 2.21   Qualitative RMS voltage versus RMS current in the case of an upstream source

measurement section while a negative slope is indicative of a flicker source that is downstream from the measurement section.

Simulation and experimental results have shown the ability to detect the source location in cases of both cyclic and random flicker. Further experimental verification should be conducted to consider the simultaneous presence of flicker sources upstream and downstream from the measurement section.

Additional methods based on voltage and current measurements have been proposed for attributing responsibilities. For example, a method was proposed in which the low-frequency fluctuations were decomposed into voltage and current phasors, and the fluctuations where the voltage and current phasors were in phase were attributed to the utility whereas the fluctuations where the voltage and current phasors were out of phase were attributed to customers. Another approach solves the problem of flicker level sharing among several lines connected to the same bus by transforming the measured load currents into voltage drops [63].

2.4.6   Power-based approaches

Two approaches will be presented within this section. The first is based on the interharmonic power direction [68] while the second introduces a new index, the flicker power [69, 71, 72]. Both approaches give information about the identification of the flicker source.

The interharmonic power direction method starts from the well-known consideration that a correlation between flicker and interharmonics exists. In fact, as is well known, if interharmonics exist, the waveform RMS can fluctuate and this can cause flicker. On the other hand, if a waveform causes flicker it can produce interharmonics. The method requires knowledge of voltage and current waveforms, but it does not require any knowledge of impedance.

The proposed algorithm for locating flicker source, in some cases, can be more robust in identifying the disturbance source than the techniques based on the harmonic active powers shown in Section 2.2.2 (for the harmonic source identification), since generally the interharmonic frequencies change as the operation of the disturbing load changes (see the interharmonics generated by variable frequency drives), so that it is uncommon for two or more interharmonic-producing loads to generate the interharmonics at the same frequencies (and, as such, be characterized by the same interharmonic power) and at the same time. This consideration does not apply to the flicker produced by arc furnaces.

The method is based on the following steps:

1. Apply the Fourier algorithm to the current and voltage waveforms at the PCC and store the interharmonic components that have the maximum magnitude.[13]

2. Calculate the active power of the stored interharmonic(s).

3. Identify the flicker source on the basis of the interharmonic power sign; a negative interharmonic power indicates an interharmonic (flicker) source since it is assumed that a customer producing interharmonic power can be identified as an interharmonic source and, consequently, as a flicker source.

---

[13] Interharmonics can appear in pairs; in such cases, store the two dominant ones.

The method has been tested, with encouraging results, using simulations on a motor with fluctuating load and using field tests on a variable frequency drive. However, the approach, as previously described, is not effective in detecting some flicker sources and might require more tests (particularly in systems with several sources of flicker), mainly to ascertain the measurement problems that may occur in the presence of low-amplitude interharmonics.

Example 2.16   In this example the flicker source was identified using the interharmonic power direction. The injection of interharmonics is due to a 4 MW fluctuating R load (FL1) connected to a 20 kV, medium voltage busbar that also feeds a constant R-L load ($P = 5$ MW, $\cos(\varphi) = 0.9$). A time window length of 1 s was used. The main interharmonics are at the frequencies $f_{ih1} = 40$ Hz and $f_{ih2} = 60$ Hz.

Table 2.15 shows the values of the powers of the interharmonics, the signs of which help to correctly locate the flicker source.

Table 2.15   Interharmonic active powers

	FL1	R-L Load
P_{ih1} [W]	−101.606	15.332
P_{ih2} [W]	−120.257	34.169

Example 2.17   In this example the simulated circuit is the same as in Example 2.16 but an additional interharmonics source due to a 3 MW fluctuating R load (FL2) has been added at the MV busbar. The main current interharmonics in the waveforms at the line feeding FL1 and at the line feeding the static load are at the frequencies $f_{ih1} = 40$ Hz and $f_{ih2} = 60$ Hz while the main current interharmonics in the waveforms at the line feeding FL2 are at the frequencies $f_{ih1} = 42$ Hz and $f_{ih2} = 58$ Hz.

Table 2.16 shows the values of the powers of the interharmonics, the signs of which help once again to correctly locate the flicker sources.

Table 2.16   Interharmonic active powers

	FL1	R-L Load	FL2
P_{ih1} [W]	−110.152	15.174	−68.502
P_{ih2} [W]	−140.106	33.79	−84.341

The second approach is based on the introduction of a new quantity: the flicker power index. This approach starts from the considerations that (i) the fluctuations in voltage and current at the PCC can be considered as amplitude modulation of the fundamental frequency component and (ii) the low-frequency fluctuations (i.e. the envelopes) of the voltage and current have opposite signs in the presence of a downstream source while, in the presence of an upstream source, the fluctuations have the same sign (Figures 2.20 (a) and (b)). The voltage and current fluctuations can be used properly to calculate the flicker power, the sign of which gives information about the location of the flicker source. (A negative sign indicates a downstream source while a positive sign an upstream source).

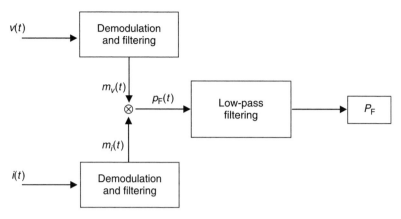

Figure 2.22   Flow diagram for the flicker power calculation

To better explain the steps of the approach, let us refer to the flow diagram in Figure 2.22.

First, the sampled input signals are demodulated (square demodulation), creating a new spectrum in which the low-frequency signal can be recovered and separated from the fundamental (e.g. 50 Hz) carrier. Two filters are then used, the first being the flicker sensitivity filter, the transfer function of which is defined in IEC 61000-4-15 [73]; the second filter is used to further attenuate the waveform components outside the flicker frequency window. The obtained filtered voltage $m_v(t)$ and current $m_i(t)$, which indicate how an 'average human' responds to flicker, are multiplied to obtain the instantaneous flicker power:

$$p_F(t) = m_v(t)\, m_i(t), \qquad (2.97)$$

then, the mean value of $p_F(t)$ is calculated to obtain the desired flicker power:

$$P_F = \frac{1}{T}\int_0^T p_F(t)\mathrm{d}t, \qquad (2.98)$$

where T is the integration window, chosen to be 1 s or 2 s.

The power sign in Equation (2.98) furnishes an indication of the direction of flicker power. A positive sign indicates a direction from the utility to the customer, and a negative sign indicates a direction from the customer to the utility. If the power is correctly evaluated, it can help to identify the dominant source.

Recently, an advanced version of the method was proposed, with the aim of improving the calculation of flicker power by introducing an envelope demodulation instead of using square demodulation; additional discussions were included concerning the summation criteria of flicker power in a power system bus and concerning the working region for estimating the flicker power. In particular, envelope demodulation is used to overcome a drawback of the square demodulator which introduces additional low-frequency components within the flicker frequency range that are not negligible in the current signal, making the flicker power superposition unreliable. Thanks to the improved demodulation process, the magnitude of flicker power can be calculated correctly so, if we consider a bus with more outgoing lines feeding flicker sources, a line with lower flicker power magnitude contributes less to the resulting actual flicker level at the considered bus.

The proposed method has been tested with simulations and in field tests. The authors who proposed this approach acknowledged that it can give false detections, for example when measuring a constant-power load. In this case, the decrease in the voltage due to an upstream flicker source will result in an increase in the value of the load current, thereby causing a misinterpretation. In addition, the authors showed that, if the fluctuating load characteristics differ, there will be a flicker power cancellation effect. In any case, we think that the link between the flicker power and the flicker emission level introduced in the previous sections should be researched and clarified.

2.4.7   A simplified approach

This method is based on the assumption that an indication of the emission level of a fluctuating load connected to the secondary side of a transformer can be given by the value of the short-term flicker severity P_{st} index calculated directly at this side [67]. The assumption is that the contribution of background flicker at the primary side to the overall flicker level at the secondary side bus can be neglected, especially when the 99th percentile of P_{st} is taken into account.

Once the secondary side (SS) flicker emission level is obtained, it must be transposed at the primary side (PS) of the transformer where the emission limits are usually specified. This can be accomplished with the following relationship:

$$P_{st}(PS) \cong \frac{S_{SC,A}}{S_{SC,B}} P_{st}(SS), \tag{2.99}$$

where $S_{SC,A}$ and $S_{SC,B}$ are the short-circuit power at the secondary and primary sides, respectively. The relationship (2.99) was also applied directly to the 99th percentile calculation.

2.5   Voltage sags

For voltage sags the research interest has concentrated on the source side identification more than on the proposal of new indices for sharing responsibility, since the probability is very low (near zero) that a disturbance occurs on both sides simultaneously; in practice, in the case of voltage sags, the responsibility assessment coincides with the side localization of the disturbance source.

Some contributions have been published with the aim of proposing methods that can be used to find the source in all electrical systems (including one-source systems) and using the data obtained at one measurement section. Criteria based on the following will be presented [74–80]:

- disturbance power and energy approach;
- slope of the system trajectory approach;

- resistance sign approach;

- real current component approach;

- distance relay approach.

Usually, these methods locate the sag upstream or downstream from the monitored bus using the pre-event power flow direction. Upstream (downstream) is the region against (following) the power flow.

2.5.1   Disturbance power and energy approach

The disturbance power $DiP(t)$ is the difference between the instantaneous power delivered to the load during the fault $p_f(t)$ and during normal steady-state conditions $p_n(t)$ [74]:

$$DiP(t) = p_F(t) - p_n(t). \tag{2.100}$$

The disturbance energy, DiE, is the integral of Equation (2.100) over the disturbance event duration. DiE indicates the change in the energy flow.

This method assumes that less power (and energy) is delivered to the load during the fault (faults act as energy sinks), so that information about changes in $DiP(t)$ and DiE can help in making a decision about the origin of the sag. This method was also proposed for capacitor switching origin detection. (When a capacitor is switched on, energy is supplied in order to charge the capacitor).

The results obtained by applying $DiP(t)$ and DiE do not always agree; obviously, this reduces the robustness of the method and introduces a degree of uncertainty in the source identification. Moreover, attention must be paid when the disturbance energy at the end of an event does not have a large enough value to detect the origin with some certainty. Then, the following 'heuristic' rules are suggested to locate the disturbance source once the disturbance power and energy have been calculated:

1. The final value of DiE is greater than or equal to 80% of the peak excursion of DiE during the event. If the final value of DiE is positive (negative) the disturbance source is located downstream (upstream). In this case, if the polarity of the initial peak of DiP agrees with the polarity of the final value of DiE, we can conclude that a high degree of confidence can be associated with the selected origin. A lesser degree of confidence arises when the polarities do not match.

2. The final value of DiE is less than 80% of the peak excursion of DiE during the event. The energy test is inconclusive and, in this case, the disturbance direction is determined on the basis of the polarity of the initial peak of DiP.

2.5.2   Slope of the system trajectory approach

Let us consider a two-source system, the equivalent circuit of which is shown in Figure 2.23.

In Figure 2.23, as an example, a downstream fault switching on the resistance R_f is simulated. $\overline{V}_{xx}$ and $\overline{I}$ are the measured voltage and current while $\overline{V}_u^T, \dot{Z}_u$ are the equivalent

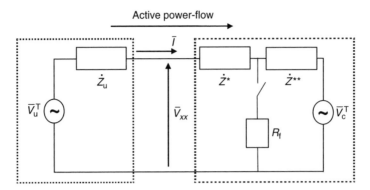

Figure 2.23   Equivalent circuit for sag source detection

parameters for the system upstream from the measurement section (with respect to the direction of the fundamental active power); the meaning of other symbols is trivial. Assuming the parameters on the left (healthy) side of the *x-x* section are unchanged before and after the switching, we can write:

$$\overline{V}_{xx} = \overline{V}_u^T - \dot{Z}_u \overline{I} = \overline{V}_u^T - (R_u + jX_u)\overline{I}. \qquad (2.101)$$

Multiplying by the complex conjugate of the current in Equation (2.101), extracting the real part and dividing for the RMS value of current $\overline{I}$, gives us:

$$V_{xx}\cos(\theta_2) = -R_u I + V_u^T \cos(\theta_1), \qquad (2.102)$$

where $\theta_2(\theta_1)$ is the phase difference between $\overline{V}_{xx}(\overline{V}_u^T)$ and $\overline{I}$.

As $\cos(\theta_2) > 0$, we have $V_{xx}\cos(\theta_2) = |V_{xx}\cos(\theta_2)|$. If $\cos(\theta_1)$ does not change greatly during the disturbance, one can imagine that the line fitting a set of points ($|V_{xx}\cos(\theta_2)|$, I) measured during the voltage sag has negative slope. On the other hand, if the fault is upstream, a relationship similar to Equation (2.102) can be derived, but one can imagine that the line fitting the points ($|V_{xx}\cos(\theta_2)|$, I) has positive slope. The above-mentioned slopes can be used in the detection of the voltage sag location.

Moreover, it has been demonstrated that the same conclusions usually arise if we leave out the assumption that $\cos(\theta_1)$ does not change much. (During the disturbance, the current argument may change significantly [75].)

This method also introduces an auxiliary rule: if a reversal in the active power direction is detected during the sag, the certain conclusion is that the voltage sag is upstream; this is an auxiliary rule to be considered together with the slope check to give more reliable sag-source detection.

However, as the authors state, the method is empirical and it is difficult to assess its application range.

2.5.3   Resistance sign approach

Let us refer once again to Figure 2.23, in which the right side represents the customer installation ($\dot{Z}^* + \dot{Z}^{**} = \dot{Z}c$) while the left side represents the utility supply system.

Before and after a fault at the customer side, we obtain:

$$\overline{V}_{xx} = \overline{V}_{u}^{T} - \dot{Z}_{u}\overline{I}_{pf,c}, \quad \overline{V}_{xx} + \Delta\overline{V}_{1} = \overline{V}_{u}^{T} - \dot{Z}_{u}(\overline{I}_{pf,c} + \Delta\overline{I}_{1}), \qquad (2.103)$$

where $\overline{I}_{pf,c}$ is the value of the current $\overline{I}$ before the fault at the customer side.
With trivial manipulation of Equation (2.103), we get:

$$\dot{Z}_{u} = -\frac{\Delta\overline{V}_{1}}{\Delta\overline{I}_{1}}. \qquad (2.104)$$

Before and after a fault at the utility side, one obtains:

$$\overline{V}_{xx} - \dot{Z}_{c}\overline{I}_{pf,u} = \overline{V}_{c}^{T}, \quad \overline{V}_{xx} + \Delta\overline{V}_{2} - \dot{Z}c(\overline{I}_{pf,u} + \Delta\overline{I}_{2}) = \overline{V}_{c}^{T}, \qquad (2.105)$$

where $\overline{I}_{pf,u}$ is the value of the current $\overline{I}$ before the fault at the utility side.
With trivial manipulation of Equation (2.105), we get:

$$\dot{Z}_{c} = \frac{\Delta\overline{V}_{2}}{\Delta\overline{I}_{2}}. \qquad (2.106)$$

Since the real parts of the impedances are always positive, and assuming:

$$\dot{Z}_{e} = \frac{\Delta\overline{V}}{\Delta\overline{I}}, \qquad (2.107)$$

we can deduce that:

- if Real $(\dot{Z}_{e}) < 0$, the impedance (multiplied by -1) is the utility impedance and the fault is on the customer side;

- if Real $(\dot{Z}_{e}) > 0$, the impedance is the customer impedance and the fault is on the utility side.

However, the authors who proposed this method reported that some false detection can occur if the during-event cycle is selected too close to or too far from the pre-event cycle [79]; this is because the estimated impedance would vary during the event. A least-squares method is proposed to obtain greater robustness. This method requires knowledge of the voltage and current during N cycles.[14] The fundamental frequency positive-sequence voltage and current are used.

The relationships linking voltages and currents, expressed in real and imaginary components, are:

$$V_{xx}^{R} + jV_{xx}^{I} = V_{u}^{T,R} + jV_{u}^{T,I} - (R_{u} + jX_{u})(I^{R} + jI^{I})$$
$$V_{xx}^{R} + jV_{xx}^{I} = V_{c}^{T,R} + jV_{c}^{T,I} + (R_{c} + jX_{c})(I^{R} + jI^{I}), \qquad (2.108)$$

with trivial meaning of symbols.

---

[14] Most PQ instruments provide three cycles or more pre-disturbance and post-disturbance data. As a result, N is typically greater than 3.

Extending Equation (2.108) to the N cycles measured during the voltage sag event, the results are:

$$
\begin{bmatrix} V_{xx}^R(1) \\ \vdots \\ \vdots \\ V_{xx}^R(N) \end{bmatrix} = \begin{bmatrix} I^R(1) & I^I(1) & 1 \\ \vdots & \vdots & \vdots \\ \vdots & \vdots & \vdots \\ I^R(N) & I^I(N) & 1 \end{bmatrix} \begin{bmatrix} -R_u \\ X_u \\ V_u^{T,R} \end{bmatrix}
$$

$$
\begin{bmatrix} V_{xx}^I(1) \\ \vdots \\ \vdots \\ V_{xx}^I(N) \end{bmatrix} = \begin{bmatrix} I^I(1) & I^R(1) & 1 \\ \vdots & \vdots & \vdots \\ \vdots & \vdots & \vdots \\ I^I(N) & I^R(N) & 1 \end{bmatrix} \begin{bmatrix} -R_u \\ -X_u \\ V_u^{T,I} \end{bmatrix}
$$

$$
\begin{bmatrix} V_{xx}^R(1) \\ \vdots \\ \vdots \\ V_{xx}^R(N) \end{bmatrix} = \begin{bmatrix} I^R(1) & I^I(1) & 1 \\ \vdots & \vdots & \vdots \\ \vdots & \vdots & \vdots \\ I^R(N) & I^I(N) & 1 \end{bmatrix} \begin{bmatrix} R_c \\ -X_c \\ V_c^{T,R} \end{bmatrix}
$$

$$
\begin{bmatrix} V_{xx}^I(1) \\ \vdots \\ \vdots \\ V_{xx}^I(N) \end{bmatrix} = \begin{bmatrix} I^I(1) & I^R(1) & 1 \\ \vdots & \vdots & \vdots \\ \vdots & \vdots & \vdots \\ I^I(N) & I^R(N) & 1 \end{bmatrix} \begin{bmatrix} R_c \\ X_c \\ V_c^{T,I} \end{bmatrix}
$$

(2.109)

By solving Equation (2.109) using a pseudo-inverse, greater robustness can be obtained. The following procedure is then suggested to detect the sag source:

1. Measure the fundamental frequency positive-sequence voltage and current at the PCC for N cycles.

2. Solve:

$$
\begin{bmatrix} V_{xx}^R(1) \\ \cdot \\ \cdot \\ V_{xx}^R(N) \end{bmatrix} = \begin{bmatrix} I^R(1) & I^I(1) & 1 \\ \cdot & \cdot & \cdot \\ \cdot & \cdot & \cdot \\ I^R(N) & I^I(N) & 1 \end{bmatrix} \begin{bmatrix} R_e \\ X_e \\ V_e^R \end{bmatrix}
$$

$$
\begin{bmatrix} V_{xx}^I(1) \\ \cdot \\ \cdot \\ V_{xx}^I(N) \end{bmatrix} = \begin{bmatrix} I^I(1) & I^R(1) & 1 \\ \cdot & \cdot & \cdot \\ \cdot & \cdot & \cdot \\ I^I(N) & I^R(N) & 1 \end{bmatrix} \begin{bmatrix} R_e \\ X_e \\ V_e^I \end{bmatrix}
$$

(2.110)

3. If the resistance signs in both parts of Equation (2.110) are coherent, a positive (negative) sign indicates the source to be on the utility (customer) side. No conclusion can be drawn if the signs are opposite.

The authors who proposed this method discussed at length the influence on the reliability of the identification process of those loads with quite different responses to linear loads [79]. For such cases the method was improved by taking into account the transient characteristics of the loads. The active power change ΔP with respect to its pre-fault disturbance value is monitored, so that the voltage and current to be used in Equation (2.110) are selected properly: if ΔP has only one sign then the cycles are selected on the basis of the whole sag duration; if ΔP changes sign, we select the cycles until the ΔP change happens.

The method has been tested with simulation results, field waveforms and experimental results. A simplification of the method, which leaves the reactances in Equation (2.110) from the unknowns has also been proposed; the resulting formulation is similar to that obtained in the method based on the slopes of lines, shown previously.

2.5.4   Real current component approach

This method investigates the phase angle difference θ_2 between the voltage and current at the measuring section when the sag begins. If the real current component $I\cos(\theta_2) > 0$ at the beginning of the sag, then the fault is assumed to be located downstream; if $I\cos(\theta_2) < 0$, then the fault is assumed to be located upstream [77]. The method has been applied for symmetrical and asymmetrical faults and for one-source and two-source systems.

2.5.5   Distance relay approach

In this method the detection of the voltage sag source is accomplished by using the observed impedance and its angle before and after the sag [78]. These quantities are estimated using voltage and current phasors at the measuring section.

The key idea of this method is the consideration that, for an upstream fault, the current direction is reversed and the resultant observed impedance changes in both amplitude and angle. The following rule is then applied to locate the source: if the amplitude of the impedance during the sag is less than the pre-sag amplitude and simultaneously the impedance sag angle is positive, then the sag source is downstream (in front of the measurement section), otherwise the sag is located upstream. This method has been applied for various systems.

It should be noted that the methods presented in Sections 2.5.1 to 2.5.5 have been compared in a paper using a simulated network including radial and meshed systems (see Example 2.18) [80].

Example 2.18   This example is adapted from information given in [80] and the authors thank the IEEE for granting permission to use this material. The example has the aim of comparing the results obtained from the application of all the methods presented for voltage sag detection. The monitored section is at the boundary between a 220 kV transmission system and a radial 138 kV subtransmission system. Six fault points (FP1 to FP6) have been considered, i.e. one at the subtransmission system to simulate a downstream event and five at the transmission level to simulate upstream events. Both three-phase and single-phase short circuits have been simulated. The loads are represented as constant impedances.

Tables 2.17 and 2.18 show the results obtained for symmetrical and asymmetrical faults, respectively.

Generally, the methods perform better in the case of three-phase faults; less accurate behaviour appears in the case of single-phase short circuits. In [80], the methods have been compared using different measurement sections, confirming that, in general, their performance is accurate for symmetrical faults (87% of successful detections) while less accurate behaviour was observed in the case of asymmetrical faults (65% of successful detections, with the method of distance relay showing the best performance); on the other hand, asymmetrical faults are more difficult to locate because each phase can exhibit different behaviour. In any case, a more in-depth analysis and more comparisons of field results are needed to be conclusive on the best-performing method.

Table 2.17 Fault location for three-phase short circuits. DS: downstream; US: upstream; NC: not conclusive; DPE: disturbance power and energy; SST: slope of system trajectory; RCC: real current component; RS: resistance sign; DR: distance relay

Fault point	Sag origin	DPE	SST	RCC	RS	DR
FP1	US	US	US	US	US	US
FP2	US	US	US	US	US	US
FP3	DS	DS	DS	DS	NC	DS
FP4	US	US	US	US	US	US
FP5	US	US	US	US	US	US
FP6	US	US	US	US	NC	US

Table 2.18 Fault location for single-phase short circuits. DS: downstream; US: upstream; NC: not conclusive; DPE: disturbance power and energy; SST: slope of system trajectory; RCC: real current component; RS: resistance sign; DR: distance relay.

Fault point	Sag origin	DPE	SST	RCC	RS	DR
FP1	US	DS	NC	NC	US	US
FP2	DS	DS	NC	DS	NC	DS
FP3	US	US	US	US	NC	US
FP4	US	US	US	US	US	US
FP5	US	DS	US	NC	NC	US
FP6	US	DS	US	NC	US	DS

In addition, it should be noted that a method that locates the sag source according to the causes of the event (e.g. faults, induction motor starting or transformer saturation) has been proposed [76]. However, this method requires that some threshold values be set to be conclusive on the source detection; these threshold values must be carefully tuned but this tuning process is not clearly understood.

A method that requires estimates of the voltage sag amplitude and the phase-angle jump has also been proposed; for a typical industrial installation, it assumes that the sags generated at the utility and at the customer sides are characterized by different phase-angle jumps versus sag amplitudes.

2.6 Voltage transients

The voltage transients associated with a capacitor switching are considered in this section. Various publications have dealt with the problem of identifying the position of the switched capacitor in an electrical power system (mainly in a distribution system), and interesting techniques in both time and frequency domains have been proposed [81–85]. For example, techniques that identify the capacitor position on a radial distribution system using a transient frequency and equivalent impedance seen by the customer have been published. Procedures based on a voltage-disturbance index and branch current phase-angle variation have recently been discussed. Autonomous expert systems have also been applied.

In this section we analyze more specifically only one contribution which deals explicitly with the problem of discriminating whether a transient is located upstream or downstream from the monitoring section x-x at the PCC between the utility and the customer installation [86]. This method expands the concept of disturbance power and energy introduced in Section 2.5.1, focusing on capacitor-switching transients.

A set of four rules for the disturbance location is proposed, and the rules are compared on the basis of various disturbances recorded in an actual distribution system. The rules are briefly described below.

- **Rule 1**

This rule is based on the consideration that when the capacitor is switched on it can act like a sink for energy during the disturbance (similar to the case of a fault): the energy flow is visible as a change in the disturbance energy. The following change in the disturbance energy can be calculated:

$$\Delta DiE = DiE_F - DiE_0 \tag{2.111}$$

where DiE_F (DiE_0) is the value of the disturbance energy DiE at the end (beginning) of the disturbance.

The rule to be applied is:

- if ΔDiE is positive, the disturbance is downstream (in front of) the metering section;

- if ΔDiE is negative, the disturbance is upstream (behind) the metering section.

- **Rule 2**

This rule is based on the analysis of the following index:

$$I_R = \left| \frac{DiE^-}{\Delta DiE} \right| \tag{2.112}$$

with:

$$DiE^- = DiE_{min} - DiE_0,$$

where DiE_{min} is the minimum value of DiE during the disturbance.

The rule to be applied requires the introduction of an empirical threshold level TR_L so that:

- if $I_R < TR_L$, the disturbance is downstream from the metering section;
- if $I_R \geq TR_L$, the disturbance is upstream from the metering section.

A threshold value of 0.25 is adopted for numerical purposes.

- **Rule 3**

This rule is based on the sign of the initial peak of the disturbance power DiP and it states:

- if the sign is positive, the disturbance is downstream from the metering section;
- if the sign is negative, the disturbance is upstream from the metering section.

- **Rule 4**

This rule is based on the empirical observation of the disturbance power DiP characteristics during the transient. The rule to be applied is:

- if the sign of the maximum peak of DiP is positive, the disturbance is downstream from the metering section;
- if the sign of the maximum peak of DiP is negative, the disturbance is upstream from the metering section.

A majority-voting scheme which employs the results of the four above-mentioned rules has also been considered. In practice, the decision about the location is made on the basis of a majority of the four rules; in the case of a 50:50 split (two rules indicating one location and the other two indicating the other), the test is not conclusive. With this technique, good decisions were made on 87% of recorded data.

2.7   Conclusions

In this chapter the problem of assessing responsibility for PQ disturbances between a utility and customers has been analyzed. Indices and methodologies for waveform distortions, voltage unbalances, voltage fluctuations, voltage sags and transients have been shown and some numerical examples provided in order to evidence the difficulties that one can meet in their application.

The main conclusion of the chapter is that the problem of assessing responsibility is a difficult, yet open, field of research. Further efforts are required to obtain solutions that are accepted all over the world. These efforts will be most welcome, because the correct assessment of responsibility is a mandatory step toward the regulation of interactions between utilities and both sensitive customers, requiring specific voltage quality characteristics, and disturbing loads, requiring control of emission levels.

References

[1] Gretsch, R. and Gunselmann, W. (1988) Disturbances on a Medium-voltage Supply System Caused by Harmonics-Measurement, Computer Simulation and Remedial Measures, *International Wroclaw Symposium on Electromagnetic Compatibility EMC'88*, Wroclaw (Poland), June.

[2] Girgis, A.A. and McManis, R.B. (1989) 'Frequency Domain Techniques for Modeling Distribution or Transmission Networks Using Capacitor Switching Induced Transients', *IEEE Transactions on Power Delivery*, **4**(3), 1882–1890.

[3] de Oliveira, A., de Oliveira, J.C., Miskulin, M.S. and Resende, J.W. (1991) 'Practical Approaches for AC System Harmonic Impedance Measurements', *IEEE Transactions on Power Delivery*, **6**(4), 1721–1726.

[4] Yang, H., Pirotte, P. and Robert, A. (1994) Assessing the Harmonic Emission Levels from One Particular Customer, *Power Quality Application Conference*, Amsterdam (The Netherlands), October.

[5] Yang, H., Pirotte, P. and Robert, A. (1995) Comparison between Emission Limit and Emission Level after Commissioning of a Nonlinear Load, *International Conference CIRED'95*, paper 2.18, Brussels (Belgium), May.

[6] CIGRE JTF 36.05.02/14.03.03 (1996) 'AC System Modelling for AC Filter Design: an Overview of Impedance Modelling', *Electra*, **164**, 132–151.

[7] Czarnecki, L.S. and Staroszczyk, Z. (1996) 'On-Line Measurement of Equivalent Parameters for Harmonic Frequencies of a Power Distribution System and Load', *IEEE Transactions on Instrumentation and Measurement*, **45**(2), 467–472.

[8] Deflandre, T. and Robert, A. (1997) Guide for Assessing the Network Harmonic Impedance, *International Conference CIRED'97*, Birmingham (UK), June.

[9] Borloo, G., De Jaeger, E., Dussart, M. and Robert, A. (1998) Practical Implementation of IEC Publications 61000-3-6 and 61000-3-7. Experiences in Belgium, *Power Quality Application Conference*, Cape Town (South Africa), June.

[10] Tsukamoto, M., Kouda, I., Natsuda, Y., Minowa, Y. and Nishimura, S. (1998) Advanced Method to Identify Harmonics Characteristic between Utility Grid and Harmonic Current Sources, *IEEE/PES International Conference on Harmonics and Quality of Power ICHQP'98*, Athens (Greece), October.

[11] CIGRE' 36.05/CIRED 2 Joint WG CC02 (Voltage Quality) (1999) *Review of Methods for Measurement and Evaluation of the Harmonic Emission Level from an Individual Distorting Load*, January.

[12] Thunberg, E. and Soder, L. (1999) 'A Norton Approach to Distribution Network Modeling for Harmonic Studies', *IEEE Transactions on Power Delivery*, **14**(1), 272–277.

[13] Xu, W. and Liu, Y. (2000) 'A Method for Determining Customer and Utility Harmonic Contributions at the Point of Common Coupling', *IEEE Transactions on Power Delivery*, **15**(2), 804–811.

[14] Gonbeau, O. Berthet, L., Javerzac, J.L. and Boudou, D. (2003) Method to Determine Contribution of the Customer and the Power System to the Harmonic Disturbance, *International Conference CIRED'2003*, paper n. 32, Barcelona (Spain), May.

[15] Chen, C., Liu, X., Koval, D., Xu, W. and Tayjasanant, T. (2004) 'Critical Impedance Method – A New Detecting Harmonic Source Method in Distribution Systems', *IEEE Transactions on Power Delivery*, **19**(1), 288–297.

[16] Wilkosz, K. and Pyzalsky, T. (2006) Localization of Sources of Current Harmonics in a Power System: Comparison of Methods Using the Voltage Rate, *MEPS'06 International Conference*, Wroclaw (Poland), September.

[17] Blažič, B., Papič, I. and Pfajfar, T. (2006) Harmonic Current Vector Method with Reference Impedances, *International Conference on Harmonics and Quality of Power*, Cascais (Portugal), October.

[18] Czarnecki, L.S. and Swietlicki, T. (1990) 'Powers in Nonsinusoidal Networks: Their Interpretation, Analysis, and Measurement', *IEEE Transactions on Instrumentation and Measurement*, **39**(2), 340–345.

[19] Ferrero, A. and Superti Furga, G. (1991) 'A New Approach to the Definition of Power Components in Three-phase Systems under Nonsinusoidal Conditions', *IEEE Transactions on Instrumentation and Measurement*, **40**(3), 568–577.

[20] Cristaldi, L. and Ferrero, A. (1994) 'A Digital Method for the Identification of the Source of Distortion in Electric Power Systems', *IEEE Transactions on Instrumentation and Measurement*, **44**(1), 14–18.

[21] Emanuel, A.E. (1995) 'On the Assessment of Harmonic Pollution', *IEEE Transactions on Power Delivery*, **10**(3), 1693–1698.

[22] IEEE Working Group on Nonsinusoidal Situations' Effects on Meter Performance and Definitions of Power (1996) 'Practical Definitions for Powers in Systems with Nonsinusoidal Waveforms and Unbalanced Loads: A Discussion', *IEEE Transactions on Power Delivery*, **11**(1), 79–99.

[23] Czarnecki, L.S. (1996) 'Comments on Active Power Flow and Energy Accounts in Electrical Systems with Nonsinusoidal Waveforms and Asymmetry', *IEEE Transactions on Power Delivery*, **11**(3), 1244–1250.

[24] Tugulea, P. (1996) 'Criteria for the Definition of the Electric Power Quality and its Measurement Systems', *European Transactions on Electrical Power (ETEP)*, **6**(5), 357–363.

[25] Swart, P.H., van Wyk, J.D. and Case, M.J. (1996) 'On Techniques for Localization of Sources Producing Distortion in Three-phase Networks', *European Transactions on Electrical Power (ETEP)*, **6**(6), 391–396.

[26] Ferrero, A., Menchetti, A. and Sasdelli, R. (1996) 'Measurement of the Electric Power Quality and Related Problems', *European Transactions on Electrical Power (ETEP)*, **6**(6), 401–406.

[27] Muscas, G. (1998) 'Assessment of Electric Power Quality: Indices for Identifying Disturbing Loads', *European Transactions on Electrical Power (ETEP)*, **8**(4), 287–292.

[28] Sasdelli, R., Muscas, C. and Peretto, L. (1998) 'A VI-based Measurement System for Sharing the Customer and Supply Responsibility for Harmonic Distortion', *IEEE Transactions on Instrumentation and Measurement*, **47**(5), 1335–1340.

[29] Rens, A.P.J. and Swart, P.H. (2001) 'On Techniques for the Localization of Multiple Distortion Sources in Three-phase Networks: Time-Domain Verification', *European Transactions on Electrical Power (ETEP)*, **11**(5), 317–321.

[30] Cegielski, M., Pyzalski, T. and Wilkosz, K. (2002) Identification of Harmonic Sources in a Power System: Survey of Methods, *International Conference MEPS'02*, Wroclaw (Poland), 11–13 September.

[31] Xu, W., Liu, X. and Liu, Y. (2003) 'An Investigation on the Validity of the Power-Direction Method for Harmonic Source Determination', *IEEE Transactions on Power Delivery*, **18**(1), 214–219.

[32] Li, C., Xu, W. and Tayjasanant, T. (2004) 'A Critical Impedance Based Method for Identifying Harmonic Sources', *IEEE Transactions on Power Delivery*, **19**(2), 671–678.

[33] Locci, N., Muscas, C. and Sulis, S. (2004) 'Investigation on the Accuracy of Harmonic Pollution Metering Techniques', *IEEE Transactions on Instrumentation and Measurement*, **53**(4), 1140–1145.

[34] Locci, N., Muscas, C. and Sulis, S. (2005) On the Measurement of Power Quality Indexes for Harmonic Distortion in the Presence of Capacitors, *Instrumentation and Measurement Technology Conference IMTC 2005*, Ottawa (Canada), May.

[35] Ortiz, A., Gherasim, C., Manana, M., Renedo, C.J., Eguìluz, L.I. and Belmans, R.J.M. (2005) 'Total Harmonic Distortion Decomposition Depending on Distortion Origin', *IEEE Transactions on Power Delivery*, **20**(4), 2651–2656.

[36] Barbaro, P.V., Cataliotti, A., Cosentino, V. and Nuccio, S. (2007) 'A Novel Approach Based on Nonactive Power for the Identification of Disturbing Loads in Power Systems', *IEEE Transactions on Power Systems*, (QU: Delivery?) **22**(3), 1782–1789.

[37] Srinivasan, K. (1996) 'On Separating Customer and Supply Side Harmonic Contributions', *IEEE Transactions on Power Delivery*, **11**(2), 1003–1012.

[38] Srinivasan, K. (1998) 'Conforming and Non-conforming Current for Attributing Steady State Power Quality Problems', *IEEE Transactions on Power Delivery*, **13**(1), 212–217.

[39] Srinivasan, K. (1998) 'Attributing Harmonics in Private Power Production', *IEEE Transactions on Industry Applications*, **34**(5), 887–892.

[40] Kandil, M.S., Farghal, S.A. and Elmitwally, A. (2001) 'Refined Power Quality Indices', *IEE Proceedings Generation, Transmission Distribution*, **148**(6), 590–596.

[41] Chandra, A., Mbang, M., Srinivasan, K., Singh, B.N. and Rastgoufard, P. (2003) A Method of Implementation of Separating Customer and Supply Side Harmonic Contributions Using an Active Filter, *International Conference CCECE – CCGEI 2003*, Montreal (Canada), May.

[42] Dell'Aquila, A., Marinelli, M., Monopoli, V.G. and Zanchetta, P. (2004) 'New Power Quality Assessment Criteria for Supply Systems under Unbalanced and Nonsinusoidal Conditions', *IEEE Transactions on Power Delivery*, **19**(3), 1284–1290.

[43] Marinelli, M., Zanchetta, P. and Dell'Aquila, A. (2004) A New Performance Factor for Weighted Harmonic Distortion Evaluation, *11th International Conference on Harmonics and Quality of Power 2004*, Lake Placid (USA), September.

[44] Ippolito, M.G., Morana, G. and Russo, F. (2005) A Contribution to Solve the Problem of Attributing Harmonic Distortion Responsibility, *International Conference CIRED'2005*, Turin (Italy), June.

[45] Dell'Aquila, A., Marinelli, M. and Zanchetta, P. (2005) New Assessing Criteria for the Contribution of Nonlinear Loads to the Supply of Voltage Harmonic Distortion, *EPE'2005*, Dresden (Germany), September.

[46] IEC 61000-3-6 (2008) *Assessment of Emission Limits for the Connection of Distorting Installations to MV, HV and EHV Power Systems*, Edition 2.0.

[47] IEC 61000-3-7 (2008) *Assessment of Emission Limits for the Connection of Fluctuating Installations to MV, HV and EHV Power Systems*, Edition 2.0.

[48] Balci, M.E. and Hocoaglu, M.H. (2008) 'Quantitative Comparison of Power Decompositions', *Electric Power System Research*, **78**(3), 318–329.

[49] Cristaldi, L., Ferrero, A. and Salicone, S. (2002) 'A Distributed System for Electric Power Quality Measurement', *IEEE Transactions on Instrumentation and Measurement*, **51**(4), 776–781.

[50] Castaldo, D., Ferrero, A., Salicone, S. and Testa, A. (2004) 'An Index for Assessing the Responsibility for Injecting Periodic Disturbances', *L'Energia Elettrica*, **81**, 134–143.

[51] Davis, E.J., Emanuel, A.E. and Pileggi, D.J. (2000) 'Evaluation of Single-Point Measurements Method for Harmonic Pollution Cost Allocation', *IEEE Transactions on Power Delivery*, **15**(1), 14–18.

[52] Davis, E.J., Emanuel, A.E. and Pileggi, D.J. (2000) 'Harmonic Pollution Metering: Theoretical Considerations', *IEEE Transactions on Power Delivery*, **15**(1), 19–23.

[53] Ferrero, A. (1998) 'Definitions of Electrical Quantities Commonly used in Non-sinusoidal Conditions', *European Transactions on Electrical Power (ETEP)*, **8**(4), 235–240.

[54] Muscas, C., Peretto, L., Sulis, S. and Tinarelli, R. (2004) Implementation of Multi-point Measurement Techniques for PQ Monitoring, *Instrumentation and Measurement Technology Conference IMTC 2004*, Como (Italy), 18–20 May.

[55] Muscas, C., Peretto, L., Sulis, S. and Tinarelli, R. (2004) Effects of Load Unbalance on Multi-point Measurement Techniques for Assessing the Responsibility for PQ Degradation, *11th International Conference on Harmonics and Quality of Power 2004*, Lake Placid (USA), September.

[56] Bergeron, R. and Slimani, K. (1999) A Method for the Determination of the Customers' Share of the Contribution to the Level of Harmonic Voltage on an Electric Network, *PES Summer Meeting*, Edmonton (Canada), July.

[57] Bergeron, R. and Slimani, K. (1999) Method for an Equitable Allocation of the Cost of Harmonics in an Electrical Network, *PES Summer Meeting*, Edmonton (Canada), July.

[58] Couvreur, M. and Robert, A. (1994) Recent Experience of Connection of Big Arc Furnaces with Reference to Flicker Level, *1994 CIGRE' Conference*, Paris (France), August/September.

[59] Srinivasan, K. (1995) RMS Fluctuations Attributable to a Single Customer, *Eight International Power Quality Solutions*, Long Beach (California), September.

[60] Sakulin, M. (1996) Flicker, State of the Art, Measuring Techniques and Evaluation, *Jornada sobre la Calidad de la Onda Electrica*, Barcelona (Spain), November.

[61] Key, T. and Sakulin, M. (1997) UIE/IEC Flicker Standard for Use in North America. Measuring Techniques and Practical Applications, *North America Conference PQA'97*, Columbus (USA), March.

[62] Couvreur, M., De Jaeger, E. and Robert, A. (1998) Short-circuit Level and Voltage Fluctuations, *Power Quality Application Conference*, Cape Town (South Africa), June.

[63] Dan, A.M. (1998) Identification of Flicker Sources, *IEEE/PES International Conference on Harmonics and Quality of Power ICHQP'98*, Athens (Greece), October.

[64] Neilson, J.B. and Hughes, M.B. (1999) *Enhanced Flickermeter*, Power Point Presentation for IEEE Task Force on Light Flicker (CC02 IWD 9924), January.

[65] De Jaeger, E. (2000) *Measurement and Evaluation of the Flicker Emission Level from a Particular Fluctuating Load*, Paper prepared on request of CIGRE/CIRED WG CC02, 2nd draft, June.

[66] Couvreur, M., De Jaeger, E. and Robert, A. (2000) Voltage Fluctuations and the Concept of Short-circuit Power, *2000 CIGRE' Conference*, Paris (France), August – September.

[67] Couvreur, M., De Jaeger, E., Goossens, P. and Robert, A. (2001) The Concept of Short-Circuit Power and the Assessment of the Flicker Emission Level, *International Conference CIRED'2001*, Amsterdam (The Netherlands), June.

[68] Nassif, A., Zhang, D. and Xu, W. (2005) Flicker Source Identification by Interharmonic Power Direction, *IEEE CCECE/CCGEI'2005 International Conference*, Saskatoon (Canada), May.

[69] Axelberg, P.G.V., Bollen, M.H.J. and Gu, I.Y.H. (2005) A Measurement Method for Determining the Direction of Propagation of Flicker and for Tracing a Flicker Source, *International Conference CIRED'2005*, Turin (Italy), June.

[70] Nassif, A., Nino, E.E. and Xu, W. (2005) A V-I Slope Based Method for Flicker Source Detection, *Annual North American Power Symposium 2005*, Edmonton (Canada), October.

[71] Axelberg, P.G.V. and Bollen, M.H.J. (2006) 'An Algorithm for Determining the Direction to a Flicker Source', *IEEE Transactions on Power Delivery*, **21**(2), 755–760.

[72] Axelberg, P.G.V., Bollen, M.H.J. and Gu, I.Y. (2008) 'Trace of Flicker Sources Using the Quantity of Flicker Power', *IEEE Transactions on Power Delivery*, **23**(1), 465–471.

[73] IEC 61000-4-15 (2003) *Electromagnetic Compatibility (EMC) – Part 4: Testing and Measurement Techniques – Section 15: Flickermeter – Functional and Design Specifications*, Edition 1.1.

[74] Parsons, A.C., Mack Grady, W., Powers, E.J. and Soward, J.C. (2000) 'A Direction Finder for Power Quality Disturbances Based upon Disturbance Power and Energy', *IEEE Transactions on Power Delivery*, **15**(3), 1081–1086.

[75] Li, C., Tayjasanant, T. and Xu, W. (2003) 'Method for Voltage-Sag-Source Detection by Investigating Slope of the System Trajectory', *IEE Proceedings Generation, Transmission, Distribution*, **150**(3), 367–372.

[76] Ahn, S., Won, D., Chung, I. and Moon, S. (2004) Determination of the Relative Location of Voltage Sag Source According to Event Cause, *PES General Meeting*, Denver (USA), June.

[77] Hamzah, N., Mohamed, A. and Hussain, A. (2004) 'A New Approach to Locate the Voltage Sag Source Using Real Current Component', *Electric Power System Research*, **72**(2), 113–123.

[78] Pradhan, A.K. and Routray, A. (2005) 'Applying Distance Relay for Voltage sag Source Detection', *IEEE Transactions on Power Delivery*, **20**(1), 529–531.

[79] Tayjasanant, T., Li, C. and Xu, W. (2005) 'A Resistance Sign-based Method for Voltage Sag Source Detection', *IEEE Transactions on Power Delivery*, **20**(4), 2544–2551.

[80] Leborgne, R.C., Karlsson, D. and Daalder, J. (2006) Voltage Sag Location Methods Performance under Symmetrical and Asymmetrical Fault Conditions, *TDC'2006: Latin America*, Caracas (Venezuela), August.

[81] Sochuliakova, D., Niebur, D., Nwankpa, C.O., Fischl, R. and Richardson, D. (1999) 'Identification of Capacitor Position in a Radial System', *IEEE Transactions on Power Delivery*, **14**(4), 1368–1373.

[82] Kim, J., Mack Grady, W., Arapostathis, A., Soward, J.C. and Bhatt, S.C. (2000) A Frequency Domain Procedure for Locating Switched Capacitors in Power Distribution Systems, *PES Summer Meeting*, Seattle (USA), July.

[83] Niebur, D. and Pericolo, P.P. (2001) Discrimination of Capacitor Transients for Position Identification, *PES Winter Meeting*, Columbus (USA), February.

[84] Kim, J., Mack Grady, W., Arapostathis, A., Soward, J.C. and Bhatt, S.C. (2002) 'A Time-Domain Procedure for Locating Switched Capacitors in Power Distribution Systems', *IEEE Transactions on Power Delivery*, **17**(4), 1044–1049.

[85] Chang, G.W., Shih, M.H., Chu, S.Y. and Thallam, R. (2006) 'An Efficient Approach for Tracking Transients Generated by Utility Shunt Capacitor Switching', *IEEE Transactions on Power Delivery*, **21**(1), 510–512.

[86] Parsons, A.C., Mack Grady, W., Powers, E.J. and Soward, J.C. (1999) Rules for Locating the Source of Capacitor Switching Disturbances, *PES Summer Meeting*, Edmonton (Canada), July.

3

Advanced methods and nonstationary waveforms

3.1 Introduction

As was pointed out in Chapter 1, PQ disturbances can be present in every waveform cycle or they can appear as isolated and independent events; they can be characterized by small deviations from normal or desired waveforms or by larger deviations.

The waveforms can be stationary (when they are statistically time invariant, e.g. their mean and variance do not change with time) or nonstationary (when they are statistically varying with time). Even though strictly stationary waveforms rarely exist in real power systems, this hypothesis is frequently accepted when the variations in the waveforms are small enough that the changes that occur can be considered statistically the same at any instant in time.[1]

In all cases, difficulties can arise in waveform analysis and, subsequently, in the calculation of traditional PQ indices, since the employed techniques are based on hypotheses that are not always verified for actual power systems; consequently, non-negligible inaccuracies can occur and, in some cases, new indices have to be introduced. It is enough to consider, as examples, the spectral leakage problems that arise in the calculation of indices for waveform distortions when deviations of fundamental frequency exist, or the difficulties linked to spectral components that are continuously varying in amplitude and/or frequency with time. In addition, well-known problems arise in the analysis of strictly nonstationary waveforms; in fact, all the necessary information cannot be deduced easily from just the analyses of their spectral components, since their location in time must also be known.

---

[1] The terms stationary and nonstationary are commonly used in signal processing and will be used extensively in this chapter. We note that some authors use terms such as normal operation, steady-state operation and stationary signals with similar, but not identical, meaning to characterize PQ variations. Analogously, since the statistical time-variant characteristics of nonstationary disturbance signals are often caused by an underlying change in the power system state, the term nonstationary and non-steady-state are often interchanged for such signals [1].

Power Quality Indices in Liberalized Markets   Pierluigi Caramia, Guido Carpinelli and Paola Verde
© 2009 John Wiley & Sons, Ltd

The aforementioned difficulties have to be considered both in the case of calculating traditional indices (e.g., the IEC groupings) and when new indices are introduced (e.g., to characterize nonstationary signals). DFT-based advanced methods and high-performance signal-analysis techniques have been proposed to overcome the problems; this chapter deals with some of these methods and their application to the calculation and proposal of PQ indices. First, some issues concerning the sampling and windowing of power system waveforms are considered; the problem of spectral leakage in the assessment of waveform distortion indices is discussed and some DFT-based advanced methods that have been proposed to solve this problem (e.g., Hanning windowing and result interpolation) are illustrated. Then, high-performance signal-analysis techniques are shown, which make it possible to calculate PQ indices based on the knowledge of what spectral components are present in the signal and where they are located in time (time–frequency representations). The short-time Fourier transform, wavelet transform, parametric high-resolution methods (such as Prony's method, ESPRIT and root-MUSIC methods) and time–frequency distributions are illustrated.

For each of the aforementioned methods, a brief summary of the theoretical background is presented, followed by the presentation of PQ indices based on their use. We demonstrate that, in some cases, these methods are used only to calculate traditional indices (e.g., total harmonic distortion or IEC groupings) in a different way; in other cases, they are the basis for introducing new PQ indices.

A further section of this chapter addresses new indices that have been proposed for harmonic bursts, which are short-duration waveform distortions that last for a very short period of time compared to the duration of the observation interval; these indices require a stochastic approach.

3.2   Discrete time waveforms and windowing

As is well-known, the Fourier transform (FT) of a signal $x(t)$ with $t \in (-\infty, +\infty)$ is defined by Equation (3.1) below [2–4]:

$$\text{FT}(f) = \int_{-\infty}^{\infty} x(t)e^{-j2\pi ft}dt \qquad f \in (-\infty, +\infty). \tag{3.1}$$

In the discrete time domain, the Fourier transform of a discrete sampled waveform $x(n)^2$ is defined as follows:

$$\text{FT}(f) = \sum_{n=-\infty}^{\infty} x(n)e^{-j2\pi fnT_s} \qquad f \in (-\infty, +\infty). \tag{3.2}$$

In practical applications, time domain waveforms are finite sequences of samples. If a finite sequence of L samples of the waveform $x(n)$ (L-point sequence $x(n)$ $n = 0, 1, \ldots, L-1$) is considered, its L-point DFT is given by:

---

[2] The samples $x(nT_s)$ obtained by sampling the continuous waveform $x(t)$ using a constant sampling time ($T_s = 1/f_s$) are indicated more simply as $x(n)$; f_s is the sampling frequency. The sampling frequency should be greater than twice the maximum frequency of the signal due to the Shannon sampling theorem.

$$\text{DFT}(k) = \sum_{n=0}^{L-1} x(n)e^{-j2\pi\frac{k}{L}n} \quad k = 0,\ 1,\ldots,L-1. \tag{3.3}$$

In Equation (3.3), the L-point DFT(k/LT_s) is indicated more simply as DFT(k); this simplification is also used for the other quantities in this section.

From the analysis of Equations (3.2) and (3.3), it can be observed that the DFT of an L-point sequence $x(n)$ is equal to the Fourier transform of the discrete signal $x(n)$ at discrete frequencies $f = \frac{k}{LT_s}$ $(\text{DFT}(k) = \text{FT}(f)|_{f=\frac{k}{LT_s}})$ with $k = 0,\ 1,\ldots,L-1$; hence, the DFT of an L-point sequence $x(n)$ corresponds to a uniform sampling of the FT of the discrete signal $x(n)$ with samples at distance $\Delta f = \frac{1}{LT_s} = \frac{1}{T_w}$ (frequency resolution), where T_w is the L-point sequence time duration ($T_w = LT_s$).

Usually, the DFT is calculated by using the fast Fourier transform (FFT), a class of algorithms that is very efficient for the computation of all DFT values. The class of FFT algorithms is applied to L-sequences with L values that are integer powers of 2; this choice leads to particularly efficient algorithms that have simple structures.

In general, the L-point finite sequence of samples of the waveform to be analyzed is obtained by multiplying the sampled waveform by a finite-length window function that is zero outside the window (windowing process) [5]. The rectangular window is the most frequently used for this purpose.

However, as is true for all windows, the use of a rectangular window can cause some problems. And one should always be aware that:

- The spectrum is composed of components at integer multiples of the reciprocal of the time window length (frequency resolution). This is analogous to looking at the true spectrum through a sort of 'picket fence', i.e. we observe the exact behaviour only at discrete points (picket-fence effect). The properties of the original signal are scrambled with the properties of the window function, so that a proper interpretation of the results has to be determined.

- The time window length T_w should be an exact integer multiple of the Fourier fundamental period (synchronization condition) to avoid spectral leakage problems; the Fourier fundamental period is the reciprocal of the Fourier fundamental frequency, i.e. the greatest common divisor of all the spectral components' frequencies contained in the signal. In practical cases, it may be that T_w reaches an enormous value due to a very small value of the Fourier fundamental frequency; in such cases, during such a long period of time, the amplitudes, frequencies and arguments of the signal components (harmonics and interharmonics) can vary.

To better analyze the problem deriving from an incorrect synchronization condition, let us consider a sampled and windowed single spectral component given by:

$$s_w(n) = x(n)w(n) = A\sin(2\pi f_a nT_s + \varphi)\,w(n) \quad \text{for } n = 0, 1, \ldots\ldots, L-1, \tag{3.4}$$

where A, f_a, and φ are the amplitude, frequency and argument of the considered single spectral component signal, respectively. In Equation (3.4) $w(n)$ (for $n = 0, 1, \ldots\ldots, L-1$) is a rectangular time window of L samples and amplitude equal to 1 obtained with the sampling frequency f_s.

Neglecting the negative frequencies, it can be shown that the spectrum of Equation (3.4) is given by:[3]

$$S_w(k) = \frac{A}{2j}\exp(j\varphi)W\left(\frac{k}{LT_s} - f_a\right) = \frac{A}{2j}\exp(j\varphi)W(k\Delta f - f_a) \quad k = 0, 1, \ldots, L/2 - 1,$$

$$(3.5)$$

where W is the spectrum of the rectangular window (the sinc function, the absolute value of which is shown in Figure 3.1[4]).

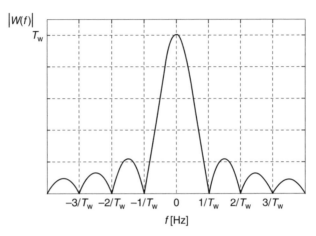

Figure 3.1   Absolute value of the spectrum of an RW of amplitude 1 in the interval $[-T_w/2,\ T_w/2]$

Then, the amplitude spectrum of the signal given in Equation (3.4) calculated by the DFT for L points and neglecting negative frequencies is given by:

$$|S_w(k)| = \frac{A}{2}|W(k\Delta f - f_a)| \quad k = 0, 1, \ldots, L/2 - 1. \quad (3.6)$$

Equation (3.6) shows that the amplitude spectrum of the signal in Equation (3.4) is linked to $L/2$ samples of the absolute value of the continuous sinc function centred at $f = f_a$; the samples are located at a distance that corresponds to the frequency resolution $\Delta f = 1/(LT_s) = 1/T_w$. Two different cases are analyzed in order to better understand where these samples are located and whether the original single spectral component signal is correctly represented in the frequency domain:

---

[3] As is well known, multiplication in the time domain corresponds to a convolution in the frequency domain, so that the spectrum of the waveform in Equation (3.4) can be calculated as the convolution of the spectrum of the signal $x(n)$ with the spectrum of the rectangular time window $w(n)$ (the sinc function).
[4] The FT of a continuous rectangular window of length T_w is a sinc function that assumes zero amplitudes at frequency multiples of $1/T_w$, except for the frequency of the central lobe.

Utility	Customer

Section x-x

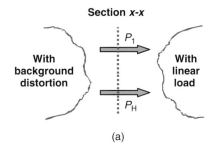

With background distortion

P_1

P_H

With linear load

(a)

Section x-x

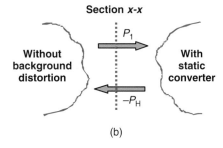

Without background distortion

P_1

$-P_H$

With static converter

(b)

Section x-x

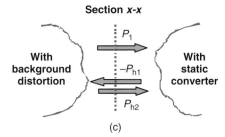

With background distortion

P_1

$-P_{h1}$

P_{h2}

With static converter

(c)

Plate 1 Active powers in electrical systems with waveform distortions (see Figure 2.12)

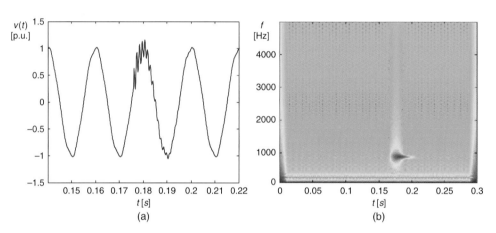

Plate 2 (a) Test signal in the time domain; (b) its TFD (see Figure 3.16)

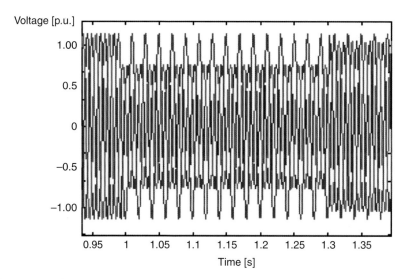

Plate 3 Zoom of the three-phase voltage waveforms corresponding to the sag presence (see Figure 4.3)

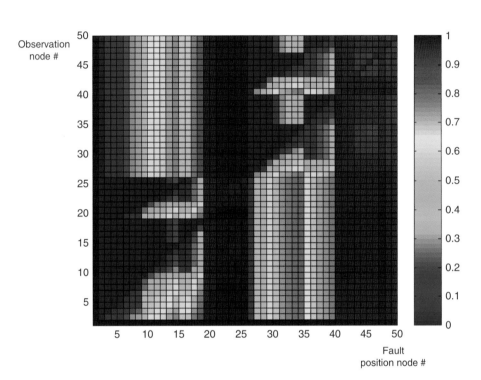

Plate 4 Graphical representation of the during-fault matrix in a distribution system (see Figure 6.1)

1. the frequency of the signal component is an integer multiple of the frequency resolution, i.e. $f_a = r\Delta f$, where r is a positive integer;

2. the frequency of the signal component is not an integer multiple of the frequency resolution, i.e. $f_a = r\Delta f + \delta$, where δ is a positive real number with $\delta < \Delta f$.

If case 1 is verified, Equation (3.6) becomes:

$$|S_w(k)| = \frac{A}{2}|W(k\Delta f - r\Delta f)| = \frac{A}{2}\left|W\left[\frac{(k-r)}{T_w}\right]\right| \qquad k = 0, 1, \ldots, L/2 - 1 \qquad (3.7)$$

and the samples of the amplitude spectrum are located at frequencies that are integer multiples of the frequency resolution ($1/T_w$), specifically:

- if $k \neq r$ the samples are located at frequencies corresponding to the frequencies of the zeros of the absolute value of the sinc function (see Figure 3.1);

- only when $k = r$ is the sample located at a frequency corresponding to the centre of the main lobe of the sinc function, whose absolute value is different from zero (see Figure 3.1.)

In case 1, neglecting negative frequencies, the amplitude spectrum (divided by $T_w/2$) correctly reproduces the original single spectral component signal in the frequency domain, as shown in Figure 3.2 (a). (The spectrum is composed of only one component at a frequency of f_a.)

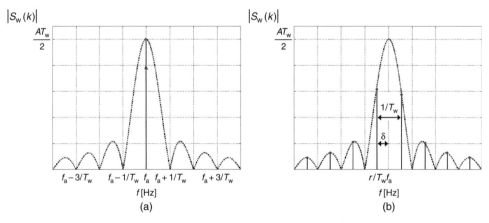

Figure 3.2 Examples of (a) synchronized spectrum; (b) nonsynchronized spectrum of the signal given by Equation (3.4)

If case 2 is verified, then Equation (3.6) becomes:

$$|S_w(k)| = \frac{A}{2}|W(k\Delta f - r\Delta f - \delta)| = \frac{A}{2}\left|W\left[\frac{(k-r)}{T_w} - \delta\right]\right| \qquad k = 0, 1, \ldots, L/2 - 1 \qquad (3.8)$$

and, due to the presence of the δ quantity, the samples of the amplitude spectrum are located at frequencies that are not integer multiples of the frequency resolution ($1/T_w$), and, for each value of k, the samples are located at frequencies that correspond to the absolute values of the sinc function that are different from zero (see Figure 3.1).

In case 2, neglecting negative frequencies, the amplitude spectrum does not correctly reproduce the original single spectral component signal in the frequency domain, as is shown in Figure 3.2 (b). (The spectrum is composed of components different from zero and none at frequency f_a.)

From the aforementioned considerations, it can be concluded that, in order to obtain a spectrum that correctly reproduces a single spectral component, the condition $f_a = r\Delta f$, with r being a positive integer, must be verified. In the time domain this condition implies that:

$$T_{\mathrm{w}} = \frac{1}{\Delta f} = \frac{r}{f_a} = rT_a. \tag{3.9}$$

Equation (3.9) means that 'the time window duration T_{W} must be exactly an integer multiple of the spectral component period T_a' (synchronization condition). In fact, in this case only one of the $L/2$ spectral components is centred at frequency f_a and corresponds to the amplitude of the original single spectral component signal (divided by $T_{\mathrm{w}}/2$) while the other components are equal to zero and each is shifted from f_a by steps equal to $(1/T_{\mathrm{w}})$, where the zeros of the absolute value of the sinc function are located (Figure 3.2 (a)).

Contrastingly, if the synchronization condition is not verified (Figure 3.2 (b)), the amplitude spectrum includes many components that are different from zero and the result is an inaccurate evaluation of the signal being studied. (Spectral leakage phenomenon.)

Since DFT is a linear transformation, the spectrum of a waveform composed of several components is equal to the sum of the spectrum of each individual component, so that the aforementioned considerations on the spectral leakage phenomenon can be extended to more general waveforms. In more detail, in the presence of several components, the lack of the synchronization condition can cause 'interference' among tones due to the spectral leakages caused by each component. This is why the rule to be followed is that the time window length T_{w} must be exactly an integer multiple of the Fourier fundamental period to avoid spectral leakage problems.

Obviously, a priori knowledge of the signal characteristics is required to satisfy the synchronization condition. This is generally impossible in real measurements, especially in the presence of interharmonics; only an approximation of the synchronization condition can be obtained using suitable tools such as a phase locked loop (PLL).

In more detail, spectral leakage problems originate in power systems primarily because of two problems [6]:

1. errors in synchronizing fundamental and harmonics due to deviations in the fundamental frequency; and

2. the presence of nonsynchronized interharmonics.

It has been shown in the literature [6, 7] that problem 1 causes the most remarkable errors, while the desynchronization of interharmonics may give minor effects. Obviously, other desynchronization issues are related to the instrumentation adopted. For instance, the frequency resolution of the sampling clock is finite and, therefore, it may be difficult, from a practical perspective, to synchronize the clock with the fundamental frequency value, even when it is stable.

Several advanced approaches have been proposed to improve the accuracy of estimating the spectra of power system waveforms for distortion assessment and, in particular, for IEC groupings calculation; some DFT-based advanced methods [7–11], such as Hanning windowing and result interpolation, will be briefly discussed below while others will be discussed in Section 3.5.

3.2.1   Hanning windowing

Spectral leakage depends on the characteristics of the time window used, and an appropriate choice of time window can reduce the problem. Different windows have been applied. In particular, the Hanning window (HW) has been proposed as an alternative to the rectangular window (RW) [11]. The HW has been considered attractive because it is characterized in the frequency domain by an amplitude spectrum with side lobes that are very low and that quickly decay (Figure 3.3), thereby limiting interference conditions. The HW amplitude spectrum has a main lobe exactly double that of the RW. The RW has the narrowest main lobe and, as a result, the best resolution among tones that are close in frequency; this characteristic of the HW main lobe guarantees acceptable resolution.

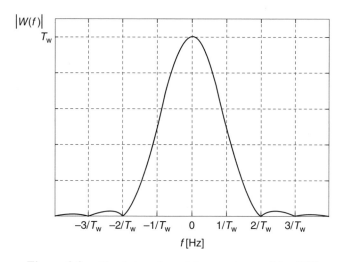

Figure 3.3   Absolute value of the spectrum of the HW

It should be noted that the use of the HW instead of the RW requires minor changes in the IEC procedure for grouping evaluation; in practice, it can be shown that one simply multiplies the IEC group values by a factor of $(2/3)^{1/2}$. It should also be noted that, in some cases (for example in the case of fluctuating harmonics or in the case of interharmonic components present in the proximity of harmonic groups), the performance of the RW may make it more appropriate for a measurement instrument in the IEC standard framework.

3.2.2   Result interpolation

Result interpolation enables the estimation of the amplitudes, frequencies and phase angles of signal components with a high degree of accuracy, starting from the results of a DFT performed at a given frequency resolution (e.g. 5 Hz).

The interpolation of a given tone is based on the assumption that the spectral leakage effects caused by the negative frequency replica and the other harmonic and interharmonic tones are negligible. These conditions occur with a good approximation if a proper window is used; the HW has been proposed because of its spectral characteristics and the simplicity of the interpolation formulas. In particular, in the presence of desynchronization, the actual amplitude, phase angle and frequency are calculated, starting from the two DFT components nearest to the actual frequency [10].

Let us consider once again the single spectral component present in Equation (3.4). Using the Hanning window, the following approximated expressions for the interpolated tone of amplitude $\hat{A}$, frequency $\hat{f}_a$ and phase angle $\hat{\varphi}$ can be used:

$$\hat{A} = \pi |S_w(M)| \frac{\hat{\delta}(1 - \hat{\delta}^2)}{\sin(\pi\hat{\delta})}$$

$$\hat{f}_a = \frac{M}{LT_s} + \delta$$

$$\hat{\varphi} = \frac{\pi}{2} - M\pi\hat{\delta} + \angle S_w(M) \tag{3.10}$$

where:

$$\hat{\delta} = \frac{2 - \alpha}{1 + \alpha}$$

$$\alpha = \frac{|S_w(M)|}{\left|S_w[M + \text{sign}(\hat{\delta})]\right|}$$

with $\text{sign}\left(\hat{\delta}\right) = \text{sign}[|S_w(M + 1)| - |S_w(M - 1)|]$.

In Equation (3.10) M is the order of the DFT component nearest to $f_a T_w$, $S_w(M)$ is the corresponding DFT spectral component and $\hat{\delta}$ is the estimation of the desynchronization error (in Hz).

3.2.3 Synchronized processing

As previously shown, the correct synchronization of the time window used reduces interference. If we initially refer to the time window width suggested by IEC standards for grouping calculation, the following sampling frequency f_s should be used to obtain L samples:

$$f_s = \frac{L}{10T_{1a}}, \tag{3.11}$$

where T_{1a} is the actual system fundamental period.

In practice, T_{1a} is only known through an estimate $\hat{T}_{1a}$, and the number L of samples should be a power of 2 to make use of the FFT algorithm. Then, the sampling frequency is:

$$f_s = \frac{2^n}{10\hat{T}_{1a}}. \tag{3.12}$$

A direct approach for obtaining an approximation of synchronization is to use the analogical phase locked loop.

Alternatively, other synchronized processing techniques can be used [10]; one possible approach consists of the following three steps:

1. the system fundamental period is estimated;

2. a nonsynchronized frequency f_s^* is chosen to obtain the first sequence of L^* samples;

3. a resampling algorithm is applied to obtain the required 2^n samples.

The system fundamental period is estimated by direct or indirect methods. Direct methods are based on algorithms which, at a fixed sampling frequency, search the samples that come closest to satisfying the synchronization condition. Indirect methods are based on an estimation of the system fundamental frequency.

Step 2 consists of obtaining, for the chosen sampling frequency f_s^*, the number of samples L^*, which is given by:

$$L^* = f_s^* 10 \hat{T}_{1a}. \tag{3.13}$$

The resampling algorithm consists of interleaving $N^* - 1$ zeros between each pair of original samples and then applying a low-pass filter to prevent aliasing problems;[5] the sampled signal so obtained is down-sampled to obtain only N samples. In so doing, the final output signal is sampled with a frequency equivalent to $(N^*/N) f_s^*$. Clearly, to obtain 2^n samples with this equivalent sampling frequency, the following condition has to be imposed:

$$\frac{N^*}{N} = \frac{2^n}{L^*}. \tag{3.14}$$

The accuracy of the above synchronization approach depends both on the estimation of the system fundamental period $\hat{T}_{1a}$ and on the resolution of the resampling stage.

3.2.4   Desynchronized processing

An example of desynchronized processing is based on harmonic filtering before the interharmonic analysis. In practice, a double-stage process is performed [7].

In the first stage, the actual frequencies, amplitudes and frequencies of harmonics (including the fundamental) are estimated using an interpolation algorithm (see Section 3.2.2). The estimations of the amplitudes $\hat{A}_h$, frequencies $\hat{f}_h$, and phases $\hat{\varphi}_h$ of all harmonic components give an estimation of the following harmonic contribution:

$$\hat{s}_w^H(n) = \sum_h \hat{A}_h \sin\left(2\pi \hat{f}_h n T_s + \hat{\varphi}_h\right) \quad \text{with} \quad n = 0, 1, \ldots \ldots, L - 1. \tag{3.15}$$

---

[5] The term *aliasing* refers to the phenomenon that causes the distortion occurring when a continuous time signal has frequencies larger than half of the sampling frequency [3].

This contribution can be filtered from the original signal, for instance in the time domain, obtaining an estimation of the interharmonic contribution:

$$\hat{s}_{\mathrm{w}}^{\mathrm{I}}(n) = s_{\mathrm{w}}(n) - \hat{s}_{\mathrm{w}}^{\mathrm{H}}(n) \qquad \text{with} \quad n = 0, 1, \ldots\ldots, L-1, \qquad (3.16)$$

where $s_{\mathrm{w}}(n)$ is the signal.

In the second stage, the interharmonics are evaluated, minimizing spectral leakage problems due to the harmonic presence. The surviving harmonic leakage is given by:

$$\varepsilon_{\mathrm{w}}^{\mathrm{H}}(n) = s_{\mathrm{w}}^{\mathrm{H}}(n) - \hat{s}_{\mathrm{w}}^{\mathrm{H}}(n) \qquad \text{with} \quad n = 0, 1, \ldots\ldots, L-1. \qquad (3.17)$$

This is generally different from zero. The lower the value of $\varepsilon_{\mathrm{w}}^{\mathrm{H}}$, the smaller are the leakage effects.

The use of a proper window for the interharmonic analysis can reduce residual harmonic leakage problems; the choice of the window must be made by considering additional aspects such as interharmonic tone interaction and IEC grouping problems.

The accuracy of the above desynchronization method is related to the filtering accuracy, which depends on the interpolation algorithms, the number of samples analyzed and interference, such as that produced by interharmonic tones close to the harmonics, which must be estimated and filtered.

With regard to the computational burden, it is important to note that, to achieve accuracy equal to or greater than that of synchronized methods, an exact starting synchronization is not needed, due to the presence of the interpolation step. Therefore, it is possible to choose a sampling frequency that is independent of the actual supply frequency and to acquire a number of samples using the power of 2.

The technique generally implies a doubled number of FFTs. It is worth noting that by using the same window for both the first and second stages, harmonic components can be filtered directly in the frequency domain due to the DFT linearity.

Example 3.1   In this example the amplitudes and frequencies of some spectral components are calculated, with the DFT characterized by a rectangular window $T_{\mathrm{w}} = 200$ ms (DFT-RW), the interpolation technique (I-HW) applied to the components obtained by a DFT on 200 ms using the Hanning window and the desynchronized procedure (DP). These calculated spectral components are then used for the evaluation of some IEC interharmonic subgroups.

All the data used in the experimental case study were entered with the maximum allowable precision. For the sake of simplicity, the exact number of zeros after the last significant cipher is not reported.

Three test waveforms are considered; they constitute a tone of amplitude 1 p.u. at the fundamental frequency of 50 Hz with an interharmonic tone of amplitude 0.001 p.u. at frequencies of 73 Hz, 75 Hz and 78 Hz, respectively. As an example, Table 3.1 reports the results in terms of magnitude error for the interharmonic subgroup C_{isg1} obtained by using DFT-RW, I-HW and DP (see Section 1.3.1 in Chapter 1).

From the analysis of Table 3.1 the following considerations can be made:

- the errors in the DFT-RW reach the larger values at 73 Hz and 78 Hz; the error is null in the experiment characterized by an interharmonic frequency of 75 Hz, where the interharmonic is synchronized with T_{w};

- the errors in the DP assume the larger values at 73 Hz and 78 Hz with values smaller than the errors furnished by the DFT-RW (and, once again, the error is null in the experiment characterized by an interharmonic frequency of 75 Hz);

- the results obtained by using I-HW give the smallest errors except for the experiment characterized by an interharmonic frequency of 75 Hz.

Table 3.1 Interharmonic subgroup C_{isg1} magnitude error (in %) obtained by using DFT-RW, I-HW and DP

	Interharmonic subgroup C_{isg1} magnitude error [%]		
Experiments	73 [Hz]	75 [Hz]	78 [Hz]
DFT-RW	−2.912	0.000	−2.795
I-HW	1.016×10^{-4}	-1.148×10^{-5}	0.875×10^{-4}
DP	-2.885×10^{-3}	0.000	-3.594×10^{-3}

3.3   Short-time Fourier transform

In this section, first the theoretical background of the short-time Fourier transform is briefly described and then some indices based on its application are illustrated [1, 3, 12–20].

3.3.1   Theoretical background

The short-time Fourier transform (STFT) is a simple extension of the FT of a general signal $x(t)$, where the FT is evaluated repeatedly for a windowed version of the time domain signal, as shown in Figure 3.4, where the window, as an example, is rectangular. In this way, the

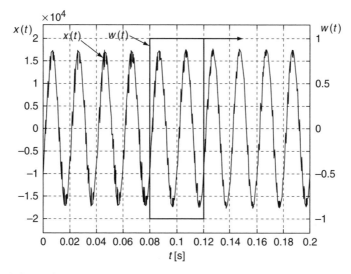

Figure 3.4   STFT: rectangular sliding window $w(t)$ and time-varying signal $x(t)$

signal is broken into small segments; each segment is then analyzed using FT to ascertain variations in the spectrum with time. In practice, each FT gives a frequency domain 'slice' associated with the time value (usually) at the centre of the window. With such an approach, the waveform can be represented in a three-dimensional plot (frequency and magnitude of the transform versus time) so it is possible to know what spectral components are present in the signal and where they are located in time (time–frequency representation). So, STFT can be used for nonstationary waveform assessment.

In the continuous time domain, STFT is defined by:

$$\text{STFT}(f, \tau) = \int_{-\infty}^{+\infty} x(t)w(t - \tau)e^{-j2\pi ft}dt, \tag{3.18}$$

where w is the sliding time window of length T_w, and τ is the time to which the spectrum refers (for example, as previously mentioned, the central time of the sliding time window).

In the discrete time domain, Equation (3.18) becomes:

$$\text{STDFT}(k, m) = \sum_{n=0}^{L-1} x(n)w(n - m)e^{-j2\pi \frac{k}{L}n} \qquad k = 0, 1, \ldots, L - 1, \tag{3.19}$$

where $x(n)$ and $w(n)$ are the sampled versions of the signal and of the sliding time window, respectively, L is the number of samples and m is related to the discrete time to which the spectrum refers.

The sliding time window length $T_w = LT_s$ fixes the resolution of the DFT spectral components in the frequency domain according to the frequency resolution $\Delta f = 1/T_w$. Moreover, since the product of the time resolution and the frequency resolution is constrained by the uncertainty principle, for a narrow window, a detailed time resolution and a coarse frequency resolution are obtained, and vice versa; problems can then arise for short-duration/high-frequency waveforms and for long-duration/low-frequency waveforms, due to the fixed duration of the time window. Analyses with variable window lengths may be required.

The shape of the sliding time window can influence the spectral component values. The choice of the window (e.g. rectangular, Gaussian, Hanning) is therefore an important issue, and this topic has been discussed extensively in the literature when power quality disturbances are dealt with [6, 10, 11, 16]. In practice, the rectangular window is used frequently when the issue of power quality indices is considered.

Finally, it should be noted that, with particular reference to the time-varying harmonic estimation, in [21] a sliding window DFT method has been presented, while in [22] a method based on the theory of multi-rate filter banks is shown; both methods are applied for time-varying harmonic decomposition and allow the extraction of each harmonic component present in the signal in the time domain.

3.3.2   STFT-based indices

First, we note that, in practice, the IEC harmonic and interharmonic groupings are calculated using the STDFT with a sliding rectangular window 10 or 12 fundamental periods long (50 Hz and 60 Hz, respectively), without overlapping between successive windows.

Moreover, other indices, such as single harmonic amplitudes, total harmonic distortion (THD), K-factor and communication interference factors, can also be considered, but one must remember to take into account the peculiar characteristics of the STDFT [15, 16, 18, 19].

For example, the short-term harmonic distortion (STHD) index has been proposed and applied. If the frequency resolution is Δf, the STHD is defined as follows:[6]

$$\text{STHD}(m) = \cfrac{\sqrt{\displaystyle\sum_{\substack{j \neq \frac{f_\theta}{\Delta f} + 1}}^{N_f} \text{STDFT}^2(j, m)}}{\text{STDFT}\left(\dfrac{f_\theta}{\Delta f} + 1, m\right)} \tag{3.20}$$

where:

N_f is the number of spectral components of a positive STDFT spectrum

STDFT(j,m) is the amplitude of the jth component of the STDFT spectrum with reference to m

f_θ is the power system frequency

Δf is the frequency resolution.

Two different STHDs have been proposed to separately account for harmonic and interharmonic contributions in arc furnace waveforms [20].

Similar extensions for other waveform distortion indices, such as the K-factor and the crest factor, have been considered.

Example 3.2   In this example we use the STDFT and analyze the voltage waveform at the MV busbar of a typical DC arc furnace. In DC arc furnaces, the presence of the AC/DC static converter and the random motion of the DC electric arc are responsible for time-varying waveform distortions. The scheme of the simulated DC arc furnace is shown in Figure 3.5, and it consists of a DC arc fed by two thyristor rectifiers connected to an MV AC busbar through a transformer with secondary Δ and Y windings. The MV busbar is connected to the HV busbar with an HV/MV transformer, the windings of which are Y-Δ connected. The data of the system are reported in [20].

Figures 3.6 (a) and (b) show the STHD values of the arc furnace supply voltage waveform obtained using a sliding window width $T_w = 0.2$ s without overlap and with overlap of 0.1 s, respectively.

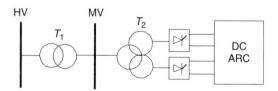

Figure 3.5   DC arc furnace plant

---

[6] It should be noted that the first spectral component of the DFT corresponds to the mean value of the signal.

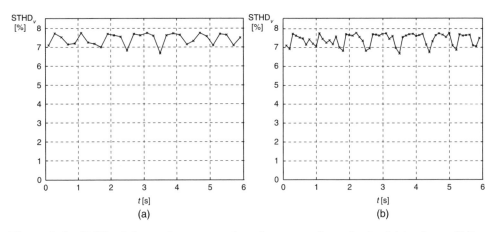

Figure 3.6 STHD of the arc furnace supply voltage waveform obtained (a) using a sliding window with $T_w = 0.2\,\text{s}$ without overlap; (b) with overlap of 0.1 s

From the analysis of Figure 3.6, the time-varying nature of the distorted waveform is clearly evident; moreover, similar values of STHD with and without overlap were determined, even though, in the presence of overlap, much more information is available, due to the greater number of STHDs performed.

Figure 3.7 shows the STHD values of the same waveform in Figure 3.6, except that they are based on the use of a sliding time window length of 0.5 s without overlap. Figure 3.7 shows a smoothing effect due to the choice of a longer time window.

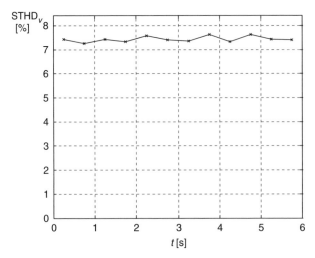

Figure 3.7 STHD of the arc furnace supply voltage waveform obtained using a sliding window with $T_w = 0.5\,\text{s}$ without overlap

3.4   Wavelet transform

In this section, first the theoretical background of the wavelet transform is briefly presented and then some indices based on its application are illustrated [1, 13, 14, 16, 17, 19, 23–28].

3.4.1   Theoretical background

Recently, the wavelet transform (WT) has been proposed more and more for processing power quality disturbances. It, like the STFT, is a form of time–frequency technique since it allows the simultaneous analysis of waveforms in both time and frequency domains. The main advantage of WT over the STFT is the possibility of conducting multi-resolution time–frequency analyses, which, as will be shown later, allows different resolutions of an analyzed spectrum to be obtained at different frequencies.

Unlike Fourier analysis, WT deals with wavelets ('small waves') that are functions with limited energy and zero mean; they have the same role as the sine and cosine functions in Fourier analysis.

In the WT, first the choice of a specific wavelet is made. This is called the *mother wavelet*. The wavelet $\psi(t)$ can be dilated (stretched) and translated (shifted in time) by adjusting the two parameters that characterize it, i.e. a = the scale parameter and b = the translation parameter:

$$\psi_{a,b}(t) = \frac{1}{\sqrt{a}} \psi\left(\frac{t-b}{a}\right), \tag{3.21}$$

with $a \, \varepsilon \, \mathrm{R}^+$ and $b \, \varepsilon \, \mathrm{R}$. A value of $a > 1$ corresponds to a more stretched basis function, while a value of $a < 1$ corresponds to a compressed basis function; Figure 3.8 shows an example of the wavelet for two different values of parameter a. It can be noted from the figure that the functions for values of $a > 1$ vary slowly, so they can better take into account the lower frequencies of the signal, while the functions for values of $a < 1$ vary more rapidly, so they can better take into account the higher frequencies. The parameter b in Equation (3.21) introduces a time translation of the wavelet by which the wavelet can be moved to various locations of the waveform: the 'small wave' moves along the time axis. The term $1/\sqrt{a}$ is a weight used for reasons of energy conservation.

As seen earlier, wavelet functions defined in Equation (3.21) can be used to obtain information at lower and higher frequencies of a signal using $a < 1$ or $a > 1$. In order to

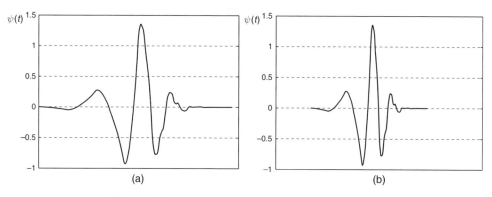

Figure 3.8   Wavelet examples (a) $a > 1$; (b) $a < 1$

obtain information at low frequencies of the signal, it is possible to introduce the so-called scaling function $\varphi(t)$, which can be considered an aggregation of the wavelets characterized by scale parameters $a > 1$. Like the wavelet function, it can be scaled and translated.

The continuous wavelet transform (CWT) of a given signal $x(t)$ with respect to the mother wavelet $\psi_{a,b}(t)$, (if $\psi_{a,b}(t)$ is a complex signal) is the following dot product:

$$\text{CWT}(a,b) = \langle x, \psi_{a,b} \rangle = \int_{-\infty}^{+\infty} x(t) \frac{1}{\sqrt{a}} \psi^* \left(\frac{t-b}{a} \right) dt, \tag{3.22}$$

where * means the complex conjugate.

For an assigned pair of parameters a and b, the coefficient obtained by Equation (3.22) represents how well the signal $x(t)$ and the scaled and translated mother wavelet match. By varying the values of a and b, we can obtain the wavelet representation of the waveform $x(t)$ in time (thanks to the b coefficient) and frequency (thanks to the a coefficient) domains with respect to the chosen mother wavelet.

The CWT is a redundant transform; in fact, it considers continuous translations and dilations of the same function, which makes the transform heavily implementable. A non-redundant version of the wavelet transform, which refers to a discrete number of expansions (a values) and translations (b values), is the discrete wavelet transform (DWT), which is often used instead of CWT in the field of power quality disturbance assessment.

In order to obtain discrete scale and time parameters of the wavelet function defined in Equation (3.21), we can choose discrete scaling and translation parameters such as $a = a_0^j$ and $b = ka_0^j b_0$ where k and j are integers and $a_0 > 1$, $b_0 > 0$ are fixed. The result is:

$$\psi_{j,k}(t) = \frac{1}{\sqrt{a_0^j}} \psi \left(\frac{t - kb_0 a_0^j}{a_0^j} \right). \tag{3.23}$$

A similar expression holds for the scaling function $\varphi(t)$, as is the case for all the following relations.

The wavelet transform of a continuous signal $x(t)$ using discrete scale and time parameters of wavelets leads to the following wavelet coefficients:

$$\text{CWT}(j,k) = \langle x, \psi_{j,k} \rangle = \frac{1}{\sqrt{a_0^j}} \int_{-\infty}^{+\infty} x(t) \psi^* \left(\frac{t - kb_0 a_0^j}{a_0^j} \right) dt. \tag{3.24}$$

The reconstruction of the signal can be made by using the inverse wavelet transform:

$$x(t) = \sum_{j=-\infty}^{\infty} \sum_{k=-\infty}^{\infty} \langle x, \psi_{j,k} \rangle \psi_{j,k}(t). \tag{3.25}$$

Now let us consider a finite sequence of L samples of the sampled waveform $x(n)$; the L-point DWT of the L-point sequence $x(n)$ $n = 0,1,\ldots,L-1$ (once again using discrete scale and time parameters) is defined by:

$$\text{DWT}(j,k) = \frac{1}{\sqrt{a_0^j}} \sum_{n=0}^{L-1} x(n) \psi^* \left(\frac{n - kb_0 a_0^j}{a_0^j} \right). \tag{3.26}$$

The right choice of a_0 and b_0 depends on the wavelet; the most frequently used is the dyadic expansion, characterized by the scale parameter values $a = 2^j$, $j = 1, 2 \ldots (a_0 = 2)$, and $b_0 = 1$.

The DWT is characterized by nonuniform time and frequency spreads across the frequency plane. In practice, both spreads vary with the scale parameter a, but in an opposite manner: the time spread is proportional to a, while the frequency spread is proportional to $1/a$. In particular, the scaling used in Equation (3.26) gives the DWT a logarithmic frequency coverage in contrast to the uniform frequency coverage of STDFT, as illustrated in Figure 3.9, where a comparison between DWT and STDFT is presented.

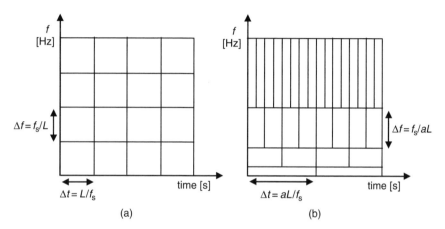

Figure 3.9 Comparison of time–frequency resolution between (a) STDFT and (b) DWT with $a = 2^j$

As discussed in Section 3.3, the time and frequency resolutions Δt and Δf in STDFT are constant (fixed square boxes in Figure 3.9 (a)). The resolution of DWT varies across the plane, and at low frequency, the time resolution is coarse while the frequency resolution is fine (dilated version of the mother wavelet, high value of scale parameter a); in the high frequency range, the time resolution is fine and the frequency resolution is coarse (contracted version of the mother wavelet, low value of scale parameter a). As stated earlier, the analysis of waveforms with different resolutions is called *multi-resolution analysis*.

In practice, the DWT can be implemented using a multi-stage filter, which is made up of a cascade of high-pass and low-pass filters with down-sampling, as shown in Figure 3.10. The high-pass and low-pass filters are linked to the chosen wavelet and scaling functions. As shown in Figure 3.10, the output of the high-pass filters D_j ($j = 1, 2, \ldots$) gives a detailed version of the high-frequency components of the decomposition, while the low-pass filters produce approximations A_j ($j = 1, 2, \ldots$) of the low-frequency components of the decomposition.

Eventually, in multi-resolution analysis, the DWT decomposes a waveform into a discrete number of frequency bands at successive decomposition levels. At Level 1, the waveform's bandwidth is split into two parts made up of high and low frequencies; only the low-frequency component is further split into low- and high-frequency parts. The process is then repeated several times. In practice, this behaviour can be implemented through successive steps of high-pass and low-pass filtering, with filters constructed from the wavelet function

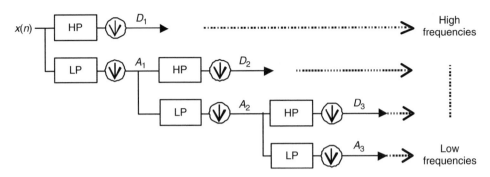

Figure 3.10   Fast DWT decomposition

characterized respectively by approximate A_j and detailed D_j coefficients of the original signal. This will lead to an array of wavelet coefficients with j frequency bands and k coefficients that will represent the original signal.

The coefficients D_j and A_j in Figure 3.10 can be computed using the following relationships:

$$D_j(k) = \frac{1}{\sqrt{a_0^j}} \sum_{n=0}^{L-1} x(n) \psi^* \left(\frac{n-k}{a_0^j} \right) \tag{3.27}$$

$$A_j(k) = \frac{1}{\sqrt{a_0^j}} \sum_{n=0}^{L-1} x(n) \varphi^* \left(\frac{n-k}{a_0^j} \right), \tag{3.28}$$

where k denotes the translation in time and scale j denotes the jth frequency bands, from high to low frequencies.

It is interesting to note that at high frequencies, the frequency bandwidths are wide, which leads to poor frequency resolution. In addition, if information about specific spectral components is needed, the DWT can fail in this objective; in fact, it furnishes only information about a frequency band, the location of which depends on the choices selected.

The most delicate steps in WT PQ applications are the choice of the mother wavelet and the choice of the number of expansion levels [28]. With reference to the choice of the mother wavelet, only knowledge of the phenomena to be analyzed and experience can help to make the best choice and often several trials are necessary to obtain the best choice. This is the reason why a wide number of mother wavelets has been applied in the field of power quality applications; the most frequently used have been the Daubechies family. Just as when making the choice of an appropriate mother wavelet, knowledge of the phenomena to be analyzed and experience are necessary in order to make the best choice of the number of decomposition levels.

3.4.2   Wavelet-based indices

As with the case of STDFT, efforts have been devoted to extending the usual indices, such as total harmonic distortion (THD), the K-factor, and communication interference factors, to take into account the peculiar characteristics of the WT. In practice, the indices are calculated by considering different frequency bands instead of different spectral components; in the case of the total harmonic distortion, it has been proposed that each decomposition frequency level should have different weights, such that the higher the level (characterized by lower frequency), the larger the weight [19].

An adapted version of the WT, called the wavelet packet transform (WPT) has been proposed for the estimation of harmonic groups [27]. The WPT has the advantage of guaranteeing the existence of similar bandwidths across the entire frequency plane, thereby allowing the necessary 5 Hz resolution required by the grouping calculation. In comparison to the decomposition shown in Figure 3.10, the high frequencies are also decomposed further at each subsequent level. The Vaidyanathan 24 and the Daubechies 20 have been proposed as wavelet functions, and the filter bank with five levels of decomposition is applied; the output of the filter bank is divided into 32 uniform bands of 25-Hz width. To make the decomposition compatible with the harmonic group definition, the outputs are grouped into 15 bands with each harmonic frequency component in the centre and with a uniform 50-Hz interval.

It should be noted that both WT and STDFT (such as the high-resolution methods analyzed in the next section) have been used widely to detect and analyze other power quality disturbances, such as voltage sags and transients. This is a very interesting topic, but it is not discussed here for the sake of the conciseness of this book. More detail on this topic can be found in [1].

3.5   Parametric methods

In the field of signal processing, there are two main approaches to spectral analysis. In the first approach, the studied signal is the input of a filter, the output of which is linked to the spectral content of the same signal; this is the 'classical approach,' and it is the basis of the methods analyzed in the previous sections. The second approach, called the 'parametric approach,' is based on the postulation of a model for the studied signal; this approach reduces the spectral estimation problem to only the estimation of the parameters in the assumed model.

The parametric approach is the basis of the parametric methods; these methods can guarantee a more accurate spectral estimation than the classical methods when the studied signals satisfy the model assumed. For example, as discussed in Section 3.2, the crucial drawback of DFT-based methods is that the length of the window constrains the frequency resolution; furthermore, to ensure the accuracy of the DFT, the sampling interval of analysis should be an exact integer multiple of the waveform's fundamental period. Some parametric methods can be applied in order to overcome these drawbacks and obtain a more accurate signal spectrum; some of these methods are analyzed in this section [1, 2, 4, 13, 29–50].

We wish to point out that the parametric methods analyzed assume that the signal model is made up of a certain number of exponential functions; this assumption leads to so-called 'discrete spectra,' which are spectra that contain a fixed number of spectral lines (the spectral components of the signal model), while the classical methods are characterized by a spectrum composed of a theoretically infinite number of spectral lines.

The parametric spectrum estimation methods considered in this section are Prony's method, the estimation of signal parameters by rotational invariance technique (ESPRIT) and the multiple signal classification (MUSIC) method.

Prony's method approximates the sampled data with a linear combination of exponentials; it is closely related to the least squares linear prediction algorithms used for estimating the auto regressive (AR) and auto regressive moving average (ARMA) parameters.[7] In practice, it allows the estimation of the model parameters via the minimization of the error between the signal samples and the approximation obtained using the assumed model.

In signal processing theory, the signals characterized by some properties can be considered elements of a linear space; in particular, power signals enjoy all of the properties required for a linear space. The ESPRIT and MUSIC methods are based on the concept of the signals' linear space. The model of the signal in this case is the sum of the exponentials in the background noise. The model parameters are estimated using the concepts of 'signal subspace' and 'noise subspace.' (These methods are called 'subspace methods.') In particular, the ESPRIT and MUSIC methods use different properties of the signal and noise subspaces for estimating the model's parameters.

Parametric methods have been applied to characterize both stationary and nonstationary waveforms. In the first case, an entire data sequence is used for the estimation of the waveform parameters. In the second case, instead of using all available data samples from the recorded data sequence, we can properly divide the data into blocks that can be either overlapped or nonoverlapped, depending on the application. As a result, we obtain a representation of each of the block parameters (for example, spectral component amplitudes and frequencies) versus time (time–frequency representation).

In the next sections, a brief discussion of the theoretical background of some of the parametric methods is presented, and the possibility of calculating waveform distortion indices with them is outlined. It should be noted that the parametric methods (in particular, Prony's method) have also been applied for calculating indices for voltage sags and transients; this topic will be analyzed in Section 3.6.2.

3.5.1   Theoretical background

In this section Prony's, ESPRIT and MUSIC methods are briefly described.

3.5.1.1   Prony's method

Prony's method allows us to obtain a spectral estimation of the signal using the assumption that the signal model is a linear combination of a finite number M of exponential functions. The parameters of the model are the amplitudes, the frequencies and the initial phases of the exponentials. These parameters are estimated by solving linear systems of equations that allow the minimization of the error between the signal samples and the approximation obtained using the assumed model.

In order to explain Prony's idea for the estimation of the model parameters, let us consider the L-point sequence $x(n)$ $n = 0, 1, \ldots, L-1$ of the investigated waveform; Prony's method approximates each sample using the following linear combination of M exponential functions:

$$\hat{x}(n) = \sum_{k=1}^{M} A_k e_k^{(\alpha_k + j\omega_k)nT_s + j\phi_k}, \qquad n = 0, 1, \ldots, L-1 \qquad (3.29)$$

---

[7] AR and ARMA are classes of signals that can be used to obtain parametric models of the signal to analyze; the least squares method is one of the methods used to obtain an estimation of the model parameters.

where A_k, α_k, ω_k and ϕ_k are the amplitude, the damping factor, the angular velocity and the initial phase of the kth exponential, respectively, and k is the exponential code. It is important to note that, in Equation (3.29), since the signal to approximate is real, the number of exponentials M is even, and the couples of exponentials have to be complex conjugated. Equation (3.29) can be rewritten introducing the variables:

$$z_k = e^{(\alpha_k + j\omega_k)T_s}, \quad h_k = A_k e^{j\phi_k};$$ (3.30)

considering only M relationships from Equation (3.29) and substituting $\hat{x}(n)$ with the actual samples, we have:

$$\begin{bmatrix} x(0) \\ x(1) \\ \vdots \\ x(M-1) \end{bmatrix} = \begin{bmatrix} z_1^0 & z_2^0 & \cdots & z_M^0 \\ z_1^1 & z_2^1 & \cdots & z_M^1 \\ \vdots & \vdots & \vdots & \vdots \\ z_1^{M-1} & z_2^{M-1} & \cdots & z_M^{M-1} \end{bmatrix} \begin{bmatrix} h_1 \\ h_2 \\ \vdots \\ h_M \end{bmatrix}.$$ (3.31)

Prony's idea is based on a method that separately calculates the variables z_k and h_k. To do this, first the following polynomial which has z_k as roots is introduced:

$$z^M + a(1)z^{M-1} + \ldots + a(M-1)z + a(M) = 0;$$ (3.32)

obviously, to calculate z_k by solving Equation (3.32), the coefficients $a(m)$, $m = 1, \ldots, M$ must be known.

It is possible to calculate the values of the coefficients $a(m)$, $m = 1, \ldots, M$ by solving the following system of linear equations obtained by means of simple manipulation of Equations (3.31) and (3.32). Assuming $L > 2M$, the result is:

$$\sum_{m=0}^{M} a(m)x(n-m) = 0, \quad n = M, M+1, \ldots, L-1.$$ (3.33)

Once the variables z_k have been calculated by solving successively Equations (3.33) and (3.32), the h_k values can be obtained by solving the system of linear equations (3.31).

Once the values of z_k and h_k are known, the parameters of exponential components for $k = 1, 2, \ldots, M$ can be calculated using the following relationships:

$$f_k = \omega_k/2\pi = tg^{-1}[\text{Im}(z_k)/\text{Re}(z_k)]/(2\pi T_s), \quad \alpha_k = \ln|z_k|/T_s,$$

$$A_k = |h_k|, \quad \phi_k = tg^{-1}[\text{Im}(h_k)/\text{Re}(h_k)].$$

As previously mentioned, Prony's method, as is the case with all the parametric methods, can be applied to an entire data sequence or to several data blocks (obtained, in practice, using a sliding window on all available samples) that can be either overlapped or nonoverlapped to obtain a representation of model parameters (typically, spectral component amplitudes and frequencies) versus time. The data blocks can be of fixed or variable size, depending on the forecast characteristics of the waveform.

We will now describe a method that has been proposed in the literature [47] to properly select contiguous time windows inside the overall waveform to be analyzed. This proposal (adaptive windows) seems to be particularly useful for detecting harmonics and interharmonics with amplitudes and frequencies that vary with time; in fact, the widths of these windows (and then the numbers of samples for each contiguous block of samples) are variable, ensuring the best fit of the waveform time variations.

Once again, let us consider the signal $x(n)$ and the L_j samples of the generic jth time window, obtained using the sampling frequency f_s. For each sample, the following estimation error can be introduced:

$$e_n = |\hat{x}(n) - x(n)|, \qquad n = 0, \ 1, \ldots, L_j - 1 \tag{3.34}$$

where $\hat{x}(n)$ is given by Equation (3.29).

Applying Equation (3.29) for all L_j samples of the jth time window, the following mean square relative error can be defined:

$$\varepsilon_{j-\text{curr}}^2 = \frac{1}{L_j} \sum_{n=0}^{L_j - 1} \frac{|\hat{x}(n) - x(n)|^2}{x(n)^2}. \tag{3.35}$$

The mean square relative error given in Equation (3.35) gives a measure of the fidelity of the considered model. By defining a threshold $\varepsilon_{\text{thr}}^2$ (an acceptable mean square relative error), it is possible to choose a length for the time window (and then a subset of the data segment length), ensuring a satisfactory approximation ($\varepsilon_{j-\text{curr}}^2 \leq \varepsilon_{\text{thr}}^2$).

It should be noted that, for the model parameter calculation in each time window, the number of components M has to be selected, since this number is an input parameter for the model given by Equation (3.29). Choosing the number of components is a well-known problem in the field of signal processing. As a general rule, it is assumed that, if we select a model with too low an order (low number of components), we obtain a highly smoothed spectrum; on the other hand, if the order selected is too high, we run the risk of introducing spurious low-level peaks in the spectrum.

Several criteria have been proposed to solve this problem, such as the final prediction error (FPE), the Akaike information criterion (AIC), the minimum description length (MDL) criterion, the autoregressive transfer criterion (CAT) and a criterion based on the eigen-decomposition of the sample autocorrelation matrix [3, 4, 29–31, 33, 34]. The MDL criterion seems an appropriate method to evaluate the optimal number of components in the case of power system waveform distortions. This method is based on the selection of the M value corresponding to the minimum of the MDL function, which is defined as follows:

$$\text{MDL}(M) = L \ln \left(\hat{\sigma}_M^2 \right) + M \ln(L), \tag{3.36}$$

where L is the number of time-window samples and $\hat{\sigma}_M^2$ is the estimated variance of the square prediction error. For more details on this topic see [48].

3.5.1.2   ESPRIT method

As mentioned earlier, the ESPRIT method is one of the so-called 'subspace methods;' these methods use the linear algebraic concept of subspace and assume, for the signal, a model that includes the sum of complex conjugated exponentials in the background noise. The background noise represents the additive observation noise, the statistical characterization of which, in terms of the covariance function, is assumed to be known. The parameters of the model are the damping factors, the amplitudes and the frequencies of the exponentials.

These parameters are estimated using the properties of linear algebraic subspaces. In particular, assuming that the vector of the signal samples belongs to a subspace and using this vector to obtain an estimation of the autocorrelation matrix, it is possible to define two subspaces, i.e. the signal subspace and the noise subspace. The original ESPRIT method was based on the naturally existing shift invariance between discrete time series, which leads to rotational invariance between the corresponding signal subspaces.

In order to explain the ESPRIT method, let us consider the assumed signal model defined by the following relationship:

$$\hat{x}(n) = \sum_{k=1}^{M} h_k e_k^{(\alpha_k + j\omega_k)n} + r(n), \qquad n = 0, 1, \ldots, L-1, \qquad (3.37)$$

where h_k was introduced in the previous section on Prony's method and $r(n)$ represents additive noise.

Starting from the model in Equation (3.37) and considering a block of $L_1 < L$ samples, the result is:

$$\mathbf{x}(n) = \mathbf{V}\mathbf{\Phi}^n\mathbf{H} + \mathbf{r}(n), \qquad (3.38)$$

where:

$$\mathbf{x}(n) = [\, x(n) \qquad x(n+1) \qquad \ldots \qquad x(n+L_1-1) \,]^T$$

$$\mathbf{V} = \begin{bmatrix} 1 & 1 & \ldots & 1 \\ e^{\alpha_1 + j\omega_1} & e^{\alpha_2 + j\omega_2} & \ldots & e^{\alpha_M + j\omega_M} \\ \vdots & \vdots & \ddots & \vdots \\ e^{(\alpha_1 + j\omega_1)(L_1-1)} & e^{(\alpha_2 + j\omega_2)(L_1-1)} & \ldots & e^{(\alpha_M + j\omega_M)(L_1-1)} \end{bmatrix}$$

$$\mathbf{H} = [\, h_1 \qquad h_2 \qquad \ldots \qquad h_M \,]^T \qquad (3.39)$$

$$\mathbf{\Phi} = \begin{bmatrix} e^{(\alpha_1 + j\omega_1)} & 0 & \ldots & 0 \\ 0 & e^{(\alpha_2 + j\omega_2)} & \ldots & 0 \\ \vdots & \vdots & \ddots & \vdots \\ 0 & 0 & \ldots & e^{(\alpha_M + j\omega_M)} \end{bmatrix}$$

$$\mathbf{r}(n) = [\, r(n) \qquad r(n+1) \qquad \ldots \qquad r(n+L_1-1) \,]^T.$$

The matrix $\mathbf{\Phi}$ contains all information about the damping factors and frequencies of the M components.

In order to obtain an estimation $\hat{\mathbf{\Phi}}$ of matrix $\mathbf{\Phi}$, let us consider an estimation $\hat{\mathbf{R}}$ of the autocorrelation matrix $\mathbf{R}$ of the signal,[8] obtained by using the samples of the signal given in Equation (3.38). The eigenvectors of the autocorrelation matrix can be separated into two sets, with one set corresponding to the M largest eigenvalues $\mathbf{S} = [\, \mathbf{s}_1 \quad \mathbf{s}_2 \ldots \mathbf{s}_M \,]$ and the other set corresponding to the remaining eigenvalues $\mathbf{E} = [\, \mathbf{e}_1 \quad \mathbf{e}_2 \ldots \mathbf{e}_{L_1-M} \,]$; these eigenvectors define two subspaces (the signal and noise subspaces). It can be noted that, using two matrices $\mathbf{\Gamma}_1$ and $\mathbf{\Gamma}_2$ which select the first L_1-1 rows and the last L_1-1 rows of $\mathbf{S}$,[9] respectively, the following matrices can be introduced:

$$\begin{aligned} \mathbf{S}_1 &= \mathbf{\Gamma}_1\mathbf{S} \\ \mathbf{S}_2 &= \mathbf{\Gamma}_2\mathbf{S}. \end{aligned} \qquad (3.40)$$

---

[8] The autocorrelation matrix is a square matrix whose elements are obtained by using the autocorrelation function, estimated using the unbiased estimator defined in Equation (2.18) in Chapter 2.
[9] $\mathbf{\Gamma}_1$ and $\mathbf{\Gamma}_2$ are constructed by applying $\mathbf{\Gamma}_1 = [\, \mathbf{I}_{L_1-1} \quad 0 \,]$ and $\mathbf{\Gamma}_2 = [\, 0 \quad \mathbf{I}_{L_1-1} \,]$, respectively where $\mathbf{I}$ is the identity matrix.

Rotational invariance theory shows that the calculation of the rotation matrix $\mathbf{\Phi}$ is equivalent to the calculation of the matrix $\mathbf{\Psi}$ that satisfies the following equation:[10]

$$S_2 = S_1\mathbf{\Psi}. \tag{3.41}$$

One of the most common methods used to estimate the matrix $\mathbf{\Psi}$ is the total least-squares approach (TLS-ESPRIT); this method uses an estimation of the matrices S_1 and S_2 ($\hat{S}_1, \hat{S}_2$) obtained by applying Equations (3.40) to find an estimation of the matrix $\mathbf{\Psi}(\hat{\mathbf{\Psi}})$ using the following relationship:[11]

$$\hat{\mathbf{\Psi}} = (\hat{S}_1^* \hat{S}_1)^{-1} \hat{S}_1^* \hat{S}_2. \tag{3.42}$$

Frequencies and damping factors can be calculated by means of simple relationships from the knowledge of the eigenvalues of matrix $\hat{\mathbf{\Psi}}$. Once frequencies and damping factors are obtained, the amplitudes of the M components can be found by solving a linear system of M equations obtained by using the estimated autocorrelation matrix $\hat{R}$.

As in the case of Prony's method, the ESPRIT method can be applied to an entire data sequence or to several data blocks, with or without overlapping. Moreover, the adaptive technique considered by Prony's method can also be applied in the ESPRIT method. However, since the ESPRIT method does not provide any phase estimation of the spectral components, the contiguous time windows are obtained by minimizing the error between the actual waveform energy content evaluated in the time domain and the estimated waveform energy content obtained by using the spectral components in the frequency domain. For more details on this topic see [49].

3.5.1.3 MUSIC method

As mentioned earlier, the MUSIC method is also a subspace method and, for the signal, it assumes the same model as the ESPRIT method; the parameters of the model are, once again, the amplitudes, the damping factors and the frequencies of the exponentials.

In order to explain the MUSIC method, let us remind the reader of the matrices of the eigenvectors of the autocorrelation matrix $\hat{R}$; as for the ESPRIT method, they can be divided into signal and noise eigenvectors.

The MUSIC method uses only the noise subspace for the estimation of frequencies of complex conjugated exponentials, while the ESPRIT method uses only the signal subspace. In particular, the eigenvectors $\mathbf{E} = \begin{bmatrix} \mathbf{e}_1 & \mathbf{e}_2 & \cdots & \mathbf{e}_{L_1-M} \end{bmatrix}$ are used to form the following polynomial in the Z domain:

$$\hat{P}(z) = \cfrac{1}{\sum\limits_{i=M+1}^{L_1} E_i(z)E_i^*\left(\dfrac{1}{z*}\right)} \tag{3.43}$$

where:

$$E_i(z) = \sum_{r=0}^{L_1-1} e_i(r)z^{-1}, \tag{3.44}$$

and $e_i(r)$ is the rth component of the ith eigenvector of the noise subspace.

---

[10] We note that their eigenvalues are the same.

[11] It should be noted that only an estimation of the autocorrelation matrix, and then of the matrices S_1 and S_2, is available.

In Equation (3.43), $\hat{P}(z)$ is a polynomial that has M double roots lying on the unit circle; it can be shown that these roots correspond to the frequencies of the signal components. This technique for finding frequencies is therefore called the root-MUSIC method.

Once frequencies are obtained, the amplitudes of the M components can be found by solving a linear system of M equations obtained by using the estimated autocorrelation matrix, as was done for the ESPRIT method.

As with the ESPRIT method, the root-MUSIC method does not provide any phase estimation of the spectral components. In the case of the adaptive technique, the contiguous time windows can be obtained by minimizing the error between the actual waveform energy content evaluated in the time domain and the estimated waveform energy content obtained using the spectral components in the frequency domain. For more details on this topic see [49].

3.5.2   Parametric method-based indices

All the previous advanced methods (Prony's and the others) permit the evaluation of the spectral components of distorted waveforms for only one block of data or for overlapped/nonoverlapped successive blocks. They can be used to calculate indices for both stationary and nonstationary waveforms. In particular, we stress once again their ability to obtain the spectral component parameters versus time, with particular reference to frequency variations in the spectral components that are not always easy to calculate.

The application of parametric methods does not require particular modifications in the calculation of the waveform distortion indices except for IEC groupings [51, 52], where some minor changes have to be made.

Specifically, in [46, 49] it was proposed that IEC groupings be calculated by applying parametric methods to contiguous nonoverlapped time windows selected on the basis of the proper criterion illustrated in Section 3.5.1 and based on an acceptable mean square relative error. The contiguous time windows (referred to as 'ultra-short contiguous time windows') are selected inside 10 or 12 cycles of the fundamental period in 50 and 60 Hz systems.

Once the spectral components of distorted waveforms inside the ultra-short contiguous time windows have been calculated, the need to define corresponding ultra-short time harmonic and interharmonic subgroups arises. These can be defined as the subgroups calculated for an ultra-short time window. With reference to the jth ultra-short time window, the squared RMS values of the nth ultra-short time subgroups are given by:

$$G^2_{\text{ussgn}}(j) = \sum_{k=1}^{M_{\text{nsg}}} C^2_k(j)$$

$$(3.45)$$

$$C^2_{\text{usisgn}}(j) = \sum_{k=1}^{M_{\text{nisg}}} C^2_k(j),$$

where C_k is the amplitude (RMS value) of the spectral components, M_{nsg} is the number of spectral components inside the frequency interval $[nf_0 - 7.5, nf_0 + 7.5]$ Hz, and M_{nisg} is the number of spectral components inside the frequency interval $[nf_0 + 7.5, (n+1)f_0 - 7.5]$ Hz (with f_0 being the power system frequency). With reference to the IEC intervals, the need to enlarge the frequency ranges for both harmonic and interharmonic grouping evaluations is derived from the absence of the DFT fixed frequency resolution in the advanced parametric method applications.

Once the ultra-short time harmonic and interharmonic subgroups for all windows inside an interval of 10 or 12 fundamental periods have been determined, the harmonic and interharmonic subgroup amplitudes can be calculated by averaging the amplitudes of all the above-mentioned ultra-short harmonic and interharmonic subgroups. This results in the following relationships:

$$
G_{\text{sgn}}^2 = \frac{\displaystyle\sum_{j=1}^{L_w} L_j G_{\text{ussgn}}^2(j)}{L_W}
$$

$$
C_{\text{isgn}}^2 = \frac{\displaystyle\sum_{j=1}^{L_w} L_j C_{\text{usisgn}}^2(j)}{L_W},
$$

(3.46)

where L_w is the number of samples inside 10 or 12 fundamental periods, and L_j is the number of samples in the jth ultra-short contiguous time window.

Finally, the results can be averaged over 15 intervals of ten fundamental periods in order to obtain the results with reference to the very short time measurements.

Example 3.3   In this example the IEC groupings are calculated by applying the IEC, adaptive root-MUSIC, adaptive ESPRIT and adaptive Prony's methods, referred to as IEC-N, ARM, AEM and APM, respectively. The window width is $T_w = 200$ ms for IEC-N. The acceptable error for all the adaptive methods is $\varepsilon_{\text{thr}} = 1.0 \times 10^{-6}$.

The considered signal constitutes a tone of amplitude 1 p.u. at a fundamental frequency of 50 Hz and an interharmonic tone of amplitude 0.01 p.u. at frequencies of 58 Hz, 60 Hz and 63 Hz. The true value of the interharmonic subgroup C_{isg1} is equal to 0.7071×10^{-2} in all three experiments.

As an example, Table 3.2 reports the magnitude error (in %) of interharmonic subgroup C_{isg1} obtained by using IEC-N and the adaptive methods in the three experiments. The errors in IEC-N are greater than 5% under the worst conditions, i.e. when the interharmonic tone is closest to the first harmonic subgroup frequency interval. The error is null in the experiments characterized by interharmonic frequencies of 60 Hz, where the interharmonic is synchronized with T_w. In general, all of the adaptive methods furnish a better approximation of the interharmonic subgroup C_{isg1} than those obtained by IEC-N.

It should be noted that all adaptive methods furnish a spectrum with only two components for all the experiments, showing the absence of the spectral leakage phenomenon. The errors in APM and AEM do not reach 2.0×10^{-5} %, and APM generally gives the best

Table 3.2   Interharmonic subgroup C_{isg1} magnitude error (in %) obtained by using IEC-N and adaptive methods

	Interharmonic subgroup C_{isg1} magnitude errors [%]		
Experiments	58 Hz	60 Hz	63 Hz
IEC-N	-5.3897	0.0000	-3.3244
APM	-4.5666×10^{-7}	2.7552×10^{-7}	-5.1622×10^{-6}
ARM	0.7481	0.0456	0.5095
AEM	1.4276×10^{-5}	1.4377×10^{-5}	1.4332×10^{-5}

performance. In addition, the results for AEM and APM are not influenced by the frequency position of the interharmonic tone. Only ARM suffers from the proximity of the interharmonic tone to the first harmonic subgroup (C_{isg1}); it gives varying errors in the interharmonic tone amplitude estimation, causing varying errors in the subgroup C_{isg1} evaluation. Nonetheless, the ARM error is lower than 0.8%.

Example 3.4   In this example the same voltage waveform considered in Example 3.2 is analyzed with the IEC method (DFT characterized by RW and $T_W = 200$ ms without overlap), a five-level wavelet packet decomposition (with a Daubechies 20 mother wavelet), the adaptive Prony's method, the adaptive ESPRIT method and the adaptive root-MUSIC method. The acceptable error for all the adaptive methods is $\varepsilon_{thr} = 1.0 \times 10^{-6}$. Figures 3.11 (a) and (b) show the 11th and 13th IEC harmonic subgroups (G_{sg11} and G_{sg13}) of voltage versus time calculated by using the IEC method. In Figure 3.12 the 11th and 13th IEC

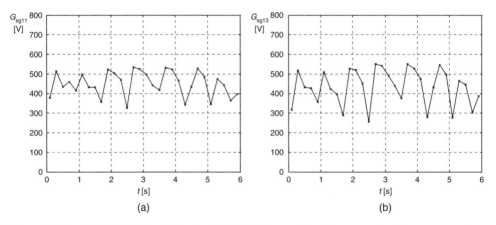

Figure 3.11   (a) The 11th and (b) the 13th IEC harmonic subgroups of voltage versus time calculated by applying the IEC method

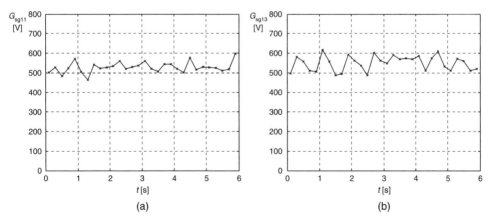

Figure 3.12   (a) The 11th and (b) the 13th IEC harmonic subgroups of voltage versus time calculated by applying five-level wavelet packet decomposition with Daubechies 20 mother wavelet

harmonic subgroups are obtained by applying five-level wavelet packet decomposition. Finally, Figures 3.13 to 3.15 show the same harmonic subgroups calculated by applying the adaptive Prony's method (Figure 3.13), the adaptive ESPRIT method (Figure 3.14) and the adaptive root-MUSIC method (Figure 3.15), respectively.

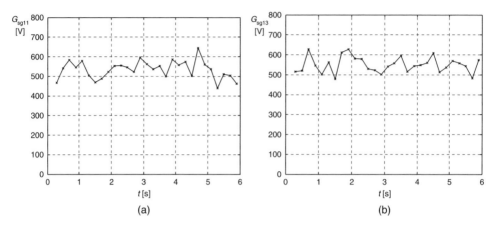

Figure 3.13   (a) The 11th and (b) the 13th IEC harmonic subgroups of voltage versus time calculated by applying the adaptive Prony's method

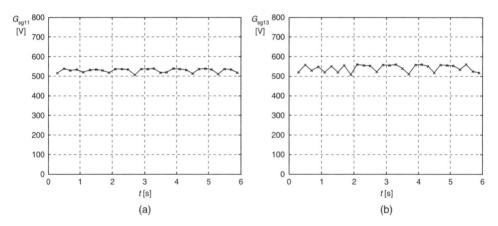

Figure 3.14   (a) The 11th and (b) the 13th IEC harmonic subgroups of voltage versus time calculated by applying the adaptive ESPRIT method

From the analysis of the figures, the following considerations arise:

- All methods furnish harmonic subgroups G_{sg11} and G_{sg13} which are significantly variable versus time but with different dynamics.

- The IEC method provides harmonic subgroup values that are generally lower than those obtained with all other methods. This is probably due to spectral leakage problems encountered with this method. In particular, part of the energy content of the V_{11} and V_{13} harmonics is shifted outside the frequency range of the harmonic subgroups G_{sg11} and G_{sg13}.

- The adaptive parametric methods furnish higher subgroup values due to the absence of spectral leakage problems in these methods.

- The adaptive ESPRIT and adaptive root-MUSIC methods show similar behaviour; this can be explained by taking into account the fact that they are based on similar techniques for spectral component calculation.

- The wavelet transform seems not to suffer from spectral leakage problems.

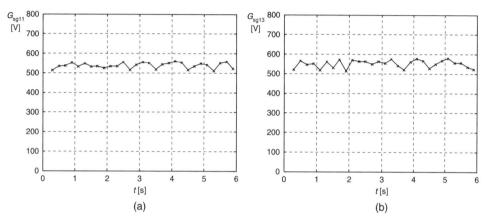

Figure 3.15 (a) The 11th and (b) the 13th IEC harmonic subgroups of voltage versus time calculated by applying the adaptive root-MUSIC method

Finally, it should be noted that the adaptive methods usually require a greater computational burden with respect to the DFT algorithm and the considered wavelet packet transform.

3.5.3 Some comparisons between DFT-based methods and parametric methods

Parametric methods have the following advantages:

- the window width is free and only linked to the waveform characteristics;

- the adaptive version ensures the best fit of waveform by a proper choice of the time window and the number of spectral components;

- the window width does not constrain the frequency resolution.

These methods can significantly reduce the inaccuracies caused by spectral leakage; some DFT-based advanced methods can also significantly reduce these inaccuracies without significantly increasing the computational burden, even though these methods, in particularly stressing situations, may give non-negligible errors. Parametric methods, however, require computational effort greater than that required by DFT-based methods.

A comprehensive comparison among the computational effort required by the parametric methods is not easy, since, for example, Prony's, the ESPRIT and the root-MUSIC methods use different models to approximate the waveforms so they can be characterized by a different number of contiguous time windows in the time observation period and each window can be characterized by a different number of spectral components to approximate the waveform.

With particular reference to the ESPRIT and the root-MUSIC methods, it can be observed that both methods are based on the same signal model, although the root-MUSIC method uses the noise subspace while the ESPRIT method uses the signal subspace. The impact on the results obtained by using the noise subspace or the signal subspace is not immediately obvious; however, the ESPRIT method is generally considered more reliable for spectral analysis [1].

It should be noted that, besides WT, STDFT and the high-resolution methods analyzed in the above section, other techniques have also been used for harmonic distortion assessment, e.g. the Kalman filtering technique. Details on this topic can be found in [1].

3.6   Time–frequency distributions

Other advanced methods for analyzing nonstationary signals (not only waveform distortions) use time–frequency distributions (TFDs); the basic objective of TFDs is to provide simultaneous time and frequency information about the energy content of the signal [12, 50, 53, 54]. In particular, TFDs can furnish information on the energy content of each signal spectral component and where this content is located in time.

In order to better understand the time–frequency representation of TFDs, let us consider a test signal that includes a spectral component of amplitude 1 p.u. at a fundamental frequency of 50 Hz, with superimposed harmonics, interharmonics and a transient disturbance. In particular, the harmonics are 5th and 7th order harmonics of amplitude 0.03 p.u. and 0.01 p.u., respectively, and the interharmonic has an amplitude of 0.002 p.u. at a frequency of 57.3 Hz; finally, a transient disturbance, which includes a sinusoidal tone at a frequency of 874.5 Hz, an initial amplitude 0.3 p.u. and a damping factor of -150, is added at $t = 0.176$ s. In Figure 3.16, the considered test signal in the time domain (Figure 3.16 (a)) and its TFD

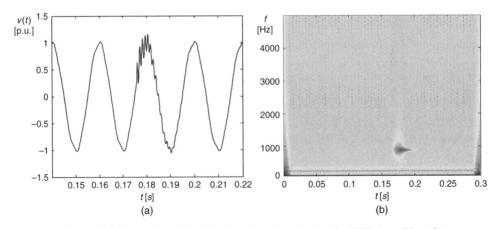

Figure 3.16   (a) Test signal in the time domain; (b) its TFD (see Plate 2)

(Figure 3.16 (b)) are shown. Figure 3.16 (b) shows the presence of the stationary spectral components at the fundamental frequency and at the 5th and 7th harmonic orders with decreasing energy content (from black to grey colours, respectively) and also shows a transient disturbance at a high frequency localized in a short time interval ([0.15–0.20] s).

In the next sections, first the theoretical background of time–frequency distributions is briefly recalled and then some indices based on their application are illustrated.

3.6.1 Theoretical background

All TFDs can be obtained from 'Cohen's class', which is a generalized formulation for the time–frequency frame.

Cohen's class of the waveform $x(t)$ is defined as:

$$\text{TFD}_x(t, \omega; \varphi) = \frac{1}{4\pi^2} \iiint z^* \left(u - \frac{\tau}{2} \right) z \left(u + \frac{\tau}{2} \right) \varphi(\theta, \tau) e^{-j(\theta t + \omega\tau - \theta u)} \, du \, d\tau \, d\theta, \qquad (3.47)$$

where the triple integration is on the interval $(-\infty, +\infty)$, $\varphi(\theta, \tau)$ is a two-dimensional function called the *kernel* of the time–frequency distribution and $z(t)$ is the complex analytic signal of the waveform $x(t)$ given by:

$$z(t) = \frac{2}{\sqrt{2\pi}} \int_{0}^{+\infty} X(\omega) e^{j\omega t} dt, \qquad (3.48)$$

where $X(\omega)$ is the Fourier transform of $x(t)$. The kernel is a function that differs for each member of Cohen's class and it can be properly selected to improve the time–frequency resolution of the time–frequency analysis.

Different types of TFDs can be obtained by selecting different kernel functions; they are characterized by special properties and have been proposed for the improvement of the time–frequency resolution. Some examples are: the spectrogram, the Wigner–Ville distribution, the Choi–Williams distribution and the reduced interference distribution (RID); Figure 3.16 (b) shows an example of a Wigner–Ville TFD. In [54], the authors have utilized a TFD analysis based on the RID to provide a unified definition of new power quality indices used for the characterization of nonstationary waveforms in power systems.

3.6.2 Time–frequency distribution-based indices

Let us consider a generic voltage waveform $v(t)$ that can be characterized by the presence of disturbances. The waveform can be decomposed into two parts: a fundamental component that is the sinusoidal component at power system frequency $v_F(t)$ and a disturbance waveform $v_D(t)$ given by:

$$v_D(t) = v(t) - v_F(t). \qquad (3.49)$$

Applying the kernel of the reduced interference distribution (RID) to Equation (3.47) and considering the waveforms $v(t)$, $v_D(t)$ and $v_F(t)$ in Equation (3.49), the following power quality indices have been introduced in [54]:

- the instantaneous distortion energy index IDE(t);
- the normalized instantaneous distortion energy index NIDE(t);

- the instantaneous frequency index IF(t);

- the instantaneous K-factor index IK(t).

IDE(t) is defined as follows:

$$IDE(t) = \left(\frac{\displaystyle\int_{-\infty}^{+\infty} TFD_{v_D}(t, \omega; \varphi)d\omega}{\displaystyle\int_{-\infty}^{+\infty} TFD_{v_F}(t, \omega; \varphi)d\omega} \right)^{1/2}. \tag{3.50}$$

The index given by Equation (3.50) can be interpreted as an extension of the well-known THD; it gives a measure of the energy content of all the spectral components included in $v_D(t)$ with respect to the energy of the fundamental component. In order to illustrate the information furnished by the IDE index, let us consider once again the test signal shown in Figure 3.16 (a); the time variation of the IDE index calculated for this signal is shown in Figure 3.17. From the analysis of Figure 3.17, it is possible to observe that the IDE values are always positive due to the presence of harmonic/interharmonic distortion and the energy content of the transient disturbance; in particular, significant values of the IDE index are observed when a transient disturbance occurs.

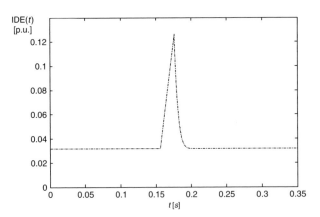

Figure 3.17   IDE index versus time

NIDE(t) is based on the same concept as IDE(t), but the energy of disturbance is referred to the energy of all the spectral components present in the voltage waveform $v(t)$, so that it is always lower than 1. NIDE(t) is defined as follows:

$$NIDE(t) = \left(\frac{\displaystyle\int_{-\infty}^{+\infty} TFD_{v_D}(t, \omega; \varphi)d\omega}{\displaystyle\int_{-\infty}^{+\infty} TFD_{v_F}(t, \omega; \varphi)d\omega + \int_{-\infty}^{+\infty} TFD_{v_D}(t, \omega; \varphi)d\omega} \right)^{1/2}. \tag{3.51}$$

The instantaneous frequency is the normalized first moment (mean value) of TFD_x in the frequency domain. It is defined by:

$$\text{IF}(t) = \frac{\displaystyle\int_{-\infty}^{+\infty} \omega \text{TFD}_v(t, \omega; \varphi) d\omega}{2\pi \displaystyle\int_{-\infty}^{+\infty} \text{TFD}_v(t, \omega; \varphi) d\omega}. \tag{3.52}$$

Observing the above definition, it is possible to note that, in the numerator, the energy density of each spectral component is weighted by its proper frequency; on the other hand, the denominator is the energy of signal $v(t)$ multiplied by 2π. In the presence of only the fundamental component, $\text{IF}(t)$ is equal to the power system fundamental frequency while it assumes increasing values in the presence of disturbances characterized by increasing frequency spectral content. In practice, $\text{IF}(t)$ provides a measure of the deviation of the frequency from the power system frequency f_0 due to the presence of spectral content at other frequencies.

In order to illustrate the information furnished by the IF index, let us consider once again the test signal shown in Figure 3.16 (a); the time variation of the IF index for this signal is shown in Figure 3.18, which shows that the IF index values are generally slightly greater than 50 Hz due to the presence of harmonic/interharmonic distortion; in addition, note that the IF index values increase significantly when a transient disturbance occurs due to the presence of the high-frequency spectral content. It should be noted that the peak value of the index during the transient is about 62 Hz, which is a significantly smaller value than the frequency of the actual high-frequency transient spectral component, which is 874.5 Hz; this is in accordance with the definition in Equation (3.52) and is due to the fact that, in Equation (3.52), the frequency 874.5 Hz is weighted with the (small) value of the energy content associated with the high-frequency transient spectral component.

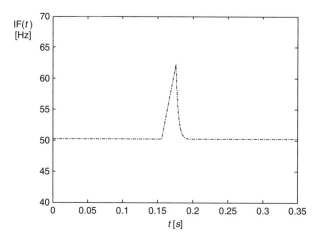

Figure 3.18   IF index versus time

The last index, IK(t), is the extension of the K-factor to nonstationary signals; in particular, it characterizes the disturbances by weighting the spectral component energy density by the square of the normalized frequency. IK(t) is defined as follows:

$$
\mathrm{IK}(t) = \frac{\displaystyle\int_{-\infty}^{+\infty} \left(\frac{\omega}{\omega_0}\right)^2 \mathrm{TFD}_v(t, \omega; \varphi)\mathrm{d}\omega}{\displaystyle\int_{-\infty}^{+\infty} \mathrm{TFD}_v(t, \omega; \varphi)\mathrm{d}\omega},
\tag{3.53}
$$

where $\omega_0 = 2\pi f_0$.

Once again, let us consider the test signal shown in Figure 3.16 (a), and the time variations of index IK (Figure 3.19). Figure 3.19 shows that IK index values are generally slightly greater than 1 due to the presence of harmonic/interharmonic distortion; also, note that the IK index values increase significantly when a transient disturbance occurs due to the presence of the high-frequency spectral content.

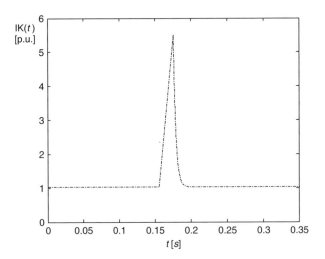

Figure 3.19   IK index versus time

Accurate evaluations of all TFD-based PQ indices are connected strictly to the accurate decomposition of the waveform $v(t)$ into fundamental and disturbance waveforms. This decomposition can be made by estimating the fundamental component $\hat{v}_F(t)$ by means of the following relation:

$$
\hat{v}_F(t) = \hat{V}_0 \cos(2\pi f_0 t + \hat{\theta}_0),
\tag{3.54}
$$

where, in [54]:

- f_0 is assumed fixed and equal to the power system frequency (for example 50 Hz);

- $\hat{V}_0$ is the amplitude of the fundamental component estimated using $\mathrm{TFD}_v(t, \omega_0; \varphi)$, that is the TFD of the voltage waveform calculated at the fundamental frequency;

- $\hat{\theta}_0$ is the initial phase of the fundamental component and is obtained by applying the following curve-fitting algorithm:

$$\hat{\theta}_0 = \arg_\theta \min |v(t) - \hat{V}_0 \cos{(2\pi f_0 t + \theta)}|^2. \tag{3.55}$$

It should be noted that, in the case of voltage sag (or swell) disturbances, the fundamental $v_F(t)$ is assumed to be the 'pre-event' voltage (the voltage waveform in the operating condition before the voltage sag or swell) so that the disturbance $v_D(t)$ is also characterized by the presence of an unavoidable spectral component at power system frequency.

In practical applications, the continuous signal $v(t)$ is sampled, resulting in a discrete sequence $v(n)$; moreover, only finite discrete sequences of samples can be analyzed. The finite discrete sequence is obtained by applying a sliding time window to the sampled waveform. Then, to calculate the aforementioned TFD-based PQ indices, a discrete version of their relationships should be applied. Using these discrete formulations, TFD-based PQ indices will be functions of time 't_m' which, for example, can be the initial time of the sliding time window used for the analysis.

With particular reference to discrete disturbances, it has been proposed in [54] that the so-called 'principal average' can be calculated to provide a single number that quantifies the disturbance in a more compact way. The principal average can be defined for each of the PQ indices defined in Equations (3.50) to (3.53). It is the mean value of the index over a time interval equal to a fundamental period ($1/f_0$); this time interval is chosen by taking into account that its central time t_{Max} is the time at which the index assumes its local maximum value. Letting TFDPQ(t) be one of the PQ indices defined in Equations (3.50) to (3.53), the principal average (PA) is defined as:

$$PA = f_0 \int_{t_{Max} - 1/(2f_0)}^{t_{Max} + 1/(2f_0)} \text{TFDPQ}(t) dt, \tag{3.56}$$

where t_{Max} is the aforementioned central time.

Recently, the use of the adaptive Prony's method (APM) outlined earlier has also been considered to redefine and calculate indices from Equations (3.50) to (3.53) [50]. The advantages of APM are that:

- the estimation of the fundamental component does not require any assumption on the fundamental frequency;

- the extraction of the actual fundamental component does not require the application of a separate fitting algorithm;

- it can be applied only to the whole waveform from which all the quantities needed for the calculation of the PQ indices are provided;

- it also provides the initial phases of each spectral component.

If we remind ourselves of Prony's model given in Section 3.5.1.1, the fundamental and disturbance waveforms can be expressed as:

$$\hat{v}(t_n) = \hat{v}_F(t_n) + \hat{v}_D(t_n) = V_{k_0} e^{j\phi_{k_0}} e^{(\alpha_{k_0} + j\omega_{k_0}) t_{n-1}}$$

$$+ \sum_{k=1 (k \neq k_0)}^{M} V_k e^{j\phi_k} e^{(\alpha_k + j\omega_k) t_{n-1}}, \tag{3.57}$$

where k_0 is the index corresponding to the fundamental component.

In order to define the APM-based PQ indices, discrete versions of the indices are considered. Finally, the indices from Equations (3.50) to (3.53) can be redefined in the discrete time domain as follows:

- The short time disturbance energy index STDE(t_m)

$$
\text{STDE}(t_m) = \left\{ \frac{\sum\limits_{k=1(k\neq k_0)}^{M} \sum\limits_{n=1}^{L} \left| V_k e^{j\phi_k} e^{(\alpha_k + j\omega_k)t_{n-1}} \right|^2}{\sum\limits_{n=1}^{L} \left| V_{k_0} e^{j\phi_{k_0}} e^{(\alpha_{k_0} + j\omega_{k_0})t_{n-1}} \right|^2} \right\}^{1/2}
\tag{3.58}
$$

- The normalized short time disturbance energy index NSTDE(t_m)

$$
\text{NSTDE}(t_m) = \left\{ \frac{\sum\limits_{k=1(k\neq k_0)}^{M} \sum\limits_{n=1}^{L} \left| V_k e^{j\phi_k} e^{(\alpha_k + j\omega_k)t_{n-1}} \right|^2}{\sum\limits_{k=1}^{M} \sum\limits_{n=1}^{L} \left| V_k e^{j\phi_k} e^{(\alpha_k + j\omega_k)t_{n-1}} \right|^2} \right\}^{1/2}
\tag{3.59}
$$

- The short time frequency deviation index STFD(t_m)

$$
\text{STFD}(t_m) = \frac{\sum\limits_{k=1}^{M} \dfrac{\omega_k}{2\pi} \sum\limits_{n=1}^{L} \left| V_k e^{j\phi_k} e^{(\alpha_k + j\omega_k)t_{n-1}} \right|^2}{\sum\limits_{k=1}^{M} \sum\limits_{n=1}^{L} \left| V_k e^{j\phi_k} e^{(\alpha_k + j\omega_k)t_{n-1}} \right|^2}
\tag{3.60}
$$

- The short time K-factor index STK(t_m)

$$
\text{STK}(t_m) = \frac{\sum\limits_{k=1}^{M} \left(\dfrac{\omega_k}{\omega_{k_0}} \right)^2 \sum\limits_{n=1}^{L} \left| V_k e^{j\phi_k} e^{(\alpha_k + j\omega_k)t_{n-1}} \right|^2}{\sum\limits_{k=1}^{M} \sum\limits_{n=1}^{L} \left| V_k e^{j\phi_k} e^{(\alpha_k + j\omega_k)t_{n-1}} \right|^2}.
\tag{3.61}
$$

In addition, it is still possible to define the APM-based principal average.

Example 3.5 In this example we consider simulated transients due to the switching of two capacitors and apply the APM to calculate the indices given by Equations (3.58) to (3.61). The scheme for this simple system is shown in Figure 3.20; the system parameters are reported in [50] except for the 5 MW load which is characterized by $\cos\varphi = 0.85$.

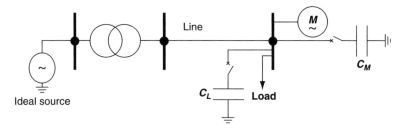

Figure 3.20   System under study

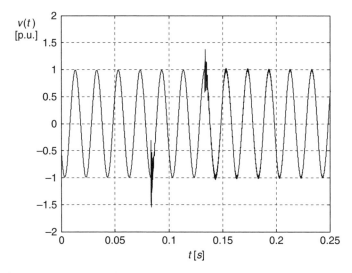

Figure 3.21   Voltage waveform versus time during the two capacitor switchings

Figure 3.21 shows the voltage at the capacitor terminals (in p.u. of peak phase voltage). The first transient (at $t = 0.083$ s) was caused by capacitor switching of the static load; the second transient (at $t = 0.133$ s) was caused by the switching of the motor capacitor.

Figure 3.22 shows the STDE (Figure 3.22 (a)) and NSTDE (Figure 3.22 (b)) indices. From the analysis of these figures, it appears that the STDE and NSTDE indices have similar time variations. In the presence of a disturbance, the value of the NSTDE index is smaller than the value of the STDE index. This is consistent with the definitions of the indices; in fact, the denominator of Equation (3.59) is greater than the denominator of Equation (3.58). Also, analyses of the STDE and NSTDE indices show that the second transient is characterized by a greater energy. Finally, the STDE and NSTDE indices have local maximum values that correspond to the initial times of the transient disturbances, i.e. 0.083 s and 0.133 s, respectively.

The STFD index values are shown in Figure 3.23 (a). It clearly appears that the frequency deviations have high values in the presence of the transient due to the high-frequency spectral content. From the analysis of Figure 3.23 (a), it also appears that the second transient disturbance causes a frequency deviation greater than the first transient because it has greater

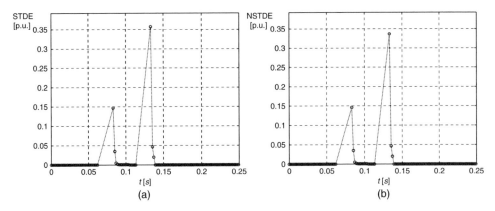

Figure 3.22   (a) STDE and (b) NSTDE index values versus time

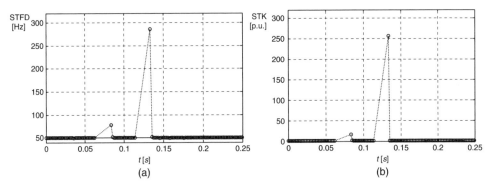

Figure 3.23   (a) STFD and (b) STK indices versus time

high-frequency spectral content. Figure 3.23 (a) shows that STFD has local maximum values that correspond to the initial times of the transient disturbances, i.e. at 0.083 s and at 0.133 s, respectively.

The same considerations made for STFD can also be made for the STK index (see Figure 3.23 (b)). It should be noted that STK has even smaller values than STFD. This is consistent with the definitions of the two indices; in fact, in the numerator of STFD (Equation (3.60)) the energy density of the spectral content is weighted by its proper frequency, while in the numerator of STK (Equation (3.61)) the energy density of the spectral content is weighted by its proper squared normalized frequency. All the indices allow us to capture the initial time of transients.

To characterize the disturbances in a more synthetic way, Table 3.3 reports the peak values, the peak times (t_{Max}) and the principal average (PA) of the indices, which were calculated with reference to the two local maximum values. As mentioned before, the central time t_{Max} in Equation (3.56) is equal to the time at which the indices reach their local maximum values. From the analysis of Table 3.3, it is confirmed that PANSTDE and PASTDE have greater values for the second transient (due to its greater energy) and PASTFD and PASTK have greater values for the second transient (due to its greater high-frequency spectral content).

Table 3.3 Principal average (PA) of APM-based PQ indices calculated with reference to the first (a) and second (b) maximum value

(a)

	t_{Max}	peak	PA
STDE [p.u.]	0.083	0.147	0.010
NSTDE [p.u.]	0.083	0.145	0.010
STFD [Hz]	0.083	77.928	50.440
STK [p.u.]	0.083	16.854	1.184

(b)

	t_{Max}	peak	PA
STDE [p.u.]	0.133	0.357	0.013
NSTDE [p.u.]	0.133	0.336	0.013
STFD [Hz]	0.133	285.745	50.493
STK [p.u.]	0.133	256.597	1.190

3.7   Transient waveform distortions (bursts)

Harmonic *bursts* are transient waveform distortions lasting a short time in comparison to the observation interval. They can be caused by nonlinear loads, subject to random and sudden shock loads (e.g. starting of drives) or by sudden variations in harmonic impedances; they can cause defective operation of control equipment resulting, for example, from the displacement of zero crossing points of the voltage waveform, and spurious tripping of control circuits.

A deterministic description of bursts is inadequate for taking into account the characteristics of these disturbances, while a description in terms of a stochastic process is more useful for providing a methodological means of identifying adequate indices [55].

3.7.1   Theoretical background

Let us consider, as an example of a waveform distortion index, the harmonic distortion factor (THD) and assume that, in the presence of bursts, the phenomenon can be described as a stochastic process $W(t)$ composed of a 'base' random process $W_0(t)$ that varies randomly around its mean value and a succession of bursts of random amplitude and duration; if a barrier value b is assigned (Figure 3.24), the THD process $W(t)$ is given by:

$$W(t) = W_0(t) + \sum_{k=1}^{N_b(t)} \varsigma_k \, \Pi_k(t - T_k) \qquad (3.62)$$

where:

- $N_b(t)$ is the number of up-crossings of barrier b in $(0,t)$;
- ς_k are the 'burst over base process' amplitudes given by $[Z_k - W_0(t)]$, with Z_k being the burst amplitude;

- Π_k represents the opportune 'pulse-shaped' functions, which are of random shape and duration, have a maximum amplitude of 1 and vanish outside $(T_k, T_k + U_k)$, where U_k is the random burst duration.

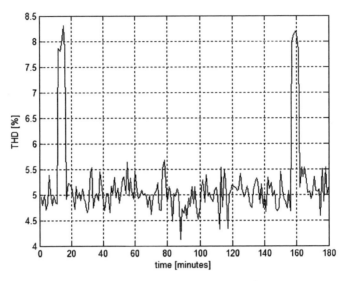

Figure 3.24   THD burst example

Equation (3.62) can be rewritten as:

$$W(t) = W_0(t) + \sum_{k=1}^{N_b(t)} V_k(t, \varsigma_k, T_k) \tag{3.63}$$

and, hence, the stochastic process $W(t)$ can be defined in terms of the discrete process $N_b(t)$ and the time random functions $V_k(t, \varsigma_k, T_k) = \varsigma_k \, \Pi_k(t - T_k)$.

It is practically impossible to derive an analytical expression for the $W(t)$ probability distribution. However, if the above functions are independent, as t diverges, for the central limit theorem, $W(t)$ approaches a Gaussian process.

3.7.2   Burst indices

The characterization of the whole process given in Equation (3.63) is frequently unnecessary, and only amplitudes Z_k and duration U_k of the bursts need be described (in the following, these variables are generically indicated as H_k); their probabilistic description can be obtained with probability density functions (pdfs) derived on the basis of available measurements or simulations.

A time-dependent index defined by the following stochastic process is introduced in [55] with regard to the 'cumulative' duration (amplitude) of bursts:

$$B_H(t) = \sum_{k=1}^{N_b(t)} H_k, \tag{3.64}$$

where H_k is, as previously defined, the kth burst duration (amplitude).

If only the mean values of the processes B_H are of interest, these statistical measures can be easily obtained under the hypothesis that the succession of the random variables Z_k or U_k is, in a broad sense, stationary. In this case, we have:

$$E[B_H(t)] = \mu_H E[N_b(t)]t, \tag{3.65}$$

where μ_H is the mean value of any single variable H_k.

In the most general case, the entire probability distributions of these two processes can be derived once the probabilistic characteristics of the discrete process $N_b(t)$ other than of the random variables Z_k and U_k are known.

On the assumptions that the mean duration of each burst U_k is much smaller than the mean time between the successive crossings τ_k and that the barrier level is high enough, the $N_b(t)$ process, here defined as the 'burst counting process', can be described, by virtue of theoretical results of the extreme values from the stochastic process, by the well-known Poisson probability law $p(k,t)$, given by:

$$p(k,t) \equiv P[N_b(t) = k] = e^{-\phi_b t} \frac{(\phi_b\ t)^k}{k!} \qquad k = 0, 1, \ldots\ldots\ldots, +\infty. \tag{3.66}$$

In Equation (3.66), ϕ_b is the mean number of up-crossings in the unit time. The mean and variance of the process $N_b(t)$ are numerically equal and are given by:

$$E[N_b(t)]\ =\ \text{Var}[N_b(t)]\ =\ \phi_b t. \tag{3.67}$$

Equations (3.66) and (3.67) strictly hold under the assumption that the process $W(t)$ is stationary, but an extension to the nonstationary case has also been derived in the literature. Moreover, in the most general case, the discrete process $N_b(t)$ can be described efficiently on the basis of numerical measurements of actual plants or simulations.

Other than the cumulative duration (amplitude) of bursts, other new indices for bursts have been proposed to characterize risk (the burst risk index) and safety (the burst safety index) conditions with respect to the disturbance effects on the system.

3.7.2.1   Burst risk index

Let us consider the following stochastic process $D(t,d)$, defined as the number of times that a characteristic parameter of the bursts exceeds a given level $d > b$:

$$D(t, d) = \sum_{k=0}^{N_b(t)} I_k, \tag{3.68}$$

where I_k is a Bernoulli random variable ($I_k = 1$ if $H_k > d$ at time T_k of the kth burst, otherwise $I_k = 0$).

The burst risk index (BRI) is defined as the mean number of burst amplitudes or durations over $(0, t)$ exceeding d; it is given by:

$$\text{BRI}(t, d)\ =\ E[D(t, d)]. \tag{3.69}$$

This index is intrinsically dynamic, allowing an efficient prediction of the performances over time. Under the hypothesis that the random variables H_k are statistically independent and identically distributed with the common, time-independent, cumulative distribution function $F(h) = F_H(h) = P(H_k \leq h)$, $\forall\ k = 1, 2, \ldots\ldots n, \ldots$, it can also be analytically calculated as:

$$\text{BRI}(t,\ d)\ =\ E[N_b(t)]\ P(H > d)\ =\ E[N_b(t)]\ [1 - F(d)]\ (t > 0). \tag{3.70}$$

It has been shown that, under the Poisson hypothesis, Equation (3.70) can be simplified [55].

It should be noted that this index can be particularly useful for establishing a component's immunity to this kind of disturbance. In fact, for the assigned process, the more sensitive a given electrical component is to the burst effect (which implies low values of the level d), the higher is its BRI value.

3.7.2.2 Burst safety index

The most natural way to identify an index with suitably robust conditions for dealing with an assigned disturbance is to introduce the following stochastic process:

$$\begin{cases} Y(t) = \max(H_1, \dots, H_{N_b}(t)) & N_b(t) = 0 \\ Y(t) = 0 & \text{otherwise.} \end{cases} \tag{3.71}$$

Once a safety level s is assigned, the following burst safety index (BSI) can be defined:

$$\text{BSI}(t) = P[Y(t) < s]. \tag{3.72}$$

Naturally, for every assigned value n_b of $N_b(t)$, the following relationship holds:

$$[\max(H_1, \dots, H_{n_b}) < s] \quad \text{if} \quad [(H_1 < s) \cap \dots \cap (H_{nb} < s)], \tag{3.73}$$

so that BSI(t) is simply the probability that the level s is *never* exceeded over $(0, t)$.

By means of trivial calculations, a compact expression can be obtained using the Poisson hypothesis. In other cases, the $Y(t)$ pdf can be readily evaluated using numerical techniques or a Monte Carlo simulation, as shown in [55].

Example 3.6 Figures 3.25 and 3.26 show the burst indices given by Equations (3.69) and (3.72) as functions of the levels d and s (in %) respectively, for different time horizons ($t = 1$, 2 and 4 days). The indices refer to burst amplitude.

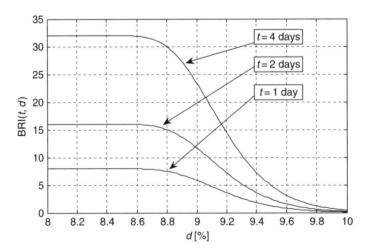

Figure 3.25 Amplitude burst risk index BRI(t, d) versus the level d for different time horizons

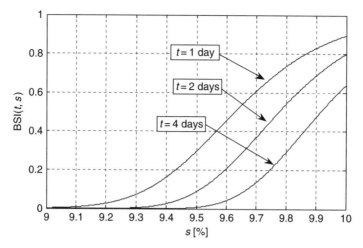

Figure 3.26 Amplitude burst safety index BSI(t, s) versus the level s for different time horizons

Figure 3.25 and Figure 3.26 refer to a test THD process and have been calculated assuming that the base random process $W_0(t)$ has a constant mean value equal to 5% and a barrier value $b = 8\%$. Moreover, it is assumed that the 'burst excess' amplitudes over the barrier, i.e. the values of the random variable $Y_k = (Z_k - b)$, with b being the barrier value, have a lognormal distribution. The following scale and shape parameters have been assumed $a = 0.1520$ and $c = 0.2462$ (corresponding to a burst excess amplitude mean value equal to 1.2 and a standard deviation equal to 0.25); a burst mean frequency ϕ_b over the barrier b equal to 0.333 hour^{-1} (corresponding to the occurrence of eight bursts, on average, per day) has been assumed.

It has been shown and stated in the relevant literature that the above indices are not dependent on the particular burst probability law [55]. For instance, it has been shown that similar results are obtained both for the lognormal distribution utilized in the example for burst-excess amplitude values and for a gamma distribution fitted to the same burst amplitude values, with equal mean and standard deviation.

3.8 Conclusions

In this chapter, some problems concerning the calculation of PQ indices have been analyzed, with particular reference to the spectral leakage problems in the waveform distortion assessment and to the analysis and characterization of nonstationary waveforms. Advanced methods as well as new indices have been shown, and numerical examples provided in order to outline the advantages that each offers.

The main conclusion of the chapter is that, even though research on these topics is ongoing and several significant contributions will soon be published, the new methods and indices considered in this chapter seem to be powerful tools for overcoming problems that have been encountered as well as those that may be anticipated.

References

[1] Bollen, M. and Yu-Hua Gu, I. (2006) *Signal Processing of Power Quality Disturbances*, Wiley-IEEE Press, New Jersey.

[2] Kay, S.M. (1988) *Modern Spectral Estimation: Theory and Application*, Prentice-Hall, Englewood Cliffs, New Jersey.

[3] Proakis, J.G. and Manolakis, D.G. (1996) *Digital Signal Processing Principles, Algorithms, and Applications*, Prentice-Hall, New Jersey.

[4] Stoica, P. and Moses, R. (1997) *Introduction to Spectral Analysis*, Prentice-Hall, New Jersey.

[5] Harris, F.J. (1978) 'On the Use of Windows for Harmonic Analysis with the Discrete Fourier Transform', *Proceedings of IEEE*, **66**(1), 51–83.

[6] Gallo, D., Langella, R. and Testa, A. (2000) On the Processing of Harmonics and Interharmonics in Electrical Power Systems, *IEEE PES Winter Meeting*, Singapore, January.

[7] Gallo, D., Langella, R. and Testa, A. (2004) 'Desynchronized Processing Technique for Harmonic and Interharmonic Analysis', *IEEE Transactions on Power Delivery*, **19**(3), 993–1001.

[8] Langlois, P. and Bergeron, R. (1992) Interharmonic Analysis by a Frequency Interpolation Method, *2nd International Conference on Power Quality*, Atlanta (USA), September.

[9] Gallo, D., Langella, R. and Testa, A. (2001) Interharmonic Analysis Utilising Optimised Harmonic Filtering, *IEEE International Symposium on Diagnostics for Electric Machines, Power Electronics and Drives (SDEMPED)*, Gorizia (Italy), September.

[10] Gallo, D., Langella, R. and Testa, A. (2002) 'A Self Tuning Harmonics and Interharmonics Processing Technique', *European Transactions on Electrical Power*, **12**(1), 25–31.

[11] Gallo, D., Langella, R. and Testa, A. (2004) 'On the Processing of Harmonics and Interharmonics: Using Hanning Window in Standard Framework', *IEEE Transactions on Power Delivery*, **19**(1), 28–34.

[12] Cohen, L. (1994) *Time–Frequency Analysis*, Prentice-Hall, New Jersey.

[13] Brouaye, F. and Meunier, M. (1998) Fourier Transform, Wavelets, Prony Analysis: Tools for Harmonics and Quality of Power, *IEEE PES 8th International Conference on Harmonics and Quality of Power (ICHQP)*, Athens (Greece), October.

[14] Poisson, O., Rioual, P. and Meunier, M. (1999) 'New Signal Processing Tools Applied to Power Quality Analysis', *IEEE Transactions on Power Delivery*, **14**(2), 561–566.

[15] Heydt, G.T. and Jewell, W.T. (1999) 'Pitfalls of Electric Power Quality Indices', *IEEE Transactions on Power Delivery*, **13**(2), 570–578.

[16] Heydt, G.T., Fjeld, P.S., Liu, C.C., Pierce, D., Tu, L. and Hensley, G. (1999) 'Applications of the Windowed FFT to Electric Power Quality Assessment', *IEEE Transactions on Power Delivery*, **14**(4), 1411–1416.

[17] Santoso, S., Powers, E.J., MackGrady, W., Lamoree, J. and Bhatt, S.C. (2000) 'Characterization of Distribution Power Quality Events with Fourier and Wavelet Transforms', *IEEE Transactions on Power Delivery*, **15**(1), 247–253.

[18] Iaramillo, S.H., Heydt, G.T., and O'Neill Carrillo, E. (2000) 'Power Quality Indices for Aperiodic Voltages and Currents', *IEEE Transactions on Power Delivery*, **15**(2), 784–790.

[19] Kandil, M.S., Farghal, S.A. and Elmitwally, A. (2001) 'Refined Power Quality Indices', *IEE Proceedings – Generation, Transmission & Distribution*, **148**(6), 590–596.

[20] Carpinelli, G., Iacovone, F., Russo, A. and Varilone, P. (2004) 'Chaos-based Modeling of DC Arc Furnaces for Power Quality Issues', *IEEE Transactions on Power Delivery*, **19**(4), 1869–1876.

[21] Duque, C., Silveira, P.M., Baldwin, T. and Ribeiro, P.F. (2008) Novel Method for Tracking Time-Varying Power Harmonic Distortions without Frequency Spillover, *IEEE PES General Meeting*, Pittsburgh (USA), July.

[22] Silveira, P.M., Duque, C.A., Baldwin, T. and Ribeiro, P.F. (2008) Time-Varying Power Harmonic Decomposition using Sliding-Window DFT, *IEEE International Conference on Harmonics and Quality of Power*, Wollongong (Australia), September/October.

[23] Santoso, S., Powers, E.J., MackGrady, W. and Hofmann, P. (1996) 'Power Quality Assessment via Wavelet Transform Analysis', *IEEE Transactions on Power Delivery*, **11**(2), 924–930.

[24] Mallat, S.G. (1998) *A Wavelet Tour of Signal Processing*, Elsevier Academic Press, California.

[25] Pham, V.L. and Wong, K. (1999) 'Wavelet-Transform-Based Algorithm for Harmonic Analysis of Power System Waveforms', *IEE Proceedings – Generation, Transmission & Distribution*, **146**(3), 249–254.

[26] Hamid, E.Y., Mardiana, R. and Kawasaki, Z.I. (2002) 'Method for RMS and Power Measurements Based on the Wavelet Packet Transform', *IEE Proceedings – Science, Measurement and Technology*, **149**(2), 60–66.

[27] Barros, J. and Diego, R.I. (2006) 'Application of the Wavelet-Packet Transform to the Estimation of Harmonic Groups in Current and Voltage Waveforms', *IEEE Transactions on Power Delivery*, **21**(1), 533–535.

[28] Chen, S. and Zhu, H.Y. (2007) 'Wavelet Transform for Processing Power Quality Disturbances', *EURASIP Journal on Advances in Signal Processing,* Article ID 47695, doi:10.1155/2007/47695.

[29] Akaike, H. (1974) 'A New Look at the Statistical Model Identification', *IEEE Transactions on Automatic Control*, **19**(6), 716–723.

[30] Kashyap, R. (1980) 'Inconsistency of the AIC Rule for Estimating the Order of Autoregressive Models', *IEEE Transactions on Automatic Control*, **25**(5), 996–998.

[31] Wax, M. and Kailath, T (1985) 'Detection of Signals by an Information Theoretic Criteria', *IEEE Transactions on Acoustics, Speech and Signal Processing*, **33**(2), 387–392.

[32] Roy, R. and Kailath, T. (1989) 'ESPRIT – Estimation of Signal Parameters via Rotational Invariance Techniques', *IEEE Transactions on Acoustics, Speech and Signal Processing*, **37**(7), 984–995.

[33] Zhang, X.D. and Zhang, Y.S. (1993) 'Determination of the MA Order of an ARMA Process Using Sample Correlations', *IEEE Transactions on Signal Processing*, **41**(6), 2277–2280.

[34] Liang, G., MitchellWilkes, D. and Cadzow, J.A. (1993) 'ARMA Model Order Estimation Based on the Eigenvalues of the Covariance Matrix', *IEEE Transactions on Signal Processing*, **41**(10), 3003–3009.

[35] Dafis, A.J., Nwankpa, C.O. and Petropulu, A. (2000) Analysis of Power System Transient Disturbances Using an ESPRIT-based Method, *IEEE PES Summer Meeting*, Seattle, July.

[36] Lobos, T., Leonowicz, Z. and Rezmer, J. (2000) Harmonics and Interharmonics Estimation Using Advanced Signal Processing Methods, *9th IEEE PES International Conference on Harmonics and Quality of Power (ICHQP)*, Orlando (USA), October.

[37] Lobos, T., Rezmer, J. and Koglin, H.J. (2001) Analysis of Power System Transients Using Wavelets and Prony Method, *IEEE Power Tech*, Porto (Portugal), September.

[38] Leonowicz, Z., Lobos, T. and Rezmer, J. (2003) 'Advanced Spectrum Estimation Methods for Signal Analysis in Power Electronics', *IEEE Transactions on Industrial Electronics*, **50**(3), 514–519.

[39] Lobos, T., Rezmer, J. and Schegner, J. (2003) Parameter Estimation of Distorted Signals Using Prony Method, *IEEE Power Tech*, Bologna (Italy), June.

[40] Bracale, A., Carpinelli, G., Lauria, D., Leonowicz, Z., Lobos, T. and Rezmer, J. (2004) On Some Spectrum Estimation Methods for Analysis of Non-Stationary Signals in Power Systems Part I: Theoretical Aspects, *IEEE PES 11th International Conference on Harmonics and Quality of Power (ICHQP)*, Lake Placid (USA), September.

[41] Bracale, A., Carpinelli, G., Lauria, D., Leonowicz, Z., Lobos, T. and Rezmer, J. (2004) On Some Spectrum Estimation Methods for Analysis of Non-Stationary Signals in Power Systems Part II: Numerical Applications, *IEEE PES 11th International Conference on Harmonics and Quality of Power (ICHQP)*, Lake Placid (USA), September.

[42] Bracale, A., Carpinelli, G., Leonowicz, Z., Lobos, T. and Rezmer, J. (2005) Waveform Distortions Due to AC/DC Converters Feeding DC Arc Furnaces, *Electrical Power Quality and Utilization (EPQU) '05 Conference*, Krakow (Poland), September.

[43] Lobos, T., Leonowicz, Z., Rezmer, J. and Schegner, P. (2006) 'High-resolution Spectrum Estimation Methods for Signal Analysis in Power Systems', *IEEE Transactions on Instrumentation and Measurement*, **55**(1), 219–225.

[44] Zhang, J., Xu, Z., Pan,W. and Chen, H. (2006) TLS-ESPRIT Based Method for Harmonics and Interharmonics Analysis in Power Systems, *12th International Conference on Harmonics and Quality of Power (ICHQP)*, Cascais (Portugal), October.

[45] Bracale, A., Carpinelli, G., Langella, R. and Testa, A. (2006) On Some Advanced Methods for Waveform Distortion Assessment in the Presence of Interharmonics, *IEEE PES General Meeting 2006*, Montreal (Canada), June.

[46] Bracale, A., Carpinelli, G., Langella, R. and Testa, A. (2007) 'Accurate Methods for Signal Processing of Distorted Waveforms in Power Systems', *EURASIP Journal on Advances in Signal Processing*, Article ID 92191, doi:10.1155/2007/92191.

[47] Bracale, A., Caramia, P. and Carpinelli, G. (2007) 'Adaptive Prony Method for Waveform Distortion Detection in Power Systems', *International Journal of Electrical Power & Energy Systems*, **29**(5), 371–379.

[48] Bracale, A., Caramia, P. and Carpinelli, G. (2007) 'Optimal Evaluation of Waveform Distortion Indices with Prony and RootMusic methods', *International Journal of Power & Energy Systems (IJPES)*, **27**(4), October/December.

[49] Bracale, A., Carpinelli, G., Leonowicz, Z., Lobos, T. and Rezmer, J. (2008) 'Measurement of IEC Groups and Subgroups using Advanced Spectrum Estimation Methods', *IEEE Transactions on Instrumentation and Measurement*, **57**(4), 672–681.

[50] Andreotti, A., Bracale, A., Caramia, P. and Carpinelli, G. (2009) 'Adaptive Prony Method for the Calculation of Power-Quality Indices in the Presence of Nonstationary Disturbance Waveforms', *IEEE Transactions On Power Delivery*, **24**(2), 874–883.

[51] IEC 61000-4-7 (2002) *Electromagnetic Compatibility (EMC) – Part 4: Testing and Measurement Techniques – Section 7: General Guide on Harmonics and Interharmonics Measurements, for Power Supply Systems and Equipment Connected Thereto*, Edition 2.

[52] IEC 61000-4-30 (2003) *Electromagnetic Compatibility (EMC) – Part 4: Testing and Measurement Techniques – Section 30: Power Quality Measurements Methods*, Edition 1.

[53] Shin, Y.J., Parsons, A., Powers, E.J. and MackGrady, W. (1999) Time–frequency Analysis of Power System Fault Signals for Power Quality, *IEEE PES Summer Meeting*, Edmonton (Canada), July.

[54] Shin, Y., Powers, E.J., Mack Grady, W. and Arapostathis, A. (2006) 'Power Quality Indices for Transient Disturbances', *IEEE Transactions on Power Delivery*, **21**(1), 253–261.

[55] Carpinelli, G., Chiodo, E. and Lauria, D. (2007) 'Indices for the Characterization of Bursts of Short-duration Waveform Distortion', *IET Generation, Transmission & Distribution*, **1**(1), 170–179.

4

Quantifying the quality of the overall supply voltage

4.1 Introduction

The traditional power quality (PQ) indices, e.g. voltage and current total harmonic distortion, the unbalance factor, etc., analyze PQ disturbances separately. Global indices quantify the overall loss of PQ as a whole; as such, they can be of particular interest. In fact, in some cases, the instantaneous values of voltage, including all disturbances contemporaneously, can be the main cause of failure or malfunction of electrical components, such as static converters or insulated components. Moreover, global PQ indices can significantly reduce the amount of stored data required to cover the various types of disturbance. Finally, they allow the comparison of utilities for benchmarking purposes and easily quantify any loss of PQ, which is needed for PQ regulation, as will be shown in Chapter 6.

Although many papers in the relevant literature have analyzed the problem of the characterization of individual types of PQ disturbance, only a few such papers have dealt with global power quality indices (GPQIs).

GPQIs, finalized to quantify the overall supply voltage quality, can be classified into three subcategories:

- GPQIs based on the evaluation of the difference between an ideal voltage waveform and the actual voltage waveform;

- GPQIs based on opportune treatments of various traditional PQ indices in order to ascertain one or at most two figures for quantifying the overall quality of the supply voltage at the monitored site;

- GPQIs based on the economic impact produced by the PQ disturbances on the equipment of the customer.

This chapter presents and critically analyzes the theoretical aspects of some GPQIs that have been developed to quantify the quality of the supply voltage.

Power Quality Indices in Liberalized Markets   Pierluigi Caramia, Guido Carpinelli and Paola Verde
© 2009 John Wiley & Sons, Ltd

4.2   Global indices based on a comparison between ideal and actual voltages

These indices quantify directly the difference between the actual waveform of the power supply voltage and the ideal waveform of the power supply voltage, where the ideal supply voltage is a proper voltage characterized by the absence of PQ disturbances (Figure 4.1).

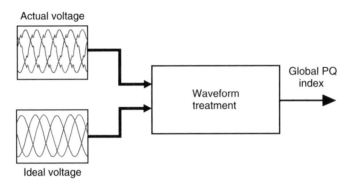

Figure 4.1   Procedure for evaluating a global power quality index based on the comparison of ideal and actual voltage waveforms

In the forthcoming sections, the following three GPQIs that have been proposed in the literature are described [1–4]:

- the normalized RMS error;
- the normalized three-phase global index;
- the voltage quality deviation factor.

4.2.1   The normalized RMS error

In Figure 4.2, the following two voltage waveforms are plotted:

- $v_a(t)$ is the actual voltage waveform characterized by the presence of PQ disturbances;
- $v_i(t)$ is a sinusoidal voltage waveform characterized by proper fixed values of amplitude and frequency (ideal supply voltage).

The root-mean-squared value of the difference between the actual and ideal voltage waveforms is defined as the RMS error (RMSE) [2]:

$$\text{RMSE} = \sqrt{\frac{1}{T}\int_T [v_a(t) - v_i(t)]^2 dt},\qquad(4.1)$$

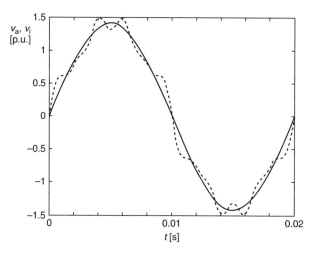

Figure 4.2   The actual v_a (dashed line) and ideal v_i (solid line) voltage waveforms

where T is the time period of the power frequency. Considering the sampled voltage waveforms, Equation (4.1) can be written as:

$$\text{RMSE} = \sqrt{\frac{\sum_{n=1}^{N}[v_a(nT_s) - v_i(nT_s)]^2}{N}}. \tag{4.2}$$

In Equation (4.2), $v_a(nT_s)$ and $v_i(nT_s)$ are the nth samples of the actual voltage and ideal voltage, respectively, N is the number of samples in a period of power frequency T and T_s is the sampling time.

The RMS error depends on the voltage level being considered; thus, to allow a meaningful comparison of PQ performance among different voltage levels, a normalized version of the RMS error (NRMSE) is considered:

$$\text{NRMSE} = \frac{\sqrt{\dfrac{\sum_{n=1}^{N}[v_a(nT_s) - v_i(nT_s)]^2}{N}}}{V_1}. \tag{4.3}$$

In Equation (4.3), V_1 is the magnitude of the actual fundamental voltage.

The NRMSE is a useful index for characterizing the whole PQ, since it considers all (both continuous and discrete) PQ disturbances, e.g. harmonics and interharmonics, voltage sags and swells and transients. If the value of NRMSE is low, the authors who proposed this index [2] suggest that no further analyses are required; otherwise, the PQ problem needs further identification to determine the nature of the power quality issue.

A crucial issue for the NRMSE index calculation is that the ideal voltage $v_i(t)$ must be known. This voltage is assumed to be an ideal sine wave at power frequency, the amplitude and argument of which are, respectively, equal to the estimated magnitude and argument of the fundamental voltage. In the case of a single event (for example, a voltage sag), the ideal voltage is assumed to be the 'pre-event' fundamental voltage, i.e. it is the fundamental voltage waveform in the operating condition before the PQ event. In both cases, the amplitude and argument of the ideal voltage can be derived by using a curve-fitting algorithm (CFA) based on the least squares estimation technique or by using the zero-crossing algorithm that requires interpolation between consecutive sample points straddling the zero crossing. Both CFA and zero-crossing techniques give good estimates of the fundamental magnitude and argument of the voltage in the presence of small background distortions. If the background distortions are not negligible, the zero-crossing method gives unacceptable results and the CFA method is the preferred method. Alternatively, the amplitude of the ideal sine wave can be assumed to be equal to the nominal value of the voltage. It should be noted that, if the amplitude of the ideal voltage coincides with the magnitude of the estimated fundamental voltage of the monitored actual voltage, the NRMSE cannot detect the presence of slow voltage variations; this disturbance can be detected if the value of the amplitude of ideal voltage is equal to the nominal amplitude.

4.2.2   The normalized three-phase global index

Unlike the NRMSE index previously introduced, the normalized three-phase global index permits the quantification of the PQ level at a three-phase busbar, considering contemporaneously all three-phase voltage waveforms.

Letting $v_{a,a}(t)$, $v_{a,b}(t)$, and $v_{a,c}(t)$ be the actual single-phase voltage waveforms monitored at the busbar of a three-phase power system, the normalized three-phase global index is defined starting from the instantaneous quality value $X(t)$, defined as:

$$X(t) = \sqrt{\frac{1}{3} \sum_{p = a,b,c} \left[v_{a,p}(t) - \sqrt{2}V_{nf} \sin\left(\omega_{nom} t + \varphi - \alpha_p\right)\right]^2}, \qquad (4.4)$$

where p is the phase code, $v_{a,p}(t)$ is the pth actual phase voltage, V_{nf} and ω_{nom} are the nominal phase voltage amplitude and angular frequency, respectively, φ is the initial argument of the first-phase voltage, α_p is equal to 0, $4\pi/3$ and $2\pi/3$ for $p = $ a, b and c, respectively. The $X(t)$ quantity is, at each instant, the root of the mean on the three phases of the squared difference between the actual single-phase voltages and the ideal single-phase voltages. In this case, the ideal single-phase voltages constitute three sinusoidal waveforms, with magnitudes and frequencies equal to the nominal values, displaced from each other by 120 degrees.

The normalized three-phase global index $\text{NGI}_{3\Phi}$ is defined as the normalized root mean square value of $X(t)$:

$$\text{NGI}_{3\Phi} = \frac{\sqrt{\frac{1}{T}\int_0^T X^2(t)\, dt}}{V_{nf}} = \frac{\sqrt{\frac{1}{T}\int_0^T \frac{1}{3}\sum_{p=a,b,c}\left[v_{a,p}(t) - \sqrt{2}V_{nf}\sin\left(\omega_{nom}t + \varphi - \alpha_p\right)\right]^2 dt}}{V_{nf}}.$$

$$(4.5)$$

The normalization at the nominal phase voltage V_{nf} is used once again to allow a meaningful comparison of PQ performance at different voltage levels.

Considering the sampled voltage waveforms, Equation (4.5) can be written as:

$$
NGI_{3\Phi} = \frac{\sqrt{\dfrac{1}{N}\displaystyle\sum_{n=1}^{N}\left(X^2(nT_s)\right)}}{V_{nf}}
$$

$$
= \frac{\sqrt{\dfrac{1}{N}\displaystyle\sum_{n=1}^{N}\dfrac{1}{3}\displaystyle\sum_{p=a,b,c}\left[v_{a,p}(nT_s) - \sqrt{2}\,V_{nf}\,\sin\left(\omega_{nom}\,n\,T_s + \varphi - \alpha_p\right)\right]^2}}{V_{nf}},
$$

(4.6)

where $v_{a,p}(nT_s)$ is the nth sample of the pth actual phase voltage.

For each busbar of the monitored system, the $NGI_{3\Phi}$ index furnishes an average indication of how close the actual three-phase voltage waveforms are to the three-phase waveforms of the ideal voltage. The $NGI_{3\Phi}$ also considers all PQ disturbances. Once again, if the value assumed by $NGI_{3\Phi}$ is low, no further analysis is required; otherwise, the PQ problem needs further assessment to determine the nature of the PQ issue.

4.2.3   The voltage quality deviation factor

The voltage quality deviation factor (VQDF) is a global PQ index that is able to quantify the deviation level of the actual voltage waveform from the ideal voltage waveform for a fixed time interval, using the energy function associated with the voltage waveform.

The voltage quality deviation factor is defined as the ratio between the voltage energy deviation and the ideal waveform's energy levels in the interval $[t, t+T]$:

$$
VQDF(t, t+T) = \frac{\displaystyle\int_{t}^{t+T} \left|v_a^2(t) - v_i^2(t)\right|dt}{\displaystyle\int_{t}^{t+T} v_i^2(t)dt}.
$$

(4.7)

The ideal voltage $v_i(t)$ is a sinusoidal voltage waveform, the magnitude of which is equal to the nominal value and the frequency of which is coincident with the fundamental frequency of the actual voltage, unlike what happens in the evaluation of the NRMSE index, where the ideal voltage is a sine wave at power frequency with amplitude equal to the estimated magnitude of the fundamental voltage. The angular frequency and argument of $v_i(t)$ can be obtained, with good precision, by using a phase-lock loop circuit. This index seems to give results similar to those produced by the instantaneous distortion energy (IDE) index for quantifying nonstationary waveforms, described in Chapter 3. The difference between these indices is the ideal voltage waveform considered; while the VQDF uses the nominal voltage as the ideal waveform, the IDE estimates the voltage component at the power frequency.

In the case of the three-phase power system, the expression for the VQDF becomes:

$$\text{VQDF}(t, t+T) = \frac{\displaystyle\int_{t}^{t+T} \sum_{p=a,b,c} \left| v_{a,p}^2(t) - v_{i,p}^2(t) \right| dt}{\displaystyle\int_{t}^{t+T} \sum_{p=a,b,c} v_{i,p}^2(t) dt}, \tag{4.8}$$

where p is the phase code, $v_{ap}(t)$ is the pth actual phase voltage and $v_{ip}(t)$ is the pth ideal phase voltage.

The index VQDF produces non-negative values; in particular, when VQDF equals zero, no deviations exist between the actual and ideal voltages and, therefore, the voltage quality is assumed to be ideal. A high VQDF value indicates a poor quality of supply voltage.

The VQDF index was also proposed to achieve a quantitative economic analysis of the effects on customers when the voltage quality is not acceptable. The authors who introduced the VQDF index [3] have proposed a formula for the energy cost paid by customers, which is a function of the extent of degradation of the power quality. In particular, the energy cost paid by the customer in the presence of PQ degradation is reduced, with respect to that paid in ideal conditions, as a function of the value assumed by the VQDF index. Considering that some deviation from the ideal voltage is permitted (due to the presence of disturbances within the fixed limits), the energy cost paid by the customer is the same as the cost in ideal conditions when the value of the VQDF index is less than a fixed threshold value. However, when the PQ deviation exceeds the fixed threshold value, the payment for energy by the customer is reduced by a factor that is proportional to the value of VQDF.

Example 4.1   In this example the performance of the GPQIs based on the evaluation of the differences between the ideal voltage waveform and the actual voltage waveform is compared using different test waveforms that represent three-phase power supply voltages character-ized by the presence of one or more disturbances. Table 4.1 shows the test waveforms considered.

Table 4.2 shows the maximum values of NRMSE, NGI$_{3\Phi}$ and VQDF indices during the analyzed time interval of one second. These indices, in fact, are calculated for each period of power frequency and, in the observed time period, their maximum values are assumed to characterize the PQ level.

Based on the analysis of the numerical results reported in Table 4.2, it is apparent that:

- All the considered indices classify test waveform 5 as the 'worst case' in terms of PQ; this waveform is characterized by the greatest number of disturbances.

- As was foreseeable, NRMSE $=$ NGI$_{3\Phi}$ and reduces to THD if only harmonics are present (waveform 1: THD $= 10.5\%$).

- In the presence of only one disturbance (waveforms 1, 2 and 3), NRMSE, NGI$_{3\Phi}$ and VQDF classify case 3 as the most critical due to the presence of a significant voltage sag.

Table 4.1   List of test signals

Code	Test signals
1	Three-phase voltages with 3rd, 5th, 7th, 9th, 11th and 13th harmonics with values equal to the limits of EN 50160 [5].
2	Three-phase unbalanced voltages with an unbalance factor equal to 3.0%.
3	Three-phase voltages with voltage sags on one phase characterized by a residual voltage equal to 0.5 p.u. and duration equal to 200 ms.
4	Three-phase voltages characterized by: • an unbalance factor equal to 3.0% • the presence of 3rd, 5th, 7th, 9th, 11th and 13th harmonics with values equal to the limits of EN 50160 [5].
5	Three-phase voltages characterized by: • an unbalance factor equal to 3.0% • the presence of 3rd, 5th, 7th, 9th, 11th and 13th harmonics with values equal to the limits of EN 50160 [5]. • the presence of a sag on one phase characterized by a residual voltage equal to 0.5 p.u. and duration equal to 200 ms

Table 4.2   Individual indicator and global index values

Test code	Global indices		
	NRMSE	$NGI_{3\Phi}$	VQDF
1	0.105	0.105	0.136
2	0.000	0.045	0.065
3	0.500	0.289	0.250
4	0.113	0.114	0.164
5	0.511	0.310	0.359

Example 4.2   In this example the GPQIs are calculated using a three-phase voltage waveform characterized by the contemporaneous presence of harmonics, unbalances and a voltage sag. The time duration of the recorded signal is about five seconds; Figure 4.3 shows a zoom of the three-phase voltage waveforms corresponding to the sag presence. The voltage waveform was obtained in the laboratory by means of a three-phase power voltage supply.

Figures 4.4 (a), (b) and (c) show the NRMSE, $NGI_{3\Phi}$ and VQDF indices versus time; in particular, the value of the NRMSE index (Figure 4.4 (a)) has been calculated for each phase.

From the analysis of the results shown in the figures, it is evident that the indices clearly identify the presence of the sag. The values of these indices present a discontinuity that corresponds to the sag.

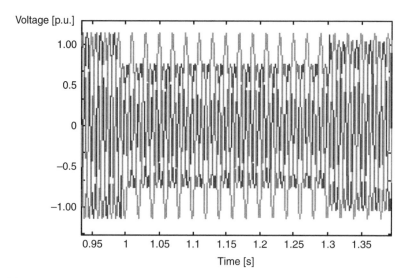

Figure 4.3   Zoom of the three-phase voltage waveforms corresponding to the sag presence (see Plate 3)

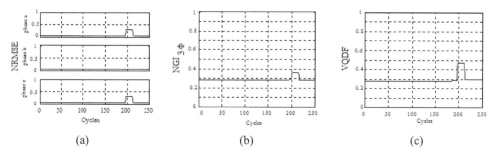

Figure 4.4   Time variations in (a) NRMSE; (b) NGI$_{3\Phi}$; (c) VQDF calculated for the signal reported in Figure 4.3

4.3   Global indices based on the treatment of traditional indices

GPQIs based on the treatment of traditional indices are able to quantify the quality of the supply voltage waveform at the busbar of an electrical power system, starting from the characterization of the single disturbance levels quantified through traditional power quality indices (Chapter 1).

The procedure for evaluating these indices is synthetically illustrated in Figure 4.5.

In this section, the following GPQIs will be described [4, 6, 7]:

- the global indicator;

- the unified power quality index.

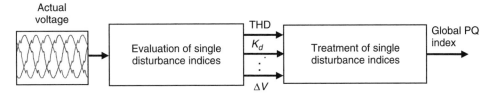

Figure 4.5   Procedure for evaluating a global power quality index based on the treatment of single disturbance indices

4.3.1   The global indicator

The global indicator I_G was designed to assist the French power transmission system operator with regard to the quality of the supply. This global index is defined as the maximum value among individual indicators relative to the slow voltage variations, harmonic distortions, voltage unbalances, voltage sags and voltage swells:

$$I_G = \max(I_T, I_H, I_{UNB}, I_C). \tag{4.9}$$

In Equation (4.9), I_T, I_H, I_{UNB} and I_C are the slow voltage variation indicator, the voltage harmonic indicator, the unbalance voltage indicator and the voltage sags and swell indicator, respectively. These indicators are measured by equipment installed at the power system busbar and configured in a cyclical manner (for example, one recording every ten minutes), for a fixed period of time (for example one week).

The slow voltage variation indicator requires the preventive definition of the variation ratio around the contractual value, $\tau_{\Delta V}$; this ratio is defined as:

$$\tau_{\Delta V} = 100 \frac{(V_a - V_C)}{V_C}, \tag{4.10}$$

where V_a is the actual RMS supply voltage and V_C is the contractual voltage, assumed to be equal to the nominal voltage or the declared supply voltage.

The slow voltage variation indicator I_T is then defined as:

$$I_T = \max(I_{TSup}, I_{TInf}), \tag{4.11}$$

where:

$$I_{TSup} = 100 \frac{\tau_{\Delta V Max}}{\tau_{\Delta V_{LimSup}}}$$
$$I_{TInf} = 100 \frac{\tau_{\Delta V Min}}{\tau_{\Delta V_{LimInf}}}; \tag{4.12}$$

$\tau_{\Delta V Max}$ and $\tau_{\Delta V Min}$ in Equation (4.12) are the maximum and minimum values of $\tau_{\Delta V}$ in the survey period of a week; $\tau_{\Delta V_{LimSup}}$ and $\tau_{\Delta V_{LimInf}}$ are the fixed contractual limits, which are values that should never be exceeded.

The voltage harmonic indicator I_H is defined as:

$$I_H = \max (I_{THDV}, F_2, F_3, F_4 \ldots), \tag{4.13}$$

where:

$$I_{THDV} = 100 \frac{THD_{Max}}{THD_{Lim}}$$

$$\left. F_h = 100 \frac{V_{h,Max}}{V_{h,Lim}} \right|_{h = 2, 3, \ldots, 25}.$$

(4.14)

In this relationship, THD_{Max} is the maximum value of total harmonic distortion at the monitored site during the survey period, THD_{Lim} is the contractual limit of total harmonic distortion, $V_{h,Max}$ is the maximum value of the hth individual voltage harmonic at the monitored site during the survey period and $V_{h,Lim}$ is the contractual limit of the hth individual voltage harmonic.

The voltage harmonic and slow voltage variation indicators are evaluated over the three phases and the maximum value is considered to quantify the slow voltage variations and the distortions at the monitored bus.

The indicator I_{UNB} relative to the voltage unbalances is based on the maximum measured value of the unbalance factor $K_{d,Max}$ at the monitored site during the survey period; in particular, it is defined as:

$$I_{UMB} = 100 \frac{K_{d,Max}}{K_{d,Lim}},$$

(4.15)

where $K_{d,Lim}$ is the contractual limit of the unbalance factor.

The evaluation of the voltage sags and swells indicator I_C is obtained by considering the voltage tolerance curves. Various voltage tolerance curves exist, but the most widely publicized are the CBEMA, ITIC and SEMI curves, as described in Section 1.3.6 in Chapter 1.

The voltage sags and swells indicator I_C refers to the ITIC curve. In particular, if N_1 is the number of events in the survey period inside the ITIC curve and N_2 is the number of events outside the curve, the indicator I_C can be calculated as follows:

$$I_C = 0 \text{ in the absence of swells and/or sags,}$$

(4.16)

$$I_C = \sqrt{\frac{\sum_{i=1}^{N_1} IPQ^2}{N_1}} \quad \text{if } N_2 = 0,$$

(4.17)

$$I_C = \sqrt{\frac{\sum_{i=1}^{N_2} IPQ^2}{N_2}} \quad \text{if } N_2 \neq 0.$$

(4.18)

where:

$$IPQ = 100 \frac{|V_C - V_{Event}|}{|V_C - V_{ITIC}|}.$$

(4.19)

In this relationship, V_{Event} is the minimum (maximum) RMS voltage during the sags (swells), V_C is the contractual voltage and V_{ITIC} is the lower (upper) boundary voltage of the ITIC curve associated with the duration of the sag (swell).[1]

Starting from the knowledge of the individual indicators, the global indicator I_G is obtained from Equation (4.9).

A value of I_G equal to 100 indicates that at least one contractual limit has been reached. A value under 100 indicates that all measured data are under the contractual limits; a value over 100 indicates that one or more contractual limits have been exceeded.

To have a comprehensive idea of the supply quality level for a given area of the power system, a global system PQ index was also defined. This global PQ system index, called the global efficiency index, is evaluated through the following relationship:

$$I_{G,\text{GEI}} = 100 \frac{N_{I_G < 100}}{N_{\text{tot}}}, \qquad (4.20)$$

where $N_{I_G < 100}$ represents the number of measuring devices for which the global indicator I_G is lower than 100, and N_{tot} is the total number of devices installed in the considered area. A value of $I_{G,\text{GEI}}$ equal to 100 indicates that, in the monitored area, the contractual limits are respected; a value of $I_{G,\text{GEI}}$ less than 100 indicates that, in one or more sites of the monitored area, the contractual limits have been exceeded.

4.3.2   The unified power quality index

The unified power quality index (UPQI) was initially defined only for continuous disturbances; this global index is obtained, as will be shown below, by properly handling the indices traditionally used to characterize single continuous disturbances. In particular, starting from the evaluation of the indices used to quantify single disturbances (e.g. the individual harmonic voltage, the total harmonic distortion factor, the unbalance factor and short- and long-term flicker indices), the value of the UPQI is obtained by performing a multi-step procedure.

The first step requires measuring the various indices as introduced in the PQ standards or recommendations. To summarize the index levels over a survey period, some statistical measures are considered; in particular, the 95th or 99th percentile values are calculated, usually over a week, during which time the values of the indices are stored every ten minutes.

In particular, the following quantities are considered:

- for harmonics, the 95th percentile value of the voltage total harmonic distortion (THD) and the 95th percentile value of each single harmonic voltage (V_h) up to the 40th order;

- for voltage unbalances, the 95th percentile value of the unbalance factor, K_d;

- for voltage fluctuation, the 99th percentile value of the short-term flicker severity (P_{st}) and the 99th percentile value of the long-term flicker severity (P_{lt}) indices (the maximum value over the three phases);

---

[1] The definition of IPQ is very similar to the PQI index introduced in Section 1.3.6 in Chapter 1. In particular, IPQ is referred to the ITIC curve, while PQI is referred to the CBEMA curve.

- for slow voltage variations, the 95th percentile value of the absolute difference between the actual RMS voltage and the permissible voltage value (the maximum value over the three phases).

For a surveying period of several weeks, the worst-case value, which is the maximum of the weekly 95th or 99th percentile values, can be considered.

During sag or swell events, the values assumed by the continuous disturbance indices are ignored. It is also recommended that the voltage total harmonic distortion and the unbalance factor be calculated relative to the nominal voltage rather than the actual voltage; otherwise, periods characterized by low voltage will lead to apparently high values for harmonic distortion and unbalances.

In the second step, all index values calculated in the first step are expressed in a per-unit base, where the limit (base) value is represented by the maximum permissible value, $I_{i,\text{lim}}$:

$$I_{i,\text{n}} = \frac{I_i}{I_{i,\text{lim}}}, \tag{4.21}$$

where I_i is the index value calculated in the first step for the ith considered disturbance type. This step is called *normalization*. The limit value $I_{i,\text{lim}}$ may be equal to the value imposed by the standards or recommendations. In this way, a value of the normalized indices equal to 1 means that the statistical measure calculated in the first step coincides with the limit (limit of acceptability); values greater than 1 represent situations in which supply quality does not satisfy the standards or recommendations, whereas values lower than 1 represent situations in which supply quality does satisfy the standards or recommendations.

In practice, the normalization step is also applied in the global indicator evaluation outlined in Section 4.3.1; in the last case, however, the limit values are assumed to be equal to the contractual values.

When a disturbance is characterized by more than one index (for example, in the case of harmonics, the 95th percentile of THD and the 95th percentile of V_h with $h = H_{\min} \div H_{\max}$), a further step is required, i.e. all the normalized values of the indices that refer to the same type of disturbance are compared to each other and, as in the case of the global indicator evaluation, only the maximum value is considered in characterizing the disturbance. This step is called *consolidation* by the authors who proposed the UPQI index [7].

These consolidated indices have the simple property that their maximum acceptable value is unity.

After applying the normalization and consolidation steps, the various indices are properly combined to give a single number. This can be accomplished in several intuitive ways, involving the average or the maximum of the consolidated indices. The average of the consolidated indices implicitly assumes that the effect of PQ is the sum of the individual disturbance types; in particular, excessive disturbances of one type can be mitigated by reducing other types of disturbance. The use of the maximum of the consolidated indices, as used by the global indicator, assumes that the effect of PQ on the customer is only linked to the maximum disturbance, independent of the magnitude of other disturbances.

In the case of the unified power quality index, an alternative way to combine the indices is adopted, involving the introduction of the concept of *exceedance*. Exceedance measures how much a disturbance index, after the normalization and consolidation steps, is over the limit. For the ith consolidated index, the exceedance value is equal to:

$$\Delta I_i = I_{i,\mathrm{n}} - 1. \tag{4.22}$$

After the calculation of the exceedance of each disturbance, the following procedure is applied to quantify the unified power quality index values:

1. If all consolidated indices are less than 1, the unified power quality index is equal to the maximum value of them.

2. If one or more of the consolidated disturbance indices exceeds 1, the unified power quality index is equal to 1 plus the sum of the exceedances. So, the value of the unified power quality index in the presence of N_c disturbances exceeding the limit is given by:

$$\mathrm{UPQI_C} = 1 + \sum_{i=1}^{Nc} \Delta I_i; \tag{4.23}$$

in Equation (4.23), suffix C means that the index refers to continuous disturbances.

One of the convenient properties of the $\mathrm{UPQI_C}$ that follows from its definition is that a value of 1 represents the limit of acceptability.

The concept of exceedance can be extended to include each type of continuous disturbance that can be represented by indices (for example, interharmonics), if normalization and consolidation are first applied. Each type of continuous disturbance can then be incorporated easily into the calculation of $\mathrm{UPQI_C}$.

The varying impacts of the disturbances on individual customers can be taken into account by imposing weight factors k_i, related to their respective exceedances:

$$\mathrm{UPQI_{C,w}} = 1 + \sum_{i=1}^{Nc} k_i \Delta I_i. \tag{4.24}$$

The coefficient k_i in Equation (4.24) can range from 0.0 to 1.0 (default value) and it could be negotiated between a utility and a customer to reflect their concerns.

An idea of the PQ level in a given area of an electrical system can be obtained using a system $\mathrm{UPQI_C}$ index, defined as the weighted average of the $\mathrm{UPQI_C}$ indices from all monitored sites in the considered area:

$$\mathrm{UPQI_{C,Sys}} = \frac{\displaystyle\sum_{j=1}^{M} w_j \mathrm{UPQI_{Cj}}}{\displaystyle\sum_{j=1}^{M} w_j}, \tag{4.25}$$

where w_j is the weighting factor of site j, $\mathrm{UPQI_{Cj}}$ is the index $\mathrm{UPQI_C}$ at the jth busbar and M is the total number of monitored sites. Weightings may be applied according to the number of customers or the maximum demand of customers, supplied by the monitored sites.

Example 4.3 In this example a comparison between the performances of indices I_G and UPQI$_C$ is effected, considering the test waveforms of Table 4.1, in which only the continuous disturbances are present (waveforms 1, 2 and 4). In Table 4.3, for different test signals, the values of the traditional PQ indices characterizing the single disturbances are summarized. The traditional indices considered are:

- the voltage total harmonic distortion THD;
- the single voltage harmonics V_h;
- the unbalance factor K_d; and
- the voltage variation $\tau_{\Delta V}$.

Table 4.3 Traditional power quality index values

Test signals code	THD [%]	h	V_h/V_n [%]	K_d [%]	$|\tau_{\Delta V}|$ [%]
1	10.5	(3, 5, 7, 9, 11, 13)	(5, 6, 5, 1.5, 3.5, 3)	0	0.6
2	0	—	—	3.0	7.5
4	11.3	(3, 5, 7, 9, 11, 13)	(5, 6, 5, 1.5, 3.5, 3)	3.0	6.9

The normalization step considers the limits proposed by EN 50160–2000; these limits are reported in Table 4.4. Finally, Table 4.5 shows the values assumed by the individual indicators and by the global indices in all considered cases.

Table 4.4 Disturbance limits by EN 50160 [5]

THD	V_3/V_n	V_5/V_n	V_7/V_n	V_9/V_n	V_{11}/V_n	V_{13}/V_n	K_d	$\tau_{\Delta V}$
8%	5%	6%	5%	1.5%	3.5%	3%	2%	±10%

Table 4.5 Individual indicator and global index values

Test code	Individual Indicators				Global Indices	
	I_H	I_{UMB}	I_T	I_C	I_G	UPQI$_C$
1	131	0	6	0	131	1.31
2	0	150	75	0	150	1.50
4	141	150	69	0	150	1.91

Based on the analysis of the numerical results reported in Table 4.5, it is apparent that:

- The I_G and UPQI$_C$ indices furnish analogous results in the presence of only one disturbance (waveforms 1 and 2).

- If the voltage waveform is characterized by the presence of more disturbances exceeding the acceptable limits, the UPQI$_C$ index assumes more severe and adequate values; the I_G index instead classifies test signal 4 (with two disturbances exceeding the limit) as equal to test signal 2 (only one disturbance exceeding the limit). In both cases (waveforms 2 and 4), in fact, the more severe disturbances, which are the only factors influencing the I_G index, are the voltage unbalances which are characterized by a K_d factor value equal to 3.0%.

The same authors who proposed the UPQI$_C$ index for continuous disturbances have also proposed indices and limits for discrete disturbances [8] that can be useful for introducing a unified power quality index for these types of disturbance.

First, to quantify discrete disturbances, they proposed an index which is practically similar to the power quality indicator (PQI) outlined in Section 1.3.6. This index, called the discrete severity indicator (DSI), is extended to all discrete disturbances (sag, swell and oscillatory and impulsive transients) and, as for the PQI, is based on voltage tolerance curves. Unlike the PQI, which refers only to the CBEMA curve, the DSI considers the ITIC curve for transients and the CBEMA curve for sags and swells.

As mentioned in Chapter 1, the voltage tolerance curves were originally developed through operating experience using mainframe computers; however, their use has been extended to provide a measure of PQ for electric drives and solid-state loads, as well as a host of wide-ranging residential, commercial and industrial loads. In particular, equipment sensitivity curves for different types of loads have similar shapes to the CBEMA/ITIC curves, but they are characterized by less restrictive values of the coupled voltage amplitude–time duration. In other words, computers are considered the most sensitive equipment, so CBEMA and ITIC curves have been assumed to represent the maximum limit of voltage quality acceptability and as a reference of the discrete disturbance severity.

Each discrete disturbance, characterized by time duration T^* and amplitude V^* (corresponding to the RMS value in the case of sags or swells and to the peak value in the case of transients), can be represented by a point in the voltage tolerance plane. This point is compared with the reference point of the CBEMA/ITIC curve with the same duration T^* to determine if the disturbances are acceptable or not. It is evident that, when the point representing the considered event is in the area of unacceptable voltage, the distance between the point and the CBEMA/ITIC curve is greater, the disturbance is more severe and, consequently, more customers will experience problems.

For any point (T^*, V^*) representing sags, swells or oscillatory transients, the DSI is defined as:

$$\text{DSI}(T^*, V^*) = \left| \frac{V^* - 1}{V_{\text{CBEMA/ITIC}}(T^*) - 1} \right|, \tag{4.26}$$

where $V_{\text{CBEMA/ITIC}}(T^*)$ is the CBEMA/ITIC voltage boundary corresponding to duration T^*. Both V^* and $V_{\text{CBEMA/ITIC}}(T^*)$ in Equation (4.26) are expressed in p.u. of the nominal value.

The evaluation of the DSI is carried out using analytical expressions of CBEMA/ITIC voltages; these expressions, obtained by a curve-fitting procedure, give the CBEMA/ITIC voltage boundary in p.u. In particular, the CBEMA fitting curves were chosen for sags and swells, while the fitted ITIC curve was used to characterize oscillatory transients (OTs); their expressions are [8]:

$$V_{\text{CBEMA-Sag}}(T) = 0.86 - \left(\frac{0.0035}{T}\right)^{\left(\frac{1}{1.22}\right)}$$

$$V_{\text{CBEMA-Swell}}(T) = 1.06 + \left(\frac{0.000295}{T}\right)^{\left(\frac{1}{1.48}\right)} \tag{4.27}$$

$$V_{\text{ITIC-OT}}(T) = 1.2 - \left(\frac{0.00076}{T}\right)^{\left(\frac{1}{1.014}\right)}.$$

For each discrete disturbance (sag, swell, oscillatory transient) characterized by amplitude V^* and time T^*, the CBEMA/ITIC voltage is calculated by Equation (4.27) with $t = T^*$ and the corresponding DSI index is obtained by applying Equation (4.26).

The DSI index is also defined for impulsive transients (ITs). However, for this type of discrete disturbance, the DSI definition is quite different and is based on the energy content associated with the disturbance $E = \int [v(t)]^2 dt$, where $v(t)$ is the transient voltage as a function of time; so, the relationship that defines the DSI for ITs is different from Equation (4.26) [9].

It is evident from Chapter 1 that, PQ monitors typically report the peak value V_{peak} and the energy content E of the transient overvoltage. From these quantities, the equivalent duration of the transient can be defined as the time interval T such that $E = V_{\text{peak}}^2 T$. In this way, each impulsive transient is characterized by the couple (V_{peak}, T), which detects a representative point on the voltage tolerance plane.

The discrete severity indicator for an impulsive transient is based on the assumption that impulsive transients that have the same energy content cause an identical customer complaint rate and are characterized by the analogous $V_{\text{peak}}^2 T$ contour on the voltage-duration plane. The relation $V_{\text{peak}}^2 T = 0.004$ (where V_{peak} is given in p.u. of the nominal voltage and T is time in seconds) corresponds to the ITIC curve that is assumed as the baseline where the equipment will begin to fail. So, the discrete severity indicator, DSI, for an impulsive transient is defined as:

$$\text{DSI}(V_{\text{peak}}, T) = \frac{V_{\text{peak}}^2 T}{0.004}. \tag{4.28}$$

From Equation (4.28), if the point representing the impulsive transient is on the ITIC curve, its coordinates (V_{peak}, T) match the relationship $V_{\text{peak}}^2 T = 0.004$, and the corresponding DSI is equal to 1. An impulsive transient characterized by DSI < 1 is represented by a point falling into the voltage-acceptable zone of the ITIC voltage-duration plane (where the energy content is less than that corresponding to the ITIC curve); an impulsive transient characterized by DSI > 1 is represented by a point falling into the voltage-unacceptable zone (where the energy content is greater than that corresponding to the ITIC curve). In Figure 4.6, together with the fitted ITIC curve, the curves corresponding to transient energy content equal to 0.5, 2, 3 and 4 times the energy content of the ITIC curve are shown.

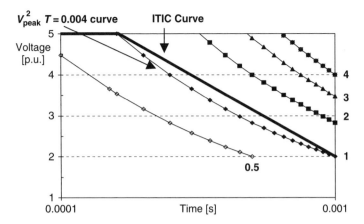

Figure 4.6 Comparison of the ITIC curve and the curves corresponding to transient energy content equal to 0.5, 2, 3 and 4 times the energy content of the ITIC curve

Generally, the DSI index values for each type of discrete disturbance give an indication of the severity levels of the disturbance. DSI values higher than unity indicate unacceptable disturbances. The greater the extent to which the DSI index exceeds unity, the greater the severity of the disturbance and, consequently, the number of customers experiencing problems increases.

The DSI indices reach the maximum value in correspondence to the highest severity of disturbances. The authors who proposed the indices calculated that [8]:

for sags	DSI = 6.9 in correspondence to $V^* = 0$ p.u. and $T^* = 3$s;
for swells	DSI = 12.9 in correspondence to $V^* = 1.8$ p.u. and $T^* = 3$s;
for oscillatory transients	DSI = 3.6 in correspondence to $V^* = 2$ p.u. and $T^* = 0.01$s;
for impulsive transients	DSI = 6.2 in correspondence to $V^* = 5$ p.u. and $T^* = 0.001$s.

Practically, it is assumed that, in correspondence to these highest values of DSI, 100% of customers experience problems, while for DSI values less than the maximum value, a linear relationship exists between the DSI and the customer complaint rate.

An important difference between the power quality indicator (PQI) and the discrete severity indicator (DSI) is the presence of a transition region that modifies the calculation of the DSI index when the point representing the discrete disturbance is close to the thresholds identifying the areas of the different disturbances in the voltage tolerance plane. The presence of the transition region is motivated by the uncertainties in detecting the disturbances that lie on the border of the set threshold of the PQ monitor. Different PQ monitors may classify events that lie on the border of the set threshold differently, even though their set threshold limits are the same.

This uncertainty can be rectified by introducing a transition region for each disturbance and by modifying the DSI index by a correction factor if the event falls into this region. For more details see [8].

Because the DSI is an event index, once again it is necessary to obtain an index that can characterize the events during the observation period at the monitored site. The site index for each discrete disturbance type is calculated as the sum of all discrete severity indicators over the specified survey period (i.e. generally one or more years):

$$\text{DSI}_{\text{Site, DD}} = \sum_{j=1}^{N_{\text{DD}}} \text{DSI}_j, \tag{4.29}$$

where DD indicates the type of discrete disturbance (sag, swell, OT, IT) and N_{DD} is the number of events of the same type that arise in the survey period.

A quantification of discrete disturbance levels in a given area of an electrical system can be obtained using a system DSI index, defined as the weighted average of the DSI site indices from all monitored sites in the considered area:

$$\text{DSI}_{\text{Sys, DD}} = \frac{\displaystyle\sum_{p=1}^{M} w_p \left(\text{DSI}_{\text{Site, DD}}\right)_p}{\displaystyle\sum_{p=1}^{M} w_p}, \tag{4.30}$$

where w_p is the weighting factor of site p, $(\text{DSI}_{\text{Site, DD}})_p$ is the site index DSI at the pth monitored site referred to DD discrete disturbance and M is the total number of monitored sites. Weightings can be applied according to the number of customers or the maximum demand of the customers supplied by the monitored sites.

The authors who proposed DSI site indices have also proposed limits for these indices [8]. The limits were obtained by applying a step procedure and using the statistical information of large-scale PQ surveys. The combined information of all these surveys gives a good comparison between different countries and regions that may be helpful in developing global limits.

The procedure used to detect the DSI site limits is illustrated in detail here with reference to voltage sags and is then extended to all discrete disturbances.

The limit for the DSI site index for voltage sags was defined using large-scale survey data derived from a measurement campaign including nine European countries. These data are reported in Table 4.6 and they distinguish between cases of underground networks (U/G) and cases of mixed networks (Mixed).

Table 4.6 Number of voltage sags that is not exceeded by 95% of the monitored sites for underground networks (U/G) and mixed networks (Mixed)

Range of residual voltage [%]	Duration [s]							
	0.01–0.1		0.1–0.3		0.3–1		1–3	
	U/G	Mixed	U/G	Mixed	U/G	Mixed	U/G	Mixed
70–90	23	61	19	68	3	12	1	6
40–70	5	8	19	38	1	4	0	1
0–40	1	2	8	20	1	4	0	2

The values reported in Table 4.6 are based on the 95% sag statistic of all surveyed sites in the nine countries. For each pre-fixed 'time duration range/residual voltage range', the number reported in Table 4.6 is the number of voltage sags (whose time durations and residual voltages are within the pre-fixed ranges) that is not exceeded by 95% of the monitored sites.

The above procedure, proposed to define the voltage sag limit, is characterized by the following steps:

1. The voltage-duration plane is segmented into a window format based on the available data, as shown in Figure 4.7, in which the windows are called A1, A2, ..., C3, C4.

2. An average disturbance index $DSI_{avg,i}$ using a fixed number of equally distributed disturbance events is calculated for each ith disturbance window (in the figure, an arbitrary number of nine is used).

3. The average disturbance index of each window $DSI_{avg,i}$ is multiplied by the respective disturbance count $N_{ev,i}$ from the available data reported in Table 4.6 to obtain the sag limit for each window WSL_i:

$$WSL_i = N_{ev,i}DSI_{avg,i}. \tag{4.31}$$

4. By summing the sag limit of each window, DSI_{sum} is calculated:

$$DSI_{sum} = \sum_{i=A_1,A_2,...,C_4} WSL_i. \tag{4.32}$$

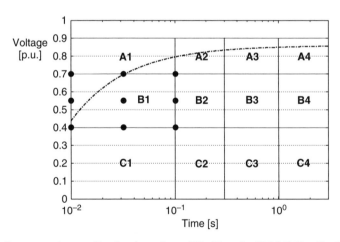

Figure 4.7 Segmented sag distribution chart [8]. Herath, H.M.S.C., Gosbell, V.J. and Perera, S. 'Power Quality (PQ) Survey Reporting: Discrete Disturbance Limits', IEEE Transactions on Power Delivery, 20(2), 851–858 © 2005 IEEE

These steps are synthesized in Table 4.7 for U/G and mixed networks. Voltage sag limits of each window are based on the 95% sag statistic of all surveyed sites in the nine countries. The authors who proposed this procedure assume that the site limit value of the voltage sag index lies between the maximum window sag limit and DSI_{sum}; in particular, a number greater than the geometric mean value between those values is suggested.

Table 4.7 Procedure to calculate the DSI_{sum} index for U/G and mixed networks [8]

Voltage sag window	$DSI_{avg,i}$	U/G networks		Mixed networks	
		$N_{ev,i}$	Window sag limit (WSL_i)	$N_{ev,i}$	Window sag limit (WSL_i)
A1	0.683	23	15.709	61	41.663
A2	1.150	19	21.850	68	78.200
A3	1.341	3	4.023	12	16.092
A4	1.392	1	1.392	6	8.352
B1	1.533	5	7.665	8	12.264
B2	2.188	19	**41.572**	38	**83.144**
B3	2.950	1	2.950	4	11.800
B4	3.048	0	0.000	1	3.048
C1	2.688	1	2.688	6	16.128
C2	4.500	8	36.000	17	76.500
C3	5.150	1	5.150	1	5.150
C4	5.394	0	0.000	3	16.182
		$DSI_{sum} = $ **138.999**		$DSI_{sum} = $ **368.523**	

For U/G networks, the geometric mean value between DSI_{sum} (138.999) and the maximum window sag limit (41.572) is equal to 76.01. While for mixed networks, the geometric mean value between DSI_{sum} (368.523) and the maximum window sag limit (83.144) is equal to 175.04. The site limit values of voltage sag for U/G and mixed networks are assumed equal to values greater than the calculated geometric mean values (76.01 and 175.04); to be exact, values of 100 and 200 are considered, respectively.

Analogous procedures can be applied for the other discrete disturbances in order to obtain the limit values. In particular, the available data used for these discrete disturbances are based on the Electric Power Research Institute's (EPRI's) DPQ project data [10].

The voltage-duration plane, including voltage swells, is segmented into a window format, as shown in Table 4.8; the windows are called P1, P2, . . . , Q3, Q4. The data from EPRI's DPQ project with reference to voltage swells falling into the aforementioned P1, P2, . . . , Q4 windows are reported in Table 4.9 (in the column 'EPRI's swell count'). This table resumes the procedure for the calculation of the DSI_{sum} index.

Table 4.8 Voltage swell windows based on EPRI's DPQ project data [8]

Range of voltage [%]	Duration [s]			
	0.01–0.1	0.1–0.3	0.3–1	1–3
110–120%	P1	P2	P3	P4
120–180%	Q1	Q2	Q3	Q4

The geometric mean value between DSI_{sum} (366.7) and the maximum window swell limit (160.7) is equal to 242.75; the limit site values of voltage swell are assumed to be 250.

Table 4.9 Procedure for calculating the DSI_{sum} index for voltage swells [8]

Voltage swell window	DSI $_{avg,i}$	EPRI's swell count	Window swell limit
P1	1.41	114	**160.7**
P2	2.00	58	116.0
P3	2.25	20	45.0
P4	2.40	12	28.8
Q1	4.76	2	9.5
Q2	6.67	1	6.7
Q3	7.46	0	0.0
Q4	7.86	0	0.0
			$DSI_{sum} = $ **366.7**

The voltage-duration plane, including voltage oscillatory transients, is segmented into four windows (called S, T, U and V), as shown in Table 4.10. The data from EPRI's DPQ project with reference to voltage oscillatory transients (OTs) are reported in the third column of Table 4.11, which illustrates the procedure for calculating the DSI_{sum} index.

Table 4.10 Voltage oscillatory transient windows based on EPRI's DPQ project data [8]

Range of voltage [%]	Duration [s]
	0.001–0.01
120–140%	S
140–160%	T
160–180%	U
180–200%	V

Table 4.11 Procedure for calculating the DSI_{sum} index for oscillatory transients [8]

Voltage OT window	$DSI_{avg,i}$	EPRI's OT count	OT window limit
S	0.70	53.01	**37.107**
T	1.16	15.39	17.852
U	1.60	2.109	3.374
V	2.05	1.151	2.360
			$DSI_{sum} = $ **60.693**

The geometric mean value between DSI_{sum} (60.693) and the maximum OT window limit (37.107) is equal to 47.46; the limit site value of oscillatory transients is assumed to be 50.

Finally, the voltage-duration plane including voltage impulsive transients (ITs) is segmented into a window format, as illustrated in Table 4.12; the windows are called Z, Y and X. The data from EPRI's DPQ project with reference to ITs are reported in the third column of Table 4.13, which illustrates the procedure for calculating the DSI_{sum} index.

Table 4.12 Voltage impulsive transient windows based on EPRI's DPQ project data [8]

Range of voltage [%]	Duration [s]
	0.0001–0.001 s
400–500%	Z
300–400%	Y
200–300%	X

Table 4.13 Procedure for calculating the DSI_{sum} index for impulsive transients [8]

Voltage IT window	$DSI_{avg,i}$	EPRI's IT count	IT window limit
Z	2.380	2.850	6.783
Y	1.450	12.350	17.908
X	0.750	139.650	**104.738**
		$DSI_{sum} =$	**129.428**

The geometric mean value between DSI_{sum} (129.428) and the maximum IT window limit (104.738) is equal to 116.43; the limit site value of oscillatory transients is assumed to be 120.

It is important to highlight that EPRI's DPQ project data were collected from a representative set of sites within continental US and Canada, and may not be as accurate as the sag limits based on the survey data from nine European countries. Therefore, a future revision of the limit based on EPRI's DPQ project data including the available data from many countries is recommended.

The procedure for detecting the limits for DSI site indices can also be used to define limits for DSI system indices. The DSI_{sum} indices give the average severity level of the different discrete disturbances in the countries that participated in the monitoring campaign. So, considering these countries as representative of the severity of the discrete disturbances, the values assumed by DSI_{sum} indices appear adequate choices as limit system values. Just as was done for the limit of the site DSI, the limit value of the system DSI can also be a value slightly higher than that assumed by the DSI_{sum} indices.

Table 4.14 presents the limit values of the site and system DSI indices obtained by applying the illustrated procedure for each type of discrete disturbance. It is useful to highlight that, in the future these limits will be updated by using the data that have resulted from

Table 4.14 Site and system DSI limit values

Type of discrete disturbance	Site DSI limit value	System DSI limit value
Sag	100 for U/G network	140 for U/G network
	200 for mixed network	370 for mixed network
Swell	250	370
Oscillatory transient	50	65
Impulsive transient	120	130

the more recent campaigns of PQ measurements. In fact, at the current time, many countries are conducting PQ surveys to define the PQ levels relative to their electrical networks, and the results of these monitoring efforts will be very useful in redefining the site and system DSI limits.

Moreover, each country will then be able to use the results of its PQ campaign of measurement to define specific limit values to consider, instead of using the values reported in Table 4.14, which might be too restrictive or inapplicable in some cases.

It is useful to show that the step procedure used to define the limit values could also be applied for the calculation of system DSI indices when only the number of events falling into pre-fixed intervals of time duration/voltage amplitude are known. In fact, often the data quantifying the severity of discrete disturbances relative to campaigns of measurement in various countries are collected in this way.

Now, having a maximum admissible limit for each index's discrete disturbance, it is possible to effect the normalization step for discrete disturbances and, consequently, to calculate the corresponding exceedances. In this way, a unified power quality index, similar to that proposed for continuous disturbances (Equation (4.23)), can also be defined for discrete disturbances:

$$\text{UPQI}_D = 1 + \sum_k \Delta D_k, \tag{4.33}$$

where ΔD_k is the ratio between the value of the (site or system) index related to the kth discrete disturbance and the corresponding limit reported in Table 4.14 or similar.

The unified power quality indices can be used in formulating voltage quality regulations; an example of this utilization will be given in Chapter 6.

4.4 Global indices based on the economic impact on the customer

Recently, the service quality index has been proposed by M. F. McGranaghan [11]; this index can be used to characterize system performance in terms of both voltage quality and continuity.

The service quality index uses the concept of economic impact on customers for different types of disturbance and combines them into a global power quality index. The objective of the service quality index is to represent the impact of the power quality on customer operations.

The key idea of the proposal is that the best measure of the impact of different disturbances is the economic impact experienced by the customer. If the economic impact for different types of customers can be estimated, these impacts can be used as a way to assign weighting factors to the different disturbance categories.

In the opinion of the author who proposed this index, once the cost of interruptions has been defined, the costs of less severe voltage quality variations and events such as voltage sags can also be estimated. The costs associated with these events can generally be expressed as some portion of the costs associated with an interruption. The weightings based on cost impact may be different from utility to utility and may even be different for different portions of the system. It would also be useful to express these costs as a probability distribution to include the variability of the costs for different customers and conditions.

Once the cost functions have been defined, the service quality index (SQI) associated with one customer or a group of customers can be calculated as:

$$SQI = P_{kW} \sum_{i=1}^{N} PQ_i \, cu_i, \tag{4.34}$$

where P_{kW} represents the active power demands of the considered customer or group of customers, PQ_i is the traditional power quality index characterizing the ith disturbance (e.g. SAIFI, SAIDI, SARFI70 or K_d) at the power supply point, cu_i is the unitary cost associated with the ith power quality disturbance and N is the total number of considered disturbances.

The economic impact experienced by the customer has to be considered in the SQI calculation only when the index characterizing the power quality disturbance has a value greater than the allowed minimum value.

In the numerical application of [11] the unitary costs are reported. These costs, which the author declared to be costs just for the purpose of illustrating the concept, are related only to interruptions in voltage and to sag events with a minimum voltage below 70%. In particular, the unitary costs are functions of the type of customer (industrial, commercial or residential), so the SQI index has first to be calculated separately for each type of customer and then referred to the considered power supply point by summing the SQI calculated for each type of customer.

4.5   Comparisons of global indices

A comparison of the indices introduced in Sections 4.2 to 4.4 leads to the following considerations.

GPQIs based on a comparison between ideal and actual voltage waveforms are not appropriate for verifying whether the supply voltage at the monitored bus is acceptable or not. In fact, these indices can be used to quantify how close the supply voltage quality at the monitored bus is to the best level of quality (supply voltage characterized by the absence of any disturbances),[2] but they are not adequate for measuring how close the supply quality is to

---

[2] In the case of the NRMSE index, we consider the ideal voltage to be equal to the nominal one.

an acceptable level of supply quality (supply voltage characterized by the presence of disturbances whose levels are lower that the standard limits).

The NRMS error and the normalized three-phase global index require the calculation of the RMS value, resulting in problems that can derive from the calculation of this mathematical feature. Moreover, the specification of the calculation procedure when using RMS is critical, mainly in the presence of short-term events.

The normalized three-phase global index and the VQDF assume, as ideal voltage, the nominal voltage value. This choice provides advantages in terms of computational effort with respect to the NRMS error, which requires the use of a curve-fitting algorithm to detect the fundamental component of the actual voltage chosen as the ideal waveform.

The NRMS error refers to only one phase of the three-phase power system, while the normalized three-phase global index averages the three phases. The latter index, then, furnishes averaged measures in the presence of disturbances characterized by different values of the three phases (e.g. a voltage sag in two phases of the power system); meanwhile, the NRMS error indicates only disturbances that occur in the monitored phase. However, clearly the NRMS error evaluation can be easily extended to all three phases of the power system.

In the absence of unbalances and when the fundamental voltage magnitude is coincident with the nominal value, the NRMS error and the normalized three-phase global index are characterized by the same values. Moreover, if only harmonics are present, Equations (4.3) and (4.6) degenerate into the well-known THD.

The NRMS error, the normalized three-phase global index and the VQDF index are calculated in each cycle of the fundamental frequency, so that a time variation in these indices can be furnished at the monitored busbar. The time values of these indices can then be grouped in order to obtain statistical measures; in particular, the 95th or 99th percentile values with reference to a fixed study period can be calculated.

Finally, the indices based on a comparison between ideal and actual voltage waveforms appear particularly adequate for detecting the presence of an event (sag, swell or transient); in fact, in these cases, the value assumed by the global indices generally presents a clear discontinuity.

The $UPQI_C$ and I_G indices, the calculations of which are based on the combination of single index disturbances, permit the verification of whether the supply voltage at the monitored bus is acceptable or not. In fact, the normalization step in their calculation procedures introduces a limit value that represents the limit of PQ acceptability. When the $UPQI_C$ is lower than 1, all disturbances are under their standard limits (limits of acceptability). A value equal to 1 indicates that at least one contractual limit is reached; a value greater than 1 indicates at least one contractual limit has been exceeded. Analogous considerations can be effected for the global indicator value, bearing in mind that, in this case, the limit value is equal to 100.

The global indicator I_G differs from the $UPQI_C$ due to the important fact that I_G is obtained using the maximum value of all the calculated consolidated indices, while $UPQI_C$ involves exceedances. In addition, another difference between $UPQI_C$ and I_G refers to the evaluation of the single disturbance indicators: in the survey period and for each indicator, I_G considers the maximum measured value while $UPQI_C$ refers to the 95th or 99th percentile values. In steady-state conditions, I_G and $UPQI_C$ indicate the same degradation of supply quality in the presence of only one disturbance exceeding the acceptable limit; in fact, if the voltage waveform is characterized by the presence of more disturbances exceeding the acceptable limits, the UPQI$_C$ index classifies the supply voltage at the monitored bus in a more severe and adequate way.

The UPQI$_C$ was initially proposed with reference to continuous disturbances only (harmonics, unbalances, and so on), while I_G includes voltage sags and swells as well. However, as illustrated in Section 4.3.2, it is possible to define an equivalent UPQI$_D$ for discrete disturbances.

The definition of the cost functions for different customer categories and different disturbances is fundamental to the application of the service quality index. The economic damage to the customers that results from PQ disturbances introduces a uniform metric to quantify the PQ level independently, based on the type of disturbance, and, consequently, permits an easy definition of a global power quality index based on these economic aspects. Nevertheless, the definition of an adequate cost function is, actually, the biggest obstacle to its practicable application.

4.6 Conclusions

Global power quality indices are important in characterizing the quality of voltage waveforms; in particular, these indices provide overall indications regarding different aspects of power quality disturbances.

In this chapter, the main global power quality indices proposed in the literature have been analyzed; in particular, the theoretical aspects have been presented and critically analyzed. Simple numerical examples have also been presented in order to compare their performance.

The main conclusion of this chapter is that global power quality indices can be very useful in characterizing the voltage quality at the power system busbars as a whole. In particular, even though they do not seem to be the best solution for regulating the interaction at the PCC between the utility and the customer installations, they can be of particular interest either in assisting the power system operators with regard to the whole quality of supply or for inclusion in a system regulator scheme finalized to guarantee acceptable average voltage quality levels for all power system customers, as will be shown in Chapter 6. In fact, employing global power quality indices for system regulation may introduce significant simplifications to the regulation mechanism.

References

[1] Muzychenko, A.D. (1986) 'The Qualitative Criteria of Electric Energy', *Electric Technology U.S.S.R.*, **1**(4), 1–9.
[2] Watson, N.R., Ying, C.K. and Arnold, C. (2000) A Global Power Quality Index for Aperiodic Waveforms, *9th International Conference on Harmonics and Quality of Power*, Orlando (USA), October.
[3] Zhao, C., Zhao, X. and Jia, X. (2004) A New Method for Power Quality Assessment Based on Energy Space, *IEEE Power Engineering Society General Meeting*, Denver (USA), June.
[4] Mastrandrea, I., Chiumeo, R. and Carpinelli, G. (2005) *Definizione di un Indice Globale per la Caratterizzazione della Qualità del Servizio*, Rapporto N. A5-032027, RdS Ricerche di sistema, available at: www.ricercadisistema.it
[5] EN 50160 (2000) *Voltage Characteristics of Electricity Supplied by Public Distribution Systems*, March.

[6] Mamo, X. and Jarzevac, J.L. (2001) Power Quality Indicators, *PowerTech 2001, Porto (Portugal),* September.

[7] Gosbell, V.J., Perera, B.S.P. and Herath, H.M.S.C. (2002) Unified Power Quality Index (UPQI) for Continuous Disturbances, *10th International Conference on Harmonics and Quality of Power*, Rio de Janeiro (Brazil), October.

[8] Herath, H.M.S.C., Gosbell, V.J. and Perera, S. (2005) 'Power Quality (PQ) Survey Reporting: Discrete Disturbance Limits', *IEEE Transactions on Power Delivery*, **20**(2), 851–858.

[9] Goedbloed, J.J. (1987) 'Transients in Low Voltage Supply Networks', *IEEE Transactions on Electromagnetic Compatibility*, **EMC-29**(2), 104–115.

[10] Dorr, D.S. (1994) Point of Utilization of Power Quality Study Results, *IEEE Industry Applications Social Annual Meeting*, Denver (USA), October.

[11] McGranaghan, M.F. (2007) 'Quantifying Reliability and Service Quality for Distribution Systems' *IEEE Transactions on Industry Applications*, **43**(1), 188–195.

5

Distribution systems with dispersed generation

5.1 Introduction

The need for energy savings and environmental impact reduction, together with technological evolution and increased customer demand for highly reliable electricity, are all pushing for the proliferation of generation units connected to distribution systems that are close to the customers. Generator ratings range from a few kilowatts to megawatts. Other important drivers leading to this proliferation are strictly linked to the liberalized electricity markets.

Such generators, with their comparatively small size, short lead times and different technologies, allow players in the electricity market (utilities, independent producers and customers) to respond in a flexible way to changing market conditions. Moreover, these generators make it possible to sell ancillary services such as reactive power and backup services.

This type of power generation is referred to as *embedded* or *distributed* generation. Sometimes it is also called *dispersed* or *decentralized* generation. In this chapter the term dispersed generation (DG) will be used.

There is no generally accepted definition of DG in the literature. Some countries define DG on the basis of the voltage level at which it is interconnected, whereas others start from the principle that DG directly supplies loads. Other countries define DG by some of its basic characteristics, e.g. using renewable sources, cogeneration and being nondispatched.

Some definitions of DG that appear in the literature are:

1. The *International Council on Large Electricity Systems* (CIGRE) considers DG to be a generation unit that is not centrally planned, not centrally dispatched and smaller than 100 MW [1].

Power Quality Indices in Liberalized Markets   Pierluigi Caramia, Guido Carpinelli and Paola Verde
© 2009 John Wiley & Sons, Ltd

2. The IEEE standard 1547–2003 defines DG, considered a subset of distributed resources, like electricity generation facilities connected to an electrical power system (EPS)[1] area through a point of common coupling [2].

3. The *International Energy Agency* (IEA) defines DG as a generating plant that provides on-site service to a customer or provides support to a distribution network that is connected to the grid at distribution-level voltages [3].

Finally, in [4], a broader definition was given, in which DG was defined as 'an electric power source connected directly to the distribution network or on the customer site of the meter.' The distinction between distribution and transmission networks is based on the legal definition that is part of the regulation of the electricity market.

No matter which definition is applied, DG has impacts on power system planning and operation, and these aspects of DG must be taken into consideration.

The growing presence of DG leads to a number of advantages. For example, since DG is often located close to the loads, both losses and voltage drops can be reduced; in addition, DG can allow the utility to postpone investments for distribution networks. Finally, the reduced size of DG units results in low financial risk.

On the other hand, DG can introduce a number of unusual effects, such as bidirectional power flows and an increase in fault current levels.

In addition, an increase in DG penetration can also have a significant impact on the power quality levels of the distribution networks [5–11]. In fact, DG may introduce several PQ disturbances, such as:

- transients, due to large current variations;

- voltage fluctuations, due to cyclic variations in generators' output power;

- long-duration voltage variations, due to variations in the generators' active and reactive power;

- unbalances, due to single-phase generators;

- voltage dips, the characteristics of which can be modified due to the increasing values of short-circuit currents.

In particular, voltage fluctuations can be caused by wind turbines and photovoltaic generators due to the fluctuations in wind speed and solar radiation. The connection and disconnection of induction generators have also been documented as potential causes of flicker.

With reference to unbalances, even though the majority of dispersed generators are three-phase, domestic combined heat and power and photovoltaic systems are becoming more and more common, meaning that the unbalance problem could become non-negligible in low-voltage distribution systems.

The use of DG can also cause waveform distortions, depending on whether the generators are connected directly or with a power electronic interface.

---

[1] EPS areas are facilities that deliver electrical power to a load (this can include generation units) that serves local EPSs. Each local EPS is contained entirely within a single premises or group of premises. The point where a local EPS is connected to the area EPS is denoted the point of common coupling.

Generators that are directly connected to the distribution system can influence the background waveform distortions. In fact, synchronous or induction generators modify the harmonic impedances and contribute to the modification of the voltage harmonic profiles at all the distribution system's busbars. In addition, the contemporaneous presence of shunt capacitors, installed to improve induction generators' power factors, can generate resonance conditions.

For generators that are connected with a power electronic interface, the power electronic interface can inject harmonic currents that lead to network voltage distortions, especially if line-commutated converters are used. The poor harmonic behaviour of line-commutated converters can be overcome by the use of switching converters with the PWM (pulse width modulation) control technique.

Finally, it should be noted that DG can also improve the PQ levels of the distribution system [6, 7, 9–12]. In fact, DG can increase the short-circuit power level. In addition, switching converters with the PWM control technique, when used as the power electronic interface, can operate at any desired power factor and, in the most general case, can be used as active filters, thereby providing an overall improvement in the power quality levels. Synchronous generators can also be used for reactive power control in distribution systems; obviously, interaction problems can arise with the existing system regulation devices, so adequate action must be taken to ensure coordination.

The assessment of PQ levels is a significant topic in the field of study related to modern distribution networks, because DG is becoming more extensively used in such networks.

There are two types of assessment of the response of power systems in terms of PQ levels for the presence of DGs, i.e. assessment prior to the installation of DG and assessment after the installation of DG.

Prior to the installation of DG, the analysis can support the distribution network manager's decision regarding size and location of DG units inside an area [13–15]. Some delicate matters, such as resonance conditions due to the presence of harmonic filters or capacitance stacks, if preliminarily ascertained, can indicate solutions for guaranteeing the safety and reliability of the system. Depending on the DG property (utility or independent producer), preliminary knowledge of ancillary services for PQ improvements can suggest management strategies for the distribution system and reduce the investment pay-back time.

After the installation of DG, the main goal of the assessment of PQ levels is to verify the effectiveness of existing limits on disturbances. Then, when PQ levels reach unacceptable values, the DG manager can correct them by taking appropriate action on the DG units or on the network, if possible.

These two analytical approaches require completely different tools and quantities for PQ assessment because the scope of their use varies widely. This chapter refers to PQ assessment prior to the installation of DG. In such a scenario, the use of indices, which are compact and practical, has been demonstrated to be the best way to assess PQ levels.

Several PQ indices have been proposed in the literature for distribution systems without DG, e.g. the traditional indices shown in Chapter 1. These indices can refer to the point of common coupling for a single customer (single site indices), to a segment of the distribution system or, more generally, to the utility's entire distribution system (system indices). Obviously, the same indices can be applied to a distribution system that includes DG.

In the recent literature, however, it has been shown that new indices which more properly account for the impact of DG can be very useful [16–18].

In this chapter, two types of index for the evaluation of PQ performance of distribution systems with DG are analyzed; the first type, PQ variation indices, can account for variations in PQ levels due to the presence of DG. The second type, the impact system index, can be useful for evaluating variations in the network performance in terms of PQ level for unit power of installed DG. These indices, calculated both for each site and for the system, offer the proper metrics for evaluating the modification of disturbances in the presence of DG.

5.2   Power quality variation indices

Let us consider a generic PQ index X, and let us assume that X_{DG} is the value of index X in the presence of DG whereas X_{NO_DG} is the value of index X in the absence of DG.

Using these assumptions, we can define the percentage variation X_V of the index X due to the introduction of DG with respect to the value without DG as:

$$X\_V = \frac{X_{NO\_DG} - X_{DG}}{X_{NO\_DG}} 100. \qquad (5.1)$$

The index given in Equation (5.1) allows us to quantify the improvement or deterioration in PQ level due to the installation of new generation units.

Starting from Equation (5.1), several variation indices can be introduced. In the next sections, some site and system indices are shown, considering, as examples, waveform distortions and voltage dip disturbances. From a theoretical point of view, it is obvious that the concept of index variations can be extended to all PQ indices, including, for example, voltage fluctuation and voltage unbalance indices.

5.2.1   Site indices

The PQ site variation indices proposed in the literature [19] for waveform distortions are:

- the individual voltage harmonic 95th percentile variation;
- the total harmonic distortion 95th percentile variation;
- the average total harmonic distortion variation.

In particular, the voltage harmonic 95th percentile variation of order h at bus j in the presence of dispersed generation ($V_h95\_V_j$) is defined as:

$$V_h95\_V_j = \frac{V_h95_{NO\_DG,j} - V_h95_{DG,j}}{V_h95_{NO\_DG,j}} 100, \qquad (5.2)$$

where $V_h95_{DG,j}$ is the voltage harmonic 95th percentile of order h at bus j in the presence of DG and $V_h95_{NO\_DG,j}$ is the voltage harmonic 95th percentile of order h at bus j in the absence of DG (for the definitions of waveform distortion traditional indices, see Section 1.3.1 in Chapter 1).

The total harmonic distortion 95th percentile variation at bus j in the presence of dispersed generation (THD95_V_j) is defined as:

$$THD95\_V_j = \frac{THD95_{NO\_DG,j} - THD95_{DG,j}}{THD95_{NO\_DG,j}} \, 100, \qquad (5.3)$$

where $THD95_{DG,j}$ is the total harmonic distortion 95th percentile at bus j in the presence of DG and $THD95_{NO\_DG,j}$ is the total harmonic distortion 95th percentile at bus j in the absence of DG.

Similar expressions can be introduced to define the average total harmonic distortion variation at bus j, ATHD_V_j.

The PQ site variation indices for voltage dips are:

- the voltage dip amplitude 95th percentile variation;
- the average voltage dip amplitude variation;
- the RMS frequency index for residual voltage X variation.

The voltage dip amplitude 95th percentile variation at bus j in the presence of DG (VDA95_V_j) is defined as:

$$VDA95\_V_j = \frac{VDA95_{NO\_DG,j} - VDA95_{DG,j}}{VDA95_{NO\_DG,j}} \, 100, \qquad (5.4)$$

where $VDA95_{DG,j}$ is the voltage dip amplitude 95th percentile at bus j in the presence of DG and $VDA95_{NO\_DG,j}$ is the voltage dip amplitude 95th percentile at bus j in the absence of DG (for the definitions of voltage dip traditional indices, see Section 1.3.6 in Chapter 1). In particular, the voltage dip amplitude at busbar j is the amplitude of the remaining voltage less than 90% of the nominal value and the 95th percentile is evaluated with respect to its probability function.

Similar expressions can be introduced to define the average voltage dip amplitude variation at bus j, AVDA_V_j, and the RMS frequency index for residual voltage X variation at bus j, RFIX_V_j.

It is interesting to note that all the above site indices are expressed in terms of probabilistic figures. This specific characteristic makes them particularly adequate for the analysis of PQ in the presence of DG. In fact, DG can introduce additional unavoidable uncertainties (due to the random nature of some renewable energy sources) to the classical uncertainties that often affect the input data in a real distribution system for changes of linear load demands, network configurations and operating modes of nonlinear loads.

The probabilistic figures that are the basis of the proposed PQ variation indices are usually the results of analytical or Monte Carlo simulation procedures used for the assessment of PQ levels in networks with DG. These methods require the statistical characterization of the random input variables introduced by the DG (e.g. wind speed and solar radiation). Usually, probability density functions (pdfs) are a good way to describe the behaviour of primary energy sources. In particular, for wind speed, a Rayleigh pdf has been frequently assumed for long-period planning; multivariate Rayleigh pdfs can describe the statistical behaviour of the wind speed at different locations in adjacent areas. Solar radiation varies randomly during the day and over the seasons. The pdf for solar radiation can be statistically derived from measured data, and many pdfs have been proposed, e.g. log-normal, beta and Weibull. In this case, the problem of correlation among adjacent areas does not generally arise. In fact, we can assume, without lack of generality, that the solar radiation pdf does not change significantly for different areas that are close to each other.

Example 5.1   In this example the PQ site variation indices have been computed to determine the influence of the installation of DG on the test system shown in Figure 5.1.

The system is characterized by the presence of nine nonlinear loads in busbars 4, 5, 6, 8, 11, 14, 15, 16 and 17. The data of the main components (transformer, lines and capacitor banks) are reported in detail in [10, 20]. Table 5.1 reports the data of loads; in particular, the

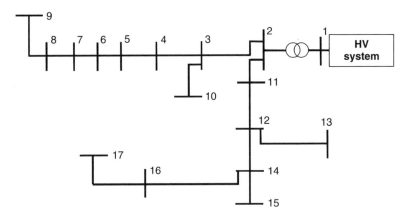

Figure 5.1   Test system [20]

Table 5.1   Linear and nonlinear load powers

Bus code	Linear loads		Nonlinear loads	Weight factors
	mean (P) [MW]	mean (Q)[MVAr]	P [MW]	[p.u.]
3	0.2	0.12	—	0.01
4	0.4	0.25	0.1	0.04
5	0.5	0.31	1.1	0.12
6	2.0	1.51	1.0	0.23
7	0.8	0.5	—	0.06
8	0.2	0.12	0.1	0.02
9	1.0	0.62	—	0.07
10	0.5	0.31	—	0.04
11	0.5	0.31	0.6	0.08
12	0.3	0.19	—	0.02
13	0.2	0.12	—	0.01
14	0.4	0.25	0.5	0.07
15	0.5	0.31	0.1	0.04
16	1.0	0.62	1.0	0.15
17	0.2	0.12	0.1	0.02

active and reactive powers of the linear loads are considered to be Gaussian-distributed random variables with the mean values reported in Table 5.1, and the standard deviations are equal to 10% of the mean values. The nonlinear loads, the nominal powers of which are also reported in Table 5.1, constitute six pulse thyristor converters with the delay angles uniformly distributed between 20° and 40°.

Table 5.1 also reports the weighted factors used to determine the weighted distributions in Example 5.2, where the system indices are calculated with reference to the same distribution system.

The presence of a wind generation unit has been considered, and the wind speed has been characterized by a Rayleigh probability density function; in all considered cases, the scale factor has been assumed equal to 8.46 m/s.

The impacts of DG on waveform distortion and voltage dip levels were assessed by evaluating the variation indices in the presence of the wind asynchronous generators that were connected to the supply system. Typical connection rules were imposed with reference to reactive power; in particular, asynchronous generators were compensated by adequate capacitor banks.

The analysis of the disturbance levels in the network was conducted by varying the bus of a 3.3-MW asynchronous generator (busbars 9, 12 and 17).

Figure 5.2 shows the total harmonic distortion 95th percentile variation at each busbar for the three different DG allocations. The greatest improvements are generally obtained when the generator is located farther from the HV/MV substation. In particular, the highest improvements are obtained when the generator is located at busbar 17.

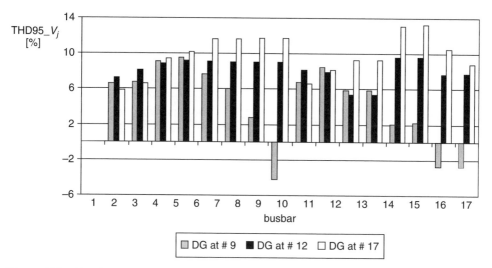

Figure 5.2 Total harmonic distortion 95th percentile variation at each busbar for the three different DG allocations

The voltage dip variation indices were calculated for the same configurations of DG considered above for waveform distortion. The method of fault position [21] was implemented, and symmetrical faults at all busbars were considered. In particular, a probabilistic analysis was performed, since the pre-fault voltage is a random variable, due to the randomness linked to loads and DG production.

In addition, in order to assess the contribution of the DG, the performance of the network was only examined during the first cycles after the occurrence of the faults, and the asynchronous generator was modelled as an impedance as suggested in [22].

In Figure 5.3, the average voltage dip amplitude variation (AVDA_V_j) at each busbar is shown, whereas Figure 5.4 shows the values of the 95th percentile of the voltage dip amplitude (VDA95_V_j).

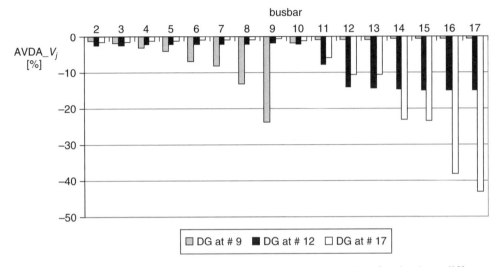

Figure 5.3 Average voltage dip amplitude variation at each busbar for the three different DG allocations

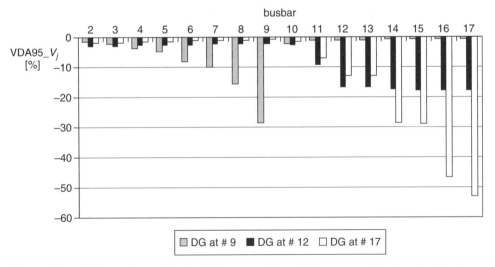

Figure 5.4 Voltage dip amplitude 95th percentile variation at each busbar for the three different DG allocations

From the results, it appears that, with reference to voltage dip disturbances, the presence of the 3.3 MW unit provides an overall improvement in network performance. In fact, the values of the variation indices reported in Figures 5.3 and 5.4 are negative, indicating that the average voltage was greater during faults, in accordance with the definition of index variation (Equation (5.1)).

The improvement was practically restricted to the feeder starting from busbar 2 and containing the asynchronous generator.

5.2.2   System indices

The PQ system variation indices proposed in the literature [16–18] for waveform distortions are:

- the system total harmonic distortion 95th percentile variation;

- the system average total harmonic distortion variation;

- the system average excessive total harmonic distortion ratio index variation.

In particular, the system total harmonic distortion 95th percentile variation in the presence of DG (STHD95_V) is defined as:

$$STHD95\_V = \frac{STHD95_{NO\_DG} - STHD95_{DG}}{STHD95_{NO\_DG}} \, 100, \qquad (5.5)$$

where $STHD95_{DG}$ is the system total harmonic distortion 95th percentile in the presence of DG and $STHD95_{NO\_DG}$ is the system total harmonic distortion 95th percentile in the absence of DG (for the definitions of waveform distortion traditional system indices see Section 1.3.1 in Chapter 1).

The $STHD95_{DG}$ and $STHD95_{NO\_DG}$ indices are defined as the 95th percentile values of a weighted distribution; this weighted distribution is obtained by computing, for each site s, the relative frequencies of the computed THD and weighting them with the ratio L_s/L_T, where L_s is the connected kVA served from the system segment where monitoring site s is, and L_T is the total connected kVA served from the system. Starting from the weighted relative frequencies at each site s, we can evaluate the system weighted relative frequencies and then the 95th percentile.

The system average total harmonic distortion variation in the presence of DG (SATHD_V) is defined as:

$$SATHD\_V = \frac{SATHD_{NO\_DG} - SATHD_{DG}}{SATHD_{NO\_DG}} \, 100, \qquad (5.6)$$

where $SATHD_{DG}$ is the system average total harmonic distortion in the presence of DG and $SATHD_{NO\_DG}$ is the system average total harmonic distortion in the absence of DG. The system average total harmonic distortion is based on the mean value of the above-mentioned weighted distributions rather than the 95th percentile value.

Since the system average total harmonic distortion is based on the mean value rather than on the 95th percentile value, it gives average indications about the system's voltage quality and, thanks to the introduced weights, allows the assignment of different levels of importance to the various sections of the entire distribution system.

The system average excessive total harmonic distortion ratio index variation in the presence of DG ($SAETHDRI_{THD*}\_V$) is defined as:

$$SAETHDRI_{THD*}\_V = \frac{SAETHDRI_{THD*,NO\_DG} - SAETHDRI_{THD*,DG}}{SAETHDRI_{THD*,NO\_DG}} \cdot 100, \qquad (5.7)$$

where $SAETHDRI_{THD*,DG}$ is the system average excessive total harmonic distortion ratio index in the presence of DG, $SAETHDRI_{THD*,\ NO\_DG}$ is the system average excessive total harmonic distortion ratio index in the absence of DG and THD* is the threshold value of the total harmonic distortion.

The system average excessive total harmonic distortion ratio index is related to the number of steady-state measurements that exhibit a value exceeding the fixed threshold. It is computed, for each monitoring site of the system, by counting the index values that exceed the threshold value THD* and normalizing this number to the total number of the index values obtained at site s. The system index is then obtained by averaging the values obtained for each monitoring site.

With reference to voltage dips, the following system indices have been introduced [18]:

- the system average voltage dip amplitude variation;

- the system average RMS frequency index for residual voltage X variation.

In particular, the system average voltage dip amplitude variation in the presence of DG ($SAVDA\_V$) is defined as:

$$SAVDA\_V = \frac{SAVDA_{NO\_DG} - SAVDA_{DG}}{SAVDA_{NO\_DG}} \; 100, \qquad (5.8)$$

where $SAVDA_{NO\_DG}$ is the system average voltage dip amplitude in the absence of DG and $SAVDA_{DG}$ is the system average voltage dip amplitude in the presence of DG (for the definitions of voltage dip traditional system indices see Section 1.3.6 in Chapter 1).

The system average voltage dip amplitude index is computed as a weighted average of the site indices $AVDA_j$. To determine the weighting factors, system and load information is needed but, often, a unity weighting factor is used for sites.

The system average RMS frequency index for residual voltage X variation in the presence of DG ($SARFIX\_V$) is given by:

$$SARFIX\_V = \frac{SARFIX_{NO\_DG} - SARFIX_{DG}}{SARFIX_{NO\_DG}} \; 100, \qquad (5.9)$$

where $SARFIX_{NO\_DG}$ is the system average RMS frequency index for residual voltage X in the absence of DG and $SARFIX_{DG}$ is the system average RMS frequency index for residual voltage X in the presence of DG.

The SARFIX indices are defined as weighted averages of the corresponding site indices $RFIX_j$.

Example 5.2   In this example, with reference to the test system in Example 5.1, the system variation indices were calculated to analyze the system performance in the presence of DG. The results are reported in Tables 5.2 and 5.3 for harmonics and voltage dips, respectively.

Table 5.2   Harmonic variation indices in the presence of DG at different busbars

Indices	DG at #9	DG at #12	DG at #17
SATHD_V [%]	6.98	10.86	15.97
STHD95_V [%]	−1.40	4.58	14.68
SAETHDRI$_{3\%}$_V [%]	0.27	100	0.66

Table 5.3   Dip voltage variation indices in the presence of DG at different busbars

Indices	DG at #9	DG at #12	DG at #17
SARFI10_V [%]	4.74	0.00	9.34
SARFI20_V [%]	0.00	6.39	8.72
SARFI30_V [%]	3.71	6.84	7.49
SARFI40_V [%]	2.07	9.45	12.41
SARFI50_V [%]	0.58	8.33	6.73
SARFI60_V [%]	1.69	4.03	3.71
SARFI70_V [%]	2.74	5.11	3.28
SARFI80_V [%]	3.59	1.90	0
SARFI90_V [%]	0	0	0
SAVDA_V [%]	−4.22	−6.29	−7.97

The system indices indicate the improvement in network performance with respect to voltage dips. When the DG unit is located at busbar 17, with reference to SARFIX_V, the results are positive for $X = 10$ to $X = 70$, whereas the results are zero for $X = 80$ and $X = 90$. The higher value for $X = 40$ shows that the presence of DG significantly affects the voltage dips characterized by this residual voltage.

5.3   Impact system indices

In the presence of DG units, it can be useful to evaluate the variation in the network's performance in terms of PQ level for installed unit power. This indication can be obtained by dividing the variations in the generic PQ system index by the total installed DG power P_{DG}:

$$ISI_X = \frac{X_{NO\_DG} - X_{DG}}{P_{DG}} \tag{5.10}$$

The impact system index ISI_X given by Equation (5.10) can be useful, for example, in understanding, at the planning stage, where the DG unit can be installed to provide the maximum reduction in disturbances; moreover, the same index can be used to compare the PQ impact due to different sizes, structures and/or connections of DG plants.

It should be noted that complete knowledge of the impacts of DG requires associating the proposed indices with the values of the traditional indices calculated in the presence of DG.

Without losing its generality, the impact system index ISI_X can be defined based on the power quality indices both for waveform distortions and for voltage dips (or for other PQ disturbances).

We note that this kind of index gives a synthetic metric of the performance of the network with respect to PQ disturbances in the presence of DG. For instance, when solving a multi-objective optimization problem for the planning of DG, the availability of this index is very useful, since it may be thought of as the objective that accounts for the sensitivity of power quality indices to the installed DG power.

Example 5.3   In this example the impact system indices ISI_X have been calculated considering both waveform distortions and voltage dips. In addition, since the usefulness of the normalized power quality variation indices is relevant when comparing the effect of different installed powers of DG, several configurations of DG have been analyzed (1.65 MW, 3.3 MW and 4.95 MW), always located at busbar 17.

Figure 5.5 shows the obtained results regarding the ISI_{SATHD} and $ISI_{SATHD95}$ indices, respectively. From the results, it clearly appears that, if we look contemporaneously at both indices, the solution with 3.3 MW of DG power installed is preferable, since it is associated with the maximum values of benefit.

Figure 5.6 shows the results obtained relative to ISI_{SAVDA}; from the analysis of Figure 5.6, a slight variation in the index with respect to the function of the installed DG power can be seen.

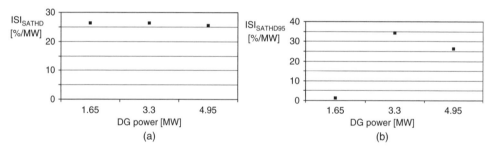

Figure 5.5   The indices (a) ISI_{SATHD} and (b) $ISI_{SATHD95}$ for three different installed DG powers at busbar 17

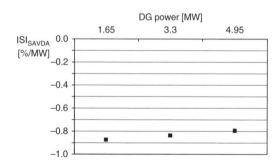

Figure 5.6 ISI_{SAVDA} for three different installed DG powers at busbar 17

Therefore, if we consider the impact of DG connected at busbar 17 on both harmonic distortion and voltage dip, the preferred solution is the installation of 3.3 MW of DG power.

5.4   Conclusions

In this chapter two types of index for the evaluation of PQ performance of distribution systems in the presence of DG have been illustrated.

PQ variation indices can account for variations in PQ levels due to the presence of DG. The impact system indices can be useful in evaluating the variation in the network performance in terms of PQ level for unit power of installed DG.

The main conclusion of this chapter is that these indices are adequate to assess the response of power systems in terms of PQ levels prior to DG installation; they represent a useful tool to support the decision of the distribution network manager regarding size and location of DG units inside an area when used together with the traditional indices calculated in the presence of DG.

References

[1] CIGRE Study Committee No 37 (1998) *Impact of Increasing Contribution of Dispersed Generation on the Power System,* Final Report of WG 37–23, September.

[2] IEEE Standard 1547 (2003) *IEEE Standard for Interconnecting Distributed Resources with Electric Power Systems,* July.

[3] The International Energy Agency (IEA) (2002) Distributed Generation in Liberalized Electricity Markets, *OECD,* Paris, May.

[4] Ackermann, T., Andersson, G. and Soder, L. (2001) 'Distributed Generation: A Definition,' *Electric Power Systems Research,* **57**(3), 195–204.

[5] Jenkins, N., Allan, R., Crossley, P., Kirschen, D. and Strbac, G. (2000) *Embedded Generation,* The Institute of Electrical Engineers, London.

[6] Barker, P.P. and De Mello, R.W. (2000) Determining the Impact of Distributed Generation on Power Systems. I. Radial Distribution Systems, *IEEE/PES Summer Meeting,* Seattle (USA), July.

[7] McDermott, T.E. and Dugan, R.C. (2002) Distributed Generation Impact on Reliability and Power Quality Indices, *IEEE 46th Annual Conference on Rural Electric Power,* Colorado Springs (USA), May.

[8] Ackermann, T. and Knyazkin, V. (2002) Interaction between Distributed Generation and the Distribution Network: Operation Aspects, *IEEE Transmission and Distribution Conference and Exhibition,* Yokohama (Japan), October.

[9] Tran-Quoc, T., Andrieu, C. and Hadjsaid, N. (2003) Technical Impacts of Small Distributed Generation Units on LV Networks, *IEEE/PES General Meeting,* Toronto (Canada), July.

[10] McDermott, T.E. and Dugan, R.C. (2003) 'PQ, Reliability, and DG', *IEEE Industry Applications Magazine,* **9**(5), 17–23.

[11] El-Samahy, E. and El-Saadany, I. (2005) The Effect of DG on Power Quality in a Deregulated Environment, *IEEE/PES General Meeting,* San Francisco (USA), June.

[12] Liang, J., Green, T.C., Weiss, G. and Zhong, Q.C. (2002) Evaluation of Repetitive Control for Power Quality Improvement of Distributed Generation, *IEEE 33rd Annual Power Electronics Specialists Conference,* Cairns (Australia), June.

[13] Carpinelli, G., Celli, G., Pilo, F. and Russo, A. (2001) Distributed Generation Siting and Sizing under Uncertainty, *Power Tech 2001,* Porto (Portugal), September.

[14] Carpinelli, G., Celli, G., Pilo, F. and Russo, A. (2003) 'Embedded Generation Planning under Uncertainty Including Power Quality Issues', *ETEP*, **13**(6), 381–389.

[15] Carpinelli, G., Celli, G., Mocci, S., Pilo, F. and Russo, A. (2005) 'Optimisation of Dispersed Generation Sizing and Siting by Using a Double Trade-off Method', *IEE Proceedings: Generation, Transmission and Distribution*, **152**(4), 503–513.

[16] Carpinelli, G., Caramia, P., Russo, A. and Verde, P. (2004) New System Harmonic Indices for Power Quality Assessment of Distribution Networks, *39th International Universities Power Engineering Conference (UPEC)*, Bristol (UK), September.

[17] Carpinelli, G., Caramia, P., Russo, A. and Verde, P. (2005) New System Power Quality Indices for Distribution Networks in the presence of Embedded Generation, *CIGRE Symposium on Power Systems with Dispersed Generation*, Athens (Greece), April.

[18] Carpinelli, G., Caramia, P., Russo, A. and Verde, P. (2006) Power Quality Assessment in Liberalized Markets: Probabilistic System Indices for Distribution Networks with Embedded Generation, *International Conference on Probabilistic Methods Applied to Power Systems*, Stockholm (Sweden), June.

[19] Bracale, A., Carpinelli, G., Di Fazio, A. and Proto, D. (2009) On the Evaluation of Power Quality Indices in Distribution Systems with Dispersed Generation, *International Conference on Renewable Energies and Power Quality (ICREPQ'09)*, Valencia (Spain), April.

[20] Chang, W.K., Mack Grady, W. and Samotj, M.J. (1994) Meeting IEEE-519 Harmonic Voltage and Voltage Distortion Constraints with an Active Power Line Conditioner, *IEEE/PES Winter Meeting*, New York (USA), February.

[21] IEEE Std 493 (1997) *Recommended Practice for the Design of Reliable Industrial and Commercial Power Systems,* December.

[22] IEEE Std 141–1993 (1993) *IEEE Recommended Practice for Electric Power Distribution for Industrial Plants,* December.

6

Economic aspects of power quality disturbances

6.1 Introduction

In the liberalized and privatized electrical power markets, several independent operators (producers, customers, utilities) tend to act independently and follow their own technical and economic objectives, which might be different and in conflict with each other. This can be to the detriment of the electrical service to customers, especially in the absence of competition, which is the case for most distribution systems operating in natural monopolies. In these cases, regulation schemes, which serve as a proxy for market competition, should be promulgated to ensure adequate levels of electrical service.

With particular reference to power quality disturbances, regulations based on economic schemes have become increasingly important due to the current capability of quantifying the economic consequences associated with disturbances in power systems.

In this chapter, Section 6.2 addresses the economic impact of PQ disturbances, focusing on the indices that are more effective in making the associated cost estimates. After this section, some economic mechanisms publicized or proposed in the literature are shown. As operational tools, the economic mechanisms use financial penalties, incentives or both.

6.2 Economic impact of power quality disturbances

The direct economic value of electric power comes from its conversion into other forms of energy, e.g. thermal energy and mechanical energy. A PQ disturbance cannot have economic relevance by itself, but the detrimental effects that it causes on the processes where this transformation takes place can have very significant economic consequences. In other words, the true economic value of PQ is linked to the effects that PQ disturbances have on equipment and other loads on the system. In the industrial sector, for example, the economic value of disturbances is actually increasing due to the extensive detrimental effects they can cause in modern, automated plants in which sensitive equipment and devices are integral components of highly complex processes. For such processes, a PQ disturbance can cause downtime that can be directly correlated with lost production and, therefore, lost revenue and profits. Additional costs that may be incurred include the costs associated with purchasing, operating and maintaining equipment for reducing the effects of disturbances, such as, for example, filters for harmonics and custom power devices for voltage dips. The economic evaluation of these devices mainly depends on their sizing and rating [1–4].

Additional economic losses associated with PQ disturbances are evident in the consequences beyond the short-term effects. For example, the increased power loss due to harmonics lowers the energy efficiency of the entire electrical system and, consequently, indirectly causes increases in environmental emissions. These secondary, long-term effects fall into the category of externalities, which are more difficult to quantify economically. Such effects are not considered in this section.

The methods available in the specialized literature for the economic evaluation of PQ disturbances refer mainly to voltage dips and harmonics. In the following, both deterministic and probabilistic methods are briefly addressed.

Deterministic methods are adequate when all the items of the analysis, e.g. the operating conditions of the system, are definitively known. This is the case, for example, with ex-post analyses performed on existing systems.

Probabilistic methods are used instead of deterministic methods when some of the problem variables are affected by uncertainty. This can happen for systems that are being installed or for existing systems for which some expansions are planned. Also, for existing systems, a degree of uncertainty is introduced in the cost estimates when technicians must estimate the costs of the future operation, because both cash flow and operating conditions of the system can vary over time. In addition, in the case of voltage dips, all pieces of equipment, even of the same type and brand, can have different sensitivities which, in turn, cause uncertain responses from the equipment and processes in terms of interruptions of service (trips). Furthermore, the actual systems experience the time-varying nature of harmonics, due to continual changes in system configurations, in linear load demands and in operating modes of nonlinear loads [5, 6]. To describe these uncertain conditions, it is useful to introduce random variables and to apply probabilistic techniques of analysis.

6.2.1   Cost of voltage dips in deterministic scenarios

The evaluation of the cost of voltage dips at a particular site in the network involves three steps:

1. Estimation of the voltage dip performance of the supply system.

2. Evaluation of the effects of voltage dips on components and equipment.

3. Economic analysis.

The performance of the supply system in terms of voltage dips can be assessed using two methods: the critical distance (CD) method and the fault position (FP) method [7]. Both methods are based on the simulation of the system in short-circuit conditions for the assigned position of fault.

The CD method allows the computation of the voltage dips in a considered node when a short circuit occurs across the system. The FP method allows derivation of the voltage dips in all the nodes of a system when short circuits occur in every node.

The CD method mainly applies to radial systems, such as distribution networks, but it has provided results close to the actual data even for transmission systems. This method assesses a node that serves a critical load and determines voltage dips that occur due to faults across the system.

The FP method mainly applies to meshed systems, such as transmission systems; it obtains the voltage dips in all the nodes of the system due to faults in every bus.

Both methods are well known and popular, but only the FP method can offer a global vision of the electrical power system's response to faults, even for radial networks. In fact, an important result of the FP method is the during-fault voltage (DFV) matrix, which is a bidimensional vector of voltages; each element (i, j) represents the RMS value of voltage at node i when a short circuit occurs at node j. To immediately capture the information about the presence and amplitude of the voltage dips, the DFV matrix can be visualized with a colour scheme. A colour with a grade proportional to the value of the during-fault voltage is assigned to every element of the DFV matrix.

Figure 6.1 gives an example of such a colour scheme for a distribution system in which three phase faults were simulated. The distribution network represents a portion of a real 20-kV distribution system supplying a mixture of industrial, commercial and residential loads. The system consists of two feeders: one feeder that is 17 km long and consists of 60% overhead lines and 40% underground cables and another feeder that is 12 km long

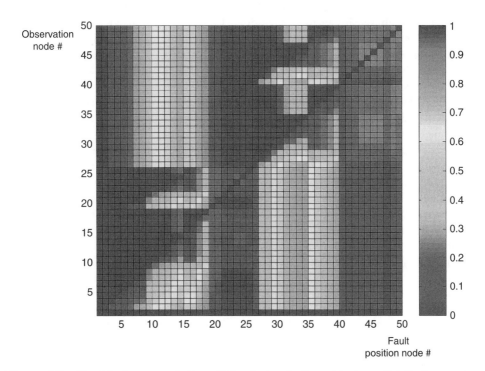

Figure 6.1 Graphical representation of the during-fault matrix in a distribution system (see Plate 4)

with 70% overhead lines and 30% underground cables. Both feeders contain both medium-voltage (MV) and low-voltage (LV) nodes at the secondary windings of the MV-LV transformers.

Figure 6.1 shows two zones (one between columns 2 and 18 and the other between columns 26 and 39) with a variation of colour covering the full scale from blue to red. These zones correspond to the MV nodes of the two feeders. The zones in which nearly all the columns are red correspond to the LV nodes. This means that faults occurring at the MV nodes affect the voltage of the other nodes with different severities, whereas faults at the low-voltage nodes do not cause voltage dips at other nodes.

The graphical depiction of the DFV matrix shows that the FP method can give a global view of the electrical power system's performance. In particular, the FP method provides information about:

- the propagation of voltage dips around the network;
- the amplitude of voltage dips for all nodes;
- the amplitude of voltage dips caused by individual nodes;
- the nodes at which the faults are more critical because of the voltage dips they cause at other nodes; and
- the nodes at which loads could experience the largest number of voltage dips.

The evaluation of the effects of a voltage dip on components and equipment requires knowledge of their sensitivity to voltage dips. The sensitivity of industrial equipment is normally expressed in terms of the magnitude and duration of the voltage dip. Different equipment has different tolerances to voltage dips. A voltage tolerance curve can be obtained either from the equipment manufacturer or from available technical documents. The Computer Business Equipment Manufacturers' Association (CBEMA), the Information Technology Industry Council (ITIC) and the Semiconductor Equipment and Materials International (SEMI) provide commonly used curves for characterizing the sensitivity of equipment to voltage dips. These curves were reported in Chapter 1.

The economic analysis of the overall effects of voltage dips requires basic knowledge of the cost of the effect of a single voltage dip. For example, for an industrial process, the costs associated with trips due to dips in the voltage supply should at least include the cost of shutting down the process, cleaning the system, restarting the process, lost production and the eventual repair or replacement of damaged equipment. These costs are real and someone must pay them. The utility's customers must decide whether they will take the risk and pay for the possible consequences or whether it would be preferable to pay the distribution company for providing higher power quality, i.e. mitigating the dips in one way or another. If the customer chooses to pay for higher power quality and the utility is unable to mitigate the dips as promised, the utility should compensate the customer for the inconvenience, costs and damage associated with voltage dips.

IEEE 1346–1998 furnishes guidelines for assessing financial losses due to voltage dips at customers' facilities, and the considered costs are related to the process disruption. With reference to industrial processes, these costs comprise downtime-related costs (lost production, idle labour, equipment damage and recovery costs), product-quality-related costs (scrap and rework costs) and other indirect costs (customer dissatisfaction, employee and customer safety, fines and penalties). Finally, the total financial losses of the facility are obtained by multiplying the total cost of process disruption by the number of disruptive dips per year. This information is obtained, at the initial stage of the evaluation, from the voltage dip performance of the utility at the point of common coupling with the industrial facility, and it is acquired either from utility data, measurements, monitoring or prediction. Using these data, the supply dip performance contours are drawn, in which each contour represents the number of voltage dips per year. Next, equipment sensitivities (voltage tolerance curves) are overlaid on the supply dip performance contours to form dip coordination charts (Figure 6.2). The sensitivity of the process is defined by the most sensitive component, with the knee point located at the upper-most, left-hand portion of the chart.

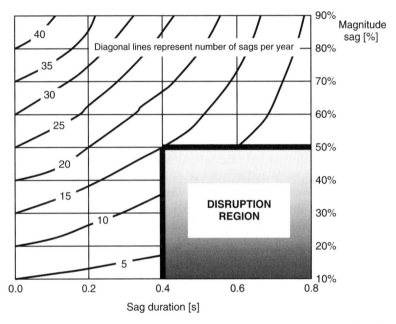

Figure 6.2 Supply sag contours and equipment sensitivity [8]. IEEE Standard 1346: "Recommended Practice for Evaluating Electric Power System Compatibility With Electronic Process Equipment" © 1998 IEEE

To facilitate the financial assessment, IEEE 1346–1998 also provides a 'standard cost of disruption evaluation form' which lists the separated costs to account for.

Other methods in the literature refer to the costs of voltage dips at utility levels that should reflect the value of PQ perceived by the utility's customers [9]. Traditionally, the utility's customers have tended to overvalue the inconvenience due to dips and interruptions, but they are not willing to pay for the measures that utilities must take to prevent them. The realistic price attributable to a single dip is somewhere between these two extreme views. The relevant value should be derived from the actual direct and indirect economic consequences of a dip. This, again, falls into the category of estimating the consequence of trips due to a voltage dip. In the category of industrial customers, the evaluation is available as previously described but, for other customer groups, the price of a sudden short interruption is commonly used.

6.2.2   Cost of harmonics in deterministic scenarios

The quantification of the economic effects of harmonics in an electrical system requires the computation of all the consequences that the harmonics of current and of voltage have on all the equipment and components [10–14]. The effects of voltage and current distortions on equipment and components fall into three main classes:

1. additional losses;

2. premature aging;

3. misoperation.

The term 'additional' means that these losses are superimposed on the fundamental losses; the term 'premature' refers to the possibility of accelerating the aging rate of equipment due to increased stress levels compared to nominal service conditions. The term 'misoperation' refers to a decrease in the equipment's performance compared to its performance at nominal conditions.

The methods first proposed for quantifying the harmonics in a system dealt with the deterministic evaluation of the costs to the electric utility to contend with the harmonics [1]. The costs included the value of total active power losses as well as the capital invested in the design, construction, operation and maintenance of filtering systems. The additional costs associated with the premature aging of the equipment were not included in the early methods. Since then, several successive studies have proposed the use of probabilistic methods to extend the costs due to harmonics to account for the premature aging of equipment [15–17]. Unfortunately, few contributions can be found regarding the economics of misoperation, which is currently approached as if it were a reliability cost [18].

6.2.2.1   Additional losses

To compute the economic value of the additional power losses arising for an operating period, the following information should be available:

- knowledge of system operating conditions during the study period, e.g. network configurations and typical duration of system states;

- knowledge of type, operating conditions and absorbed power levels of linear and nonlinear loads;

- assignment of the variation rate of the electric energy unit cost and the discount rate.

Let us initially refer to the case of a single electrical component continuously subject to H_{max} harmonics of voltage or current $G^{h_1}, \ldots, G^{H_{max}}$ in the time interval ΔT. The loss costs $(Dw_k)_{\Delta T}$ are:

$$(Dw_k)_{\Delta T} = K_w P_k\left(G^{h_1}, \ldots, G^{H_{max}}\right)\Delta T, \tag{6.1}$$

where K_w is the unit cost of electrical energy and $P_k\left(G^{h_1}, \ldots, G^{H_{max}}\right)$ represents the losses due to the harmonics $G^{h_1}, \ldots, G^{H_{max}}$ on the kth component. The loss cost $(Dw_k)_n$ of the component in a generic year n is the sum of the loss costs of all time intervals present in the considered year. Finally, the loss cost $(Dw)_n$ in the year n for the whole system, in which m components operate, can be computed as the sum of the costs associated with each component:

$$(Dw)_n = \sum_{k=1}^{m} (Dw_k)_n \tag{6.2}$$

To evaluate the loss costs of the system components with reference to more years, usually the expected life of the electrical system, it is necessary to take into account both the variation in the unit cost of electrical energy in the coming years and the present-worth value of the costs in every year of the system's life. The following relationships have been assumed for the variation in the electrical energy unit cost:

$$(K_w)_n = (K_w)_1 (1 + \beta)^{n-1} \tag{6.3}$$

and for the present worth value of the loss costs:

$$(Dw)_{n,pw} = \frac{(Dw)_n}{(1 + \alpha)^{n-1}}, \tag{6.4}$$

where β is the variation rate of the electrical energy unit cost and α is the discount rate.

Finally, the present-worth value of the total loss costs of harmonics, referred to the whole electrical system period of N_T years, is:

$$Dw = \sum_{n=1}^{N_T} (Dw)_{n,pw} = \sum_{n=1}^{N_T} \frac{(Dw)_n}{(1 + \alpha)^{n-1}}. \tag{6.5}$$

Equations (6.1) to (6.5) show that computing the economic values of losses due to harmonics requires knowledge of several quantities, among them the harmonics $G^{h_1}, \ldots, G^{H_{max}}$, which refer to currents and/or voltages for each of the components. It is worth noting that, for most real cases, the main contributions to the economic value of losses due to harmonics come from the current harmonics flowing into series components of the system, such as cables or overhead lines. However, dielectric losses linked to voltage harmonics can play a non-negligible role, for example, in medium-voltage cables.

6.2.2.2 Premature aging

To compute the economic value of replacing damaged components due to their premature aging, the following data must be known:

- system operating conditions in the study period, e.g. structure and duration of network configurations;

- types, operating conditions and absorbed power levels of linear and nonlinear loads;

- life models of equipment and components to estimate the failure times of their electrical insulation;

- purchase prices of the components, together with their variation and discount rates.

Premature aging caused by harmonics involves incremental investment costs during the observation period. Referring initially to a single component, let the aging costs $(Da_k)_{pv}$ be defined as:

$$(Da_k)_{pv} = (C_{k,ns})_{pv} - (C_{k,s})_{pv}. \qquad (6.6)$$

In Equation (6.6), $(C_{k,ns})_{pv}$ and $(C_{k,s})_{pv}$ are the present-worth values of the investment costs for buying the kth component during the system life in nonsinusoidal and sinusoidal operating conditions, respectively.

The values of $(C_{k,ns})_{pv}$ and $(C_{k,s})_{pv}$ can be evaluated when the useful lives of the component $L_{ns,k}$ and $L_{s,k}$ are known. In fact, once they are known, both the number of times the component has to be bought and the years in which the purchases will have to be made are fully estimable.

According to the cumulative damage theory [19–22], the useful lives $L_{ns,k}$ and $L_{s,k}$ can be estimated by summing the fractional losses of life, which come in succession until reaching unity.

With reference to the fractional losses of life, it is important to highlight that electrical power system components are subjected to different service stresses (e.g. electrical, thermal and mechanical), which can lead to the degradation of electrical insulation. The degradation of solid-type insulation is an irreversible process that eventually involves failure and, thus, breakdown or outage of the component.

However, electrical and thermal stresses (i.e. voltage and temperature) are, in general, the most significant stresses for insulation in MV/LV power system components. In addition, the interaction between electrical and thermal stresses can lead to a further increase in the electro-thermal aging rate with respect to the effect of these stresses applied separately, a phenomenon called *stress synergism*. The aging rate can be accelerated by a rise in stress level with respect to the nominal service conditions. This may be due just to voltage and current harmonics that may lead to increases in electrical and thermal stresses on the insulation, thus shortening the insulation's time-to-failure, i.e. the useful life of the component.

In the presence of harmonics, the life models of equipment and components can take into account either thermal stress only or electrical stresses as well, leading to a more complex life model (electro-thermal life model).

First, let us assume that the useful life of an insulated device is only linked to the thermal degradation of the insulation materials. The thermal degradation can be represented by the well-known Arrhenius reaction rate equation, in which the absolute temperature of the materials is constant.[1] From the Arrhenius relationship, it has been demonstrated that the thermal loss of life of the kth component $(\Delta L_k)_{T_c}$ in a time period T_c characterized by q different operating conditions, each at a given temperature and of a given duration, can be expressed as the summation of q fractional losses of life:

$$(\Delta L_k)_{T_c} = \sum_{i=1}^{q} \left[\frac{t_{i,k}}{\Lambda(\theta_{i,k})} \right], \qquad (6.7)$$

---

[1] The Arrhenius equation gives the dependence of the rate constant of chemical reactions on the temperature and activation energy, i.e. $K_R = A_R \exp\left(-\frac{E_a}{k\theta}\right)$, where K_R is the reaction rate, A_R is a constant, E_a is the activation energy (that is the amount of energy required to ensure that a reaction happens), θ is the absolute temperature and k is Boltzmann's constant [22].

where $t_{i,k}$ is the duration of the operating condition of the kth component at constant temperature $\theta_{i,k}$, and $\Lambda(\theta_{i,k})$ is the useful life of the kth component at temperature $\theta_{i,k}$, obtained from the Arrhenius model. The temperature of each of the insulated components $\theta_{i,k}$ can be determined by considering the heat balance relationships, in which the losses at the fundamental frequency and at the harmonics are the forcing terms.

When both thermal and electrical stresses have to be accounted for, the procedure is not modified, but the life model to be used in Equation (6.7) changes in the electro-thermal life model. In such a condition, the relative loss of life of the component in the time period T_c can again be expressed as a summation of fractional losses of life:

$$(\Delta L_k)_{T_c} = \sum_{i=1}^{q}\left[\frac{t_{i,k}}{\Lambda(E_{i,k},\theta_{i,k})}\right], \tag{6.8}$$

where $\Lambda(E_{i,k},\theta_{i,k})$ is the useful life that the kth component would experience if constant values of electrical and thermal stresses $E_{i,k}$ and $\theta_{i,k}$ were continuously applied. In the literature [17, 23, 24], electro-thermal models of the most common equipment and components of MV and LV systems can be found, i.e. the electro-thermal life models that explicitly account for voltage and current harmonics; an example will be shown in Section 6.2.4.

The present-worth value of the additional aging costs arising in the whole system for m components subjected to aging is computed as the sum of the cost of each component:

$$Da = \sum_{k=1}^{m}(Da_k)_{\text{pv}}, \tag{6.9}$$

where the value of $(Da_k)_{\text{pv}}$ is calculated using Equation (6.6), starting from the knowledge of the useful lives of the various components, obtained by applying relationships such as Equations (6.7) or (6.8).

6.2.2.3   Misoperation

The economic evaluation of misoperation is the most complex subject and, arguably, the least explored component of cost. This lack of contributions does not permit one to distinguish in a clear manner between deterministic and probabilistic methods; however, in the following we offer some considerations that would permit one to apply deterministic methods to compute the cost of misoperation.

The complexity of cost computation is strongly linked to the absence of exact knowledge of the cause–effect linkage between harmonics and the degradation of performance of equipment because of the difficulty in making a concrete determination of harmonics as the only cause of the disturbance. Assigning to harmonics the responsibility for causing several degradations in the equipment performance throughout the life of the equipment is difficult to accomplish conclusively. In the literature, however, some categories are reported for which the performance degradation due to harmonics can be more easily discriminated, i.e. electronic equipment operating with voltage zero crossing, meters and lighting devices.

Generally, the economic impact of misoperation involves financial analysis of all the effects that misoperation has on the process/activity that the equipment was a part of. Typically, the costs associated with misoperation can be estimated for existing systems that have well-known duty cycles. Unexpected tripping of protective devices, for example, can stop the entire industrial process. The cost of such an event includes several items, such as the cost of downtime, the cost of restoring/repairing and the cost of replacing the equipment, where applicable. Some interesting values can be found in the literature; for example, in Spain, extensive investigations on existing systems have been conducted among a wide range of commercial and industrial sectors [25]. The findings of the research confirm that estimating the costs of misoperation requires extensive information on the following:

- equipment malfunctioning in the presence of harmonics;
- the process or activity in which the equipment is used;
- the economic values of all the items contributing to lower productivity.

Considering the problem of evaluating misoperation costs from these perspectives, it is evident that several analogies arise with the problem of evaluating the economic effects of micro-interruptions or voltage dips. At least, for all the cases in which lower productivity is due to partial or complete stoppage of the process, the methods and the components of the financial analysis are the same.

6.2.3   Cost of voltage dips in probabilistic scenarios

The methods of assessing the economic consequences of voltage dips in probabilistic scenarios can account for various uncertainties related to equipment and process sensitivity. Two methods that were recently published in the literature are described below. The first method, called Prob-A-Sag [26], proposed:

- to represent sensitivity using both discrete states ('on' or 'off') and probabilistic values;
- to assign different cost values for voltage sag events of different characteristics;
- to account for interconnections between equipment in a probabilistic manner; and
- to include the effect of mitigation devices.

All the parameters for estimating the cost of voltage dips in a process were represented by two-dimensional arrays. For each array, the rows and columns represent sag magnitudes and sag durations, respectively. An example of an array is shown in Table 6.1.

The letter on the left side of the table describes the type of information given by the array. For instance, letter 'D' in the array shown in Table 6.1 refers to the device sensitivity, in particular, the sensitivity of a contactor. The value in the cell (i,j) is the tripping probability of the contactor when a voltage dip occurs with a magnitude and duration that correspond to the ith row and jth column. Other types of arrays with the same format were defined to furnish the other information useful for estimating the cost of voltage dips (Table 6.2).

The arrays D, S, R and E give, respectively, devices' dip sensitivity, annual dip frequency, restored annual dip frequency and event cost. These arrays represent the input data for computing the expected cost due to voltage dips.

The first step consists of calculating the process dip sensitivity array P. To obtain this array, we have to identify the critical items of process equipment and their mutual connection.

Table 6.1 Example of the array D [26]

Remaining voltage [%]		0.00	0.00	0.00	0.00	0.00
	< 100	0.00	0.00	0.00	0.00	0.00
	< 90	0.00	0.00	0.00	0.00	0.00
	< 80	0.01	0.01	0.01	0.01	0.01
	< 70	0.11	0.11	0.11	0.11	0.11
	< 60	0.40	0.40	0.40	0.40	0.40
	< 50	0.77	0.77	0.77	0.77	0.77
	< 40	0.96	0.96	0.96	0.96	0.96
	< 30	1.00	1.00	1.00	1.00	1.00
	< 20	1.00	1.00	1.00	1.00	1.00
	< 10	1.00	1.00	1.00	1.00	1.00
D		0–50	50–150	150–300	300–500	500–1000
				Sag duration [ms]		

Table 6.2 Summary of array types [26]

Array type	Definition	Cell information
D	Device sensitivity	Tripping probability of device
S	Dip frequency	Annual number of dips per phase
R	Restored dip frequency	Annual number of dips per phase with mitigation device
E	Event cost	Cost of process interruption caused by the dip
P	Process dip sensitivity	Tripping probability of process
I	Frequency of dip-oriented plant interruptions	Annual number of plant interruptions caused by voltage dips

In the general case, series and parallel connections can represent the scheme of more equipment in the process, so that the array P cells are calculated by [26]:

$$P(i,j) = 1 - \left\{ \prod_{k=1}^{q} \left[1 - \prod_{l=1}^{r_k} D_{k,l}(i,j) \right] \right\} \qquad \forall i=1\dots m, j=1\dots n, \qquad (6.10)$$

where q is the number of serial connected component groups, r_k represents the parallel components of the kth group, $D_{k,l}(i,j)$ is the probability of tripping in the lth piece of equipment in the kth serially-connected equipment group in cell (i,j) of array D and m and n are the number of rows and columns in the array, respectively.

Once the arrays P cells are known, the array I, which contains the frequency of dip-produced plant interruptions, can then be calculated using Equation (6.11) for processes without mitigation devices and Equation (6.12) for processes with mitigation devices:

$$I(i,j) = P(i,j)S(i,j) \qquad \forall i=1\dots m, j=1\dots n \qquad (6.11)$$

$$I(i,j) = P(i,j)R(i,j) \qquad\qquad \forall i = 1\ldots m, \; j = 1 \ldots n, \qquad (6.12)$$

where $I(i,j)$, $P(i,j)$, $S(i,j)$ and $R(i,j)$ are the cells (i,j) of the arrays I, P, S and R, respectively.

Finally, the financial losses due to voltage dips can be calculated by:

$$C = \sum_{i=1}^{m} \sum_{j=1}^{n} [I(i,j)E(i,j)]. \qquad (6.13)$$

The second method, proposed in [27, 28], takes into account the uncertainties associated with voltage sag calculation, the interconnection of equipment within an industrial process, the customer type and the location of the process in the network in a similar way to the preceding method. The equipment sensitivity is treated differently, as explained below.

First, the sensitive equipment is classified into various categories based on device type; in particular, four main equipment types are ascertained: personal computers (PCs), programmable logic controller (PLC), adjustable speed drives (ASDs) and AC contactors.

It can then be shown that each device, even of the same type and the same brand, responds to voltage dips in a different way, so the various sensitivities of equipment of the same type are represented in a plane (duration, amplitude of voltage) by a family of sensitivity curves inside a region of uncertainty. This approach is better than using one single curve and it allows the introduction of an area of uncertainty into the general voltage tolerance curve defined for each type of equipment. Figure 6.3 shows an example of the rectangular-type general voltage tolerance curve with shaded regions that represent the areas of uncertainty.

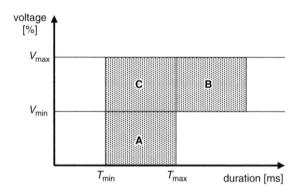

Figure 6.3 Region of uncertainty in rectangular voltage tolerance curve [27]

In the regions of uncertainty, namely A, B and C in Figure 6.3, T is the duration threshold, varying between T_{min} and T_{max}, and V is the voltage magnitude threshold, varying between V_{min} and V_{max}. Probability distribution functions for T and V are given by $p_T(T) \, p_V(V)$; they can model the uncertainty related to regions A and B, respectively. In region C, assuming the variables T and V are uncorrelated, the joint probability distribution function is given by:

$$p_{TV}(T,V) = p_T(T) \, p_V(V); \qquad T_{min} < T \le T_{max}, V_{min} < V \le V_{max} \qquad (6.14)$$

Next, to determine the equipment's response (the failure probability of equipment) to voltage dips, the equipment voltage tolerance curves are compared with the dip performance charts. The latter are prepared using the results from the voltage dip analysis performed by the fault position method; they are then characterized by the probability of occurrence using the historical fault performance of the electrical system.

The comparison between the equipment voltage tolerance curve and dip performance charts is illustrated in Figure 6.4. In the region of uncertainty, different probability density functions can be used to represent different sensitivity levels of equipment. We can see an example of such a comparison by considering the dip represented by point (2) in Figure 6.4. This point corresponds to the pair of values (T^*, V^*); therefore, the failure probability of the equipment for this dip is given by:

$$p_{TV}(T^*, V^*) = p_T(T^*)\, p_V(V^*); \qquad T_{min} \leq T \leq T^*,\ V^* \leq V \leq V_{max}. \qquad (6.15)$$

After the computation of the failure probabilities of the equipment for all the dips included in the dip performance charts, the probability of process trip, $p trip_{T,V}$, is calculated

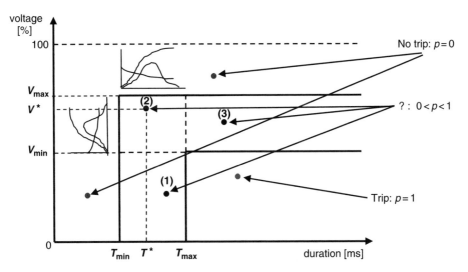

Figure 6.4 Expected behaviour of sensitive equipment experiencing voltage sags with different characteristics [27]

as in the Prob-A-Sag method previously described, accounting for the serial and/or parallel connections among the sensitive equipment. We can use Equation (6.10), properly substituting the symbols, as follows:

$$p trip_{T,V}(T, V) = 1 - \left\{ \prod_{k=1}^{q} \left[1 - \prod_{1=1}^{r_k} p_{T, V_{k,1}}(T, V) \right] \right\}; T_{min} < T \leq T_{max},\ V_{min} < V \leq V_{max}.$$

$$(6.16)$$

where q is the number of serial connected component groups, r_k represents the parallel components of the kth group, p_T, $v_{k,l}$ is the probability of tripping of the lth piece of equipment in the kth serially-connected equipment group for all the dips characterized by (T,V) values that respect the condition in Equation (6.16).

Finally, the expected total number of process trips, TN_{pt}, is determined by:

$$TN_{pt} = \sum_T \sum_V ptrip_{T,V}(T,V)\, N(T,V), \qquad (6.17)$$

where $ptrip_{T,V}(T,V)$ is the trip probability of the process defined for voltage dips with magnitude V and duration T and $N(T,V)$ is the number of such voltage dips expected at the specified site over a specified period of time.

The final economic assessment of financial losses due to voltage dips requires knowledge of the cost per process trip, which is linked to, for example, the type of industrial/commercial process and the type of customer.

6.2.4   Cost of harmonics in probabilistic scenarios

When faced with uncertainties that often unavoidably affect the input data in real systems for changes in linear load demands, in network configurations and in operating modes of nonlinear loads, the economic models must be translated to a probabilistic basis. This implies the introduction of random variables and the application of probabilistic techniques of analysis.

6.2.4.1   Additional losses

The first step in the probabilistic approach is to recognize that the output economic figures to be computed are random variables. In most cases, their probability density functions (pdfs) completely describe their statistical features. However, for the sake of estimating the economic value of losses due to harmonics, it has been considered adequate to refer to the expected value. When estimating expected values for a period of time, once again, the present-worth values must be considered.

The expected present-worth value of losses due to harmonics $E(Dw)$, with reference to the entire life of the electrical system (N_T years), is:

$$E(Dw) = \sum_{n=1}^{N_T} E(Dw)_{n,pw} = \sum_{n=1}^{N_T} \frac{E(Dw)_n}{(1+\alpha)^{n-1}} = \sum_{n=1}^{N_T} \frac{\sum_{k=1}^{m} E(Dw_k)_n}{(1+\alpha)^{n-1}}, \qquad (6.18)$$

with trivial meaning of the symbols (see Equations (6.2) and (6.5)).

It is clear from Equation (6.18) that it is necessary to compute the expected values $E(Dw_K)_n$ ($k = 1,\ldots,m$) of harmonic loss costs that occur in each year n for all the system components.

Considering each single electrical component subject to H_{max} harmonics of voltage or current $G^{h1}, .., G^{H_{max}}$ characterized in the jth time interval of the year n $\Delta T_{j,n}$ by the joint pdf $f^{\Delta T_{j,n}}_{G^{h1},..,G^{H_{max}}}$, the expected value of the harmonic loss cost $E(Dw_k)_{\Delta T_{j,n}}$ can be computed as:

$$E(Dw_k)_{\Delta T_{j,n}} = K_w \Delta T_{j,n} \int_0^\infty \int_0^\infty .. \int_0^\infty P_k(G^{h1},..,G^{H_{max}}) \, f^{\Delta T_{j,n}}_{G^{h1},..,G^{H_{max}}} \, dG^{h1}.. \, dG^{H_{max}}. \quad (6.19)$$

For the most common components of industrial energy systems, the harmonic losses $P_k(G^{h1},\ldots,G^{H_{max}})$ in Equation (6.19) can be obtained by summing separately the losses due to each harmonic, so the integral in Equation (6.19) can be significantly simplified.

The expected values $E(Dw_k)_n$ ($k=1,\ldots,m$) of harmonic loss costs incurred in each year n can be obtained by summing the contributions obtained by Equation (6.19) of all time intervals of the year n in which the pdfs are defined. For more details on this subject see [15, 17].

In spite of the apparent complexity of the procedure, it is necessary to show that the methods practically require the estimation of losses due to harmonics for each component of the system, making sure that the operating conditions are ascertained first.

The computation of losses in Equation (6.19) does not present particular difficulties, and several studies in the literature have addressed this subject for the most common components and equipment, such as transformers, cable lines and capacitors. The main difficulties arise in deriving the pdfs of voltage and current harmonics. These values can be obtained both from measurements and from simulations, using probabilistic methods of harmonic analysis [6, 29, 30].

6.2.4.2   Premature aging

The expected economic present-worth value of premature aging $E(Da)$ is evaluated by summing the expected present-worth value of the aging costs of each of the m components of the system:

$$E(Da) = \sum_{k=1}^m E(Da_k)_{pv}, \quad (6.20)$$

where the value of $E(Da_k)_{pv}$ is calculated by the relationship:

$$E(Da_k)_{pv} = E(C_{k,ns})_{pv} - E(C_{k,s})_{pv}, \quad (6.21)$$

where $E(C_{k,s})_{pv}$ and $E(C_{k,ns})_{pv}$ are the expected present-worth values of the costs of buying the kth component during the system life in sinusoidal and nonsinusoidal operating conditions, respectively.

The expected values of cost to be met for buying each component in the sinusoidal and nonsinusoidal conditions are linked to the expected values of the component life in these conditions, respectively. To estimate these figures, again, the cumulative damage theory can be applied, as in the case of the deterministic methods. In such a case, we have to refer to the expected value of the relative loss of life $E(\Delta L_k)_{\Delta T^*_{j,n}}$ in the time interval $\Delta T^*_{j,n}$, which can be calculated as:

$$E(\Delta L_k)_{\Delta T^*_{j,n}} = \Delta T^*_{j,n} \int_0^\infty \cdots \int_0^\infty \frac{f^{\Delta T^*_{j,n}}_{x_1,k x_2,k \cdots x_g,k}}{\Lambda(x_{1,k}, x_{2,k}, \ldots, x_{g,k})} \prod_1^g dx_{i,k}. \tag{6.22}$$

where $f^{\Delta T^*_{j,n}}_{x_1,k x_2,k \cdots x_g,k}$ is the joint pdf of the g random variables in the time interval $\Delta T^*_{j,n}$ on which the kth component's life depends. The successive estimation of the useful life can be conducted, as previously mentioned, by summing the expected values of the relative losses of life until reaching unity.

The main criticism of this method is the complexity of computing the dimensional integral in Equation (6.22) and, overall, by assigning the joint pdf $f^{\Delta T^*_{j,n}}_{x_1,k x_2,k \cdots x_g,k}$. Indeed, some simplifications introduced by life models of actual insulated components provide significant assistance. First of all, in most cases, it is adequate to consider electro-thermal stress models. In addition, it has been demonstrated that they can be reduced to even simpler models, such as [17, 23]:

$$\Lambda = \Lambda'_0 Kp^{-n_p} \exp(-Bc\theta), \tag{6.23}$$

where Λ'_0 is the life expectancy at nominal sinusoidal voltage and reference temperature, $c\theta = 1/\theta_0 - 1/\theta$ is the so-called conventional thermal stress (θ is the absolute temperature and θ_0 is a reference temperature) and n_p and B are model parameters. In particular, n_p is the coefficient related to the effect of the peak of the distorted voltage waveform Kp on the life expectancy. The larger this coefficient is, the stronger the influence of peak voltage is.

Using Equation (6.23), Equation (6.22) becomes:

$$E(\Delta L_k)_{\Delta T^*_{j,n}} = \Delta T^*_{j,n} \iint_{D_\theta D_{Kp}} \frac{f^{\Delta T^*_{j,n}}_{Kp_k \theta_k}}{\Lambda(Kp_k, \theta_k)} dKp d\theta, \tag{6.24}$$

where $f^{\Delta T^*_{j,n}}_{Kp_k \theta_k}$ is the joint pdf of the peak factor and temperature affecting the kth component, defined in the time interval $\Delta T^*_{j,n}$, D_{Kp} and D_θ are the variation domains of Kp_k and θ_k, respectively and $\Lambda(Kp_k, \theta_k)$ represents the equipment life model expressed by Equation (6.23).

Equation (6.24) can still present some difficulties in deriving the joint pdf of the random variables Kp_k and θ_k, which are generally not directly available. Even in the case where the statistical characterization of the variables is known, the computation of Equation (6.24) is not immediate, mainly due to the fact that Kp_k cannot be expressed in closed form as a function of voltage harmonics and fundamental components (infinite combinations of harmonic vectors can provide a given value of Kp_k). Then, the application of Equation (6.24) in actual cases requires the use of Monte Carlo simulation procedures.

Some simplifications can be pursued only in particular cases. As an example, if the goal is to highlight only the influence of voltage and current harmonics on component life, other variables, including the voltage and current at fundamental frequency, the ambient temperature and the elements of the system admittance matrices at fundamental and harmonic frequencies, can be assumed to be deterministic quantities. Under this assumption, the voltage harmonics are directly linked to the current harmonics injected by nonlinear loads via the elements of the system harmonic admittance matrices. In such a case, the expected values of the relative loss of life of the system components are functions only of the pdfs of the magnitude and phase of the current harmonics injected by nonlinear loads, thus reducing the number of random variables to be accounted for.

Further simplifications can be achieved by computing the reduction in life expectancy for the worst-case condition, i.e. the reduction that would occur when the peak voltage became the arithmetic sum of the voltage harmonic peaks. Applying this simplification, there is no need to know the pdfs of the phase of harmonic currents injected by the nonlinear loads; also, in the presence of only one group of nonlinear loads as the main cause of harmonic events, the useful life can be evaluated with a closed-form relationship, using a simplified procedure.

For more details on this subject see [17].

6.2.4.3 Misoperation

As mentioned earlier, really few contributions in the literature refer to the estimation of the misoperation cost [14, 18, 31]. The method proposed in [18] is very interesting since it provides a unified index to account for the overall equivalent cost of the reliability and power quality attributes of the system that also includes misoperation due to harmonics.

The conversion of misoperation due to harmonics into costs is based on the fact that harmonics characterized by a THD which exceeds 20% are considered interruptions. With this choice, the sector customer damage function, proposed in [32] for interruptions, can also be used for harmonics. The model is particularly suitable for distributors who, in the planning stage, can use the economic metrics to choose the best solution among future alternatives.

6.3 Some economic mechanisms for improving power quality levels

In this section, some economic mechanisms for PQ regulation that have been proposed in the literature are shown. Some of the mechanisms are actively being used and others have only been proposed and have yet to be used. The countries in which these mechanisms are in use or have been proposed, together with the authors of the proposals, are identified.

6.3.1 USA: a mechanism based on the harmonic-adjusted power factor

The first significant attempt to introduce an economic mechanism to encourage power companies to improve voltage quality was proposed in [33]. The proposal was based on the use of the harmonic-adjusted power factor.

This approach was based on the similarity between the need to apply economic disincentives for poor power factors and for generating harmonic currents. In both cases, the customer requires additional distribution capacity in a way that is not reflected in the billing structure. Consequently, the authors investigated how the economic incentives that had been adopted to improve the power factor could be extended to provide economic incentives to improve harmonic behaviour. Their intention was that the proposed incentives should not require any changes to utility billing practices and that they should be implementable using electronic revenue meters with relatively simple software.

The harmonic-adjusted power factor, hPF, is defined as:

$$hPF = \frac{P}{V_H^* I_H^*},$$
(6.25)

where P is the active power, including both fundamental and harmonic powers, and V_H^* and I_H^* are the frequency-weighted RMS voltage and current, respectively.

The values of V_H^* and I_H^* can be determined using the following relationships:

$$V_H^* = \sqrt{\sum_{h=1}^{50} C_h V_h^2}$$
(6.26)

$$I_H^* = \sqrt{\sum_{h=1}^{50} K_h I_h^2},$$
(6.27)

where C_h and K_h are weighting factors and V_h and I_h are the measured RMS voltage and current harmonics of order h, respectively.

Regarding voltage weighting factors, the weights can be set to 1 or, alternatively, they can be assigned values less than 1 or even zero, which would imply that voltage delivered at harmonic frequencies is worth less than voltage delivered at the power frequency.

Regarding current weighting factors, the following four expressions were proposed: $K_h = h$, $K_h = h^{1.333}$, $K_h = \sqrt{h}$ and $K_h = [1 + x(h^2 - 1)]$, where x is a constant value between 0.01 and 0.10. This list of possible weightings is by no means exhaustive; other possible weightings for harmonic current may be used. The last weighting factor is the current weighting factor suggested in [34], in which the calculation of the rating of transformers subjected to harmonic current is described.

In addition, it should be noted that, in the absence of harmonic pollution, the harmonic-adjusted power factor coincides with the displacement power factor[2] (dPF) while, in the presence of nondistorted voltages and currents, hPF is almost always lower than dPF. Consequently, hPF may be used to discourage harmonic pollution by requiring the payment of power factor penalties.

6.3.2   USA: a proposal based on the service quality index

In Chapter 4, the service quality index (SQI) proposed in [35] was illustrated as an example of a global power quality indicator. In the SQI calculation, the economic impact experienced by the customer is considered, and knowledge of the unitary cost function associated with each power quality disturbance is specifically required. The unitary costs associated with

---

[2] The displacement power factor is defined as the cosine of the angle between the power frequency component of the current and the power frequency component of the voltage.

disturbances may be different from one utility to another, and they may even be different for different portions of the same system.

This index, proposed by the authors to characterize system performance and as a useful tool for prioritizing investments for system improvements, is also considered useful in quantifying the economic penalties paid by the electrical distribution company to the customer as a result of poor voltage quality; the economic penalty may be chosen, for example, to be proportional to the value assumed by the SQI index.

As shown in Chapter 4, the definition of the unitary cost function associated with each type of disturbance appears to be the biggest obstacle to the practical utilization of the SQI index as the market mechanism for quality.

6.3.3   Argentina: a mechanism based on the equivalence between voltage quality and continuity

Regulation regarding voltage quality disturbances was proposed in [36] to control the level of disturbance in the networks and the injection of customer disturbances. Initially, only harmonics and flicker (voltage fluctuation) were considered.

Control of harmonic distortion and flicker is based on measured disturbance levels at the PCC. In particular, utilities perform the measurements every month at pre-fixed points in the distribution network. The measurements are conducted by registering devices that comply with requirements specified by IEC standards. Two different devices register levels of flicker and harmonics at ten-minute intervals and measure the energy supplied during the intervals. One week is the minimum time interval for a measurement period; in this case, 1008 samples result. Utilities are responsible for supplying electricity with an adequate level of quality, which is characterized by the presence of disturbance levels below regulated reference levels at least 95% of the time. When voltage quality disturbances exceed the regulated reference value for more than 5% of the time for the weekly measurement taken at ten-minute intervals, the utility will be subjected to economic penalties to compensate the customers affected by the poor quality of the voltage supply. This economic penalty is proportional to the extent to which the voltage deviates from the reference level and the energy supplied under poor voltage quality conditions.

For cases involving exceeding the flicker reference level, the penalty cost function, PCF_{Fl}, is:

$$PCF_{Fl} = \sum_{k \in \Omega_1} C_{IN} \, DPF_k^2 \, E_k + \sum_{k \in \Omega_2} C_{IN} \, E_k \qquad (6.28)$$

where E_k is the energy supplied during the interval k and DPF_k is the '*flicker distortion*' subject to penalty, given by:

$$DPF_k = \max \left[0 \, , \quad \frac{P_{st,k} - P_{st}^*}{P_{st}^*} \right], \qquad (6.29)$$

Ω_1 is the set of intervals characterized by values of $DPF_k \leq 1$ and Ω_2 is the set of intervals characterized by values of $DPF_k > 1$.

In Equation (6.29), $P_{st,k}$ is the short-term flicker severity at the kth interval measurement and P_{st}^* is the reference value of the short-term flicker severity at the considered supply point. In Argentina, the flicker reference values for all voltage levels are set to 1, and then $P_{st}^*=1$.

DPF_k is calculated for each interval of each weekly measurement period ($k = 1, 2,...,1008$) and DPF_k has a value other than zero only in the case where the reference level is exceeded.

From Equation (6.28), each k-interval registered with energy supplied during a bad voltage-quality condition will yield a unitary penalty equal to $C_{IN}\,DPF_k^2$ [US\$/kWh] in the case where $0 < DPF_k < 1$ and equal to C_{IN} [US\$/kWh] in the case where $DPF_k \geq 1$. The unitary penalty is quadratic with DPF_k, hence, the penalty levels are higher as violations of the reference level become more significant. This is the case until $DPF_k = 1$, after which the penalty is fixed at C_{IN} [US\$/kWh]. Above this level, the voltage quality is considered not acceptable and, in practice, the utility is penalized as if the customer had not been supplied; in fact, C_{IN} is the same penalty that has been proposed when interruptions occur.

For cases in which the harmonic voltage reference level is exceeded, the penalty cost function PCF_h is:

$$PCF_h = \sum_{k \in \Omega_3} C_{IN}\,DPH_k^2 E_k + \sum_{k \in \Omega_4} C_{IN}\,E_k, \qquad (6.30)$$

where DPH_k is the harmonic voltage distortion subject to penalty, given by:

$$DPH_k = \max\left[0, \quad \frac{THD_k - THD^*}{THD^*}\right] + \frac{1}{3}\sum_{i=2}^{40}\max\left[0, \quad \frac{V_{h,k} - V_h^*}{V_h^*}\right], \quad (6.31)$$

Ω_3 is the set of all intervals characterized by values of $DPH_k \leq 1$, and Ω_4 is the set of all intervals characterized by values of $DPH_k > 1$.

The THD_k value is the total harmonic distortion registered at the kth interval measurement and THD^* is the reference value for total harmonic distortion. The $V_{h,k}$ value is the short ten-minute value of the hth individual harmonic voltage measured in interval k and V_h^* is the reference value for the hth individual harmonic voltage. The reference values for THD were reported in Chapter 1; the reference values for individual harmonic voltages are different for LV, MV and HV networks.

DPH_k is calculated for each interval k ($k = 1, 2,...,1008$) of each weekly measurement period, and DPH_k is zero except in the case of a violation of reference values.

From Equation (6.30), each interval registered with energy supplied in poor voltage quality condition will yield a unitary penalty equal to $C_{IN}\,DPH_k^2$ [US\$/kWh] in the case where $0 < DPH_k < 1$ and equal to C_{IN} [US\$/kWh] in the case where $DPH_k \geq 1$. The considerations used for the unitary penalty relating to flicker can be repeated for the unitary penalty relating to harmonic distortion.

In [37], it was proposed that slow voltage variations should also be considered in the regulation regarding voltage quality disturbances. In particular, a penalty function for slow voltage variations, similar to that used for flicker and harmonics, was introduced. In cases where the acceptable limit levels are exceeded for a time greater than 3% of the survey period, the utility has to pay on the basis of the following penalty cost function:

$$\text{PCF}_V = \sum_{k \in \Omega_5} C_{\Delta V}\, E_k, \qquad (6.32)$$

where E_k is the energy supplied during the interval k and Ω_5 is the set of all intervals characterized by $|\Delta V| > \Delta V_{\text{lim}}$.

In this relationship, the summation refers to all measurement intervals characterized by voltage levels exceeding the tolerance band ΔV_{lim}:

$$|\Delta V| = \frac{|V - V_n|}{V_n} \cdot 100 > \Delta V_{\text{lim}}, \qquad (6.33)$$

where V and V_n are the supply voltage and nominal voltage, respectively. $\Delta V_{\text{lim}} = 5\%$ for HV and 8% for MV and LV networks.

The term $C_{\Delta V}$ represents the unitary energy cost. This cost increases as $|\Delta V|$ increases; for example, Table 6.3 shows the $C_{\Delta V}$ values used for the calculation of penalties in the case of LV customers.

Table 6.3   Low voltage unitary energy cost [37]

Band of voltage [%]	Unitary energy cost $C_{\Delta v}$ [US$/kWh]
$8 \le |\Delta V| < 9$	0.050
$9 \le |\Delta V| < 10$	0.140
$10 \le |\Delta V| < 11$	0.230
$11 \le |\Delta V| < 12$	0.320
$12 \le |\Delta V| < 13$	0.410
$13 \le |\Delta V| < 14$	0.500
$14 \le |\Delta V| < 15$	0.800
$15 \le |\Delta V| < 16$	1.100
$16 \le |\Delta V| < 17$	1.400
$17 \le |\Delta V| < 18$	1.700
$18 \le |\Delta V|$	2.000

The economic penalties given by Equations (6.28), (6.30) and (6.32) are maintained until the utility shows, by means of new measurements, that exceedances of the reference level have been eliminated. The penalty determined for the measurement week remains in place every week until the problem is solved.

To guarantee the reference levels at the busbars of the network, the distribution companies control customers' voltage disturbance emissions into the networks. Emission limits are fixed, both for flicker and harmonic distortions, for different types of users.[3] A particular emission limit represents the maximum disturbance level that the customer is allowed to inject into the network at the PCC. There must be a 99% probability that the emission limit will not be exceeded. If a violation of the emission limit is suspected, the distribution companies should verify the actual emission by performing a set of measurements. The required minimum measurement period is one week for checking the emission limit. The distribution company can take the following actions:

- if the reference levels have not been exceeded, the distribution company may offer an agreement to the customer to increase the assigned emission limit;

- if the reference levels have been exceeded, the customer must adopt measures to reduce the emissions;

- if the customer still does not reduce the emissions below the permissible limit in a set time period (for instance, six months), the customer is penalized economically according to the amplitude of the violation of the limits. The formulas for the economic penalties are similar to those applied to the utilities when poor-quality supply voltage is furnished. After the set period, if the customer still has not solved the emission problem, the utility may ask the regulatory authority for permission to disconnect the customer.

6.3.4   Colombia: a mechanism based on the customer's perception of the impact of each disturbance

In [38], a methodology was proposed for evaluating costs due to power quality disturbances.[4]

The penalty PCF_j^i that the utility should pay to customer j in the presence of the ith disturbance exceeding the reference value is:

$$\text{PCF}_j^i = R_V \, w_i \, E_j, \qquad (6.34)$$

where R_V is the reference value assumed equal to the cost associated with the interruptions, w_i is the weight of the voltage quality disturbance and E_j is the energy delivered to customer j during the bad-voltage-quality condition.

The weights w_i reflect the impact that the customer perceives as a result of each disturbance and, on the basis of a survey developed using distribution systems, the values reported in Table 6.4 have been suggested.

---

[3] The different types of users supplied by the Argentinean electrical distribution system are users with tariff T-1 ($P \leq 10\,\text{kW}$); users with tariff T-2 ($10\,\text{kW} < P \leq 50\,\text{kW}$); and users with tariff T-3 ($P > 50\,\text{kW}$). Users with tariffs T-1 and T-2 can only be connected to LV networks, while users with tariff T-3 can be connected to LV, MV and HV networks.

[4] It should be noted that the aim of the proposed procedure was not directly to define an economic mechanism for the regulation of voltage quality; rather, it was to estimate optimal power quality levels in Colombia. The optimal quality level is the level that minimizes the social cost of the PQ, defined as the sum of the costs supported by the electricity company in infrastructure investments, maintenance and operation of the network to improve the PQ and the costs that customers incur in the presence of a lack of PQ due to, for example, lost production.

Table 6.4   Values for disturbance weights [38]

Disturbance	Weight w_i
Interruption	1.0
Unbalance	0.1
Flicker	0.4
THD	0.3
Individual voltage harmonics	0.2

Assuming the cost of interruption as reference, the term $R_V\ w_i$ represents the unitary penalty [money/kWh] with reference to the ith voltage quality disturbance (e.g. unbalance or flicker).

If z is the number of customers for whom the ith disturbance exceeds the reference figure, the value for reimbursing the customers for the ith disturbance will be the aggregate of all customers:

$$PCF_{tot}^i = \sum_{j=1}^{z} PCF_j^i = R_V\ w_i \sum_{j=1}^{z} E_j. \tag{6.35}$$

Under conditions of uniform demand, if n is the total number of customers supplied by the monitored points, the result is:

$$\sum_{j=1}^{z} E_j = UI_i \sum_{j=1}^{n} E_j, \tag{6.36}$$

where UI_i is the fraction of customers characterized by the ith voltage disturbance exceeding the limit.

Using Equation (6.36), the total economic penalty which the utility must pay in the case of violation of the limit set for the ith voltage quality disturbance can be expressed by:

$$PCF_{tot}^i = UI_i\ R_V\ w_i \sum_{j=1}^{n} E_j. \tag{6.37}$$

Compared to the economic penalty proposed in Argentina, the penalty function proposed in Colombia is not linked to the entity of the limit violation, but it is evident that, when the percentage of customers out of limits is related directly to the limit imposed, the economic penalty paid by the utility increases with the limit restriction.

6.3.5 Iran: a proposal based on a penalty function depending on the type of customer

In Iran, distribution utilities are responsible for supplying electricity with an adequate level of quality that is characterized by the presence of disturbance levels below regulated reference levels at least 95% of the time. When the reference value of each voltage quality disturbance is exceeded in more than 5% of the ten-minute intervals of a weekly measurement period, the utilities are subjected to economic penalties to compensate the customers affected by bad-quality voltage supply [39]. The penalty cost function proposed in Iran for industrial and commercial customers is given by:

$$PCF = K_1 \text{ ESC} + K_2 \text{ IDC} + K_3 \text{ LPV} + K_4 \text{ EWC} + K_5 \text{ HWC} + K_6 \text{ HD}. \qquad (6.38)$$

where:

ESC = cost of extra shift;

IDC = cost of instrument damage;

LPV = lost production value;

EWC = extra work cost;

HWC = halt work cost;

HD = humanity damages;

K_i = weighting factors.

The economic penalty depends on the type of customer, which is taken into account using weighting factors. Table 6.5 shows the weighting factors for different types of customers (group degree), while Table 6.6 reports the classification of customers as a function of industrial processes. With reference to the type of customers, the different costs in Equation (6.38) may be present ($K_i = 1$) or absent ($K_i = 0$).

Table 6.5   Weighting factors for various load groups [39]

Group degree	K_1	K_2	K_3	K_4	K_5	K_6
A	0 or 1	1	1	0 or 1	1	1
B	0	1	0	1	1	0
C	0	1	1	1	1	0
D	1	1	1	1	1	0

To guarantee the reference limit, the distribution companies control the disturbance emissions of customers into the networks. Emission limits are fixed both for flicker and waveform distortions at different voltage levels on the networks. There must be a 99% probability that the emission limit will not be exceeded. If a violation of the emission limit is suspected, the utility should verify the actual emission by performing a set of measurements. The minimum time period required to check the emission limit is one week.

When the level of the disturbance injected by the user installation is higher than the fixed limit, solutions analogous to those used in Argentina may be applied.

Table 6.6 Industries affected by power quality issues [39]

Industry segment	Industrial process	Group degree
Continuous process	1. Paper, fibre and textile factories 2. Plastic extruding or moulding plants	D
Precision machining	1. Automotive parts manufacturing 2. Large pump forging factories	D
High technology products and research	1. Semiconductor manufacturing 2. Large particle physics research centres	B
Information technology	1. Data processing centres 2. Banks 3. Telecommunication 4. Broadcasting	C
Safety and security related	1. Hazardous processes 2. Chemical processing 3. Hospital and health-care facilities 4. Military installations	A

6.3.6 Italy: a mechanism based on the unified power quality index

A new approach to regulating voltage quality that is similar to that used in other countries for continuity regulation has been proposed by some Italian researchers [40], but it has not yet been adopted by the regulator. The proposed approach provides two levels of regulation (Figure 6.5):

1. The system level is finalized to guarantee acceptable average voltage quality levels for all customers.

2. The local regulation level takes into account the interactions between utilities and customers at the PCC.

At the system level, power quality levels are not guaranteed for each user. Instead, an adequate level of average quality for all supplied users is guaranteed. Local regulation should focus both on sensitive loads that require specific voltage quality characteristics and on disturbing loads that require control of emission levels.

In the context of the overall regulation scheme shown in Figure 6.5, the authors concentrated only on regulation at the system level; moreover, only continuous disturbances (e.g. waveform distortions, unbalances and voltage fluctuations) were considered in depth. The regulation structures proposed used global indices, such as the unified power quality index described in Chapter 4, to characterize the voltage quality, since they might be more convenient for system regulation, especially in simplifying regulatory schemes.

The proposed system voltage quality regulatory mechanism is similar to those employed in several countries for the regulation of continuity. The national territory is divided into different areas and the global index chosen to characterize voltage quality is measured separately for a pre-set number of districts for each area.

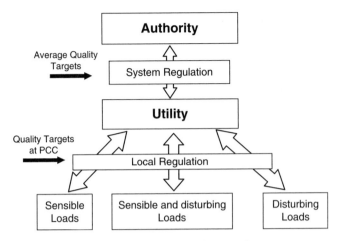

Figure 6.5   Overall regulation scheme

The regulatory scheme specifies reference levels to be obtained within a specific number of years; such levels are pursued through paths of improvement that identify the minimum improvement required for each year in each district (Figure 6.6). Also, to force the districts with more quality problems to converge towards the reference level, several paths are established based on the level of improvement required. The regulatory authority compares the actual levels with the target levels and, if the utility has improved quality more than required, it gains an incentive; otherwise, it pays a penalty based on how far it was from the objective.

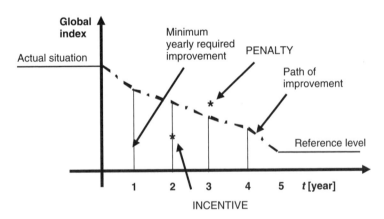

Figure 6.6   Path of improvements

The use of a scheme based on penalties and incentives for utilities is being considered since the penalties should push utilities to intervene on their own networks and with disturbing users in order to return to the levels imposed by the authority; on the other hand, the incentives should drive utilities to accelerate intervention in order to attain target levels as quickly as possible.

To apply this mechanism, once the global index of reference for the overall system has been established, the following steps must be considered:

1. Reference levels must be defined.

2. Paths to reach the reference levels (Figure 6.6) must be established, during which required minimum improvement levels should be assigned.

3. Penalties and incentives associated with the required minimum improvements must be defined.

The reference levels may be assigned by the regulatory authority by considering the limit values imposed by standards or recommendations for each index characterizing the different power quality disturbances taken into account.

Then, assuming that the nation is divided into N_{Distr} districts, for each district, the number of monitored sites, M, represents an average behaviour for the entire district; such sites divide the entire district into zones with users U_j. Also, let us assume that there is a measurement system at each monitored site and that the reference period is one week.[5]

During one week, continuous disturbances are measured at every site at intervals with duration Δt_k to obtain a number of values for each continuous disturbance equal to the number of measurement subintervals in a week.

If we consider the jth monitored site and label the limit value of the index considered for the ith continuous disturbance as $I_{i,\text{lim}}$, it is then possible to calculate the corresponding exceedance (Equations (4.21) and (4.22)) for every interval Δt_k, as:

$$\Delta I_{i,j}^{\Delta t_k} = \frac{I_{i,j}^{\Delta t_k}}{I_{i,\text{lim}}} - 1. \tag{6.39}$$

If the exceedance has a negative value, it will be considered equal to zero. Once the exceedances for the continuous disturbances under examination are known, it is possible to calculate, for the same Δt_k interval, the value of the global index UPQI_{Cj} corresponding to the jth site:

$$\left(\text{UPQI}_{Cj}\right)^{\Delta t_k} = 1 + \sum_{i=1}^{N_C} \Delta I_{i,j}^{\Delta t_k}. \tag{6.40}$$

Once we have obtained the values of the global index UPQI_{Cj} for all the Δt_k subintervals in one week, it is possible to calculate the corresponding probability function and the maximum value, $\left(\text{UPQI}_{Cj}\right)^{\text{max}}$, which is considered suitable as the reference level characterizing the site.

---

[5] This is usually the interval considered in most standards related to voltage quality continuous disturbances, such as EN 50160, as shown in Chapter 1 [41].

Once all site indices are determined $((\mathrm{UPQI}_{C,j})^{\max}, j = 1, 2, \ldots, M)$, we can calculate the values of the system index as follows:

$$(\mathrm{UPQI}_C)^{\mathrm{SYS}} = \sum_{j=1}^{M} \frac{w_j (\mathrm{UPQI}_{C,j})^{\max}}{\sum_{k=1}^{M} w_k}. \tag{6.41}$$

In Equation (6.41), w_j weights each monitored site differently, depending on the number of users or the maximum power provided to them. An alternative to Equation (6.41) is to take the maximum value of the site indices (or the 95th percentile derived from the probability function) as the system index.

Once the index for the overall system has been obtained, it can be compared with the assumed maximum permissible value, which depends on the path of improvements chosen (Figure 6.6).

If we indicate the minimum required yearly improvement as $(\mathrm{UPQI}_C)_{T_n}^{\mathrm{SYS}}$ related to the nth intermediary subperiod of the regulated period, assuming a ±5% tolerance band of the global index, we obtain the following constraints:

- if $(\mathrm{UPQI}_C)^{\mathrm{SYS}} > 1.05 \, (\mathrm{UPQI}_C)_{T_n}^{\mathrm{SYS}}$ a penalty will be assigned;
- if $(\mathrm{UPQI}_C)^{\mathrm{SYS}} < 0.95 \, (\mathrm{UPQI}_C)_{T_n}^{\mathrm{SYS}}$ an incentive will be assigned.

With reference to the problem of establishing a penalty/incentive mechanism associated with the attainment or loss of expected minimum levels, a possible penalty/incentive mechanism extends the mechanism proposed in Argentina for each PQ disturbance to the case of global indices, as shown in Figure 6.7 for the penalty function structure. Penalties are attributed proportionally to the supplied energy, for conditions of lack of voltage quality, with the penalty rising with the degree of deviation, up to an assigned threshold level. When the voltage quality exceeds the threshold level, the utility is penalized as it would be in the case of a long interruption. When the voltage quality is less than the threshold level but above the minimum required yearly improvement, the utility is penalized less, because the delivered energy quality is considered better than in the case of a long interruption.

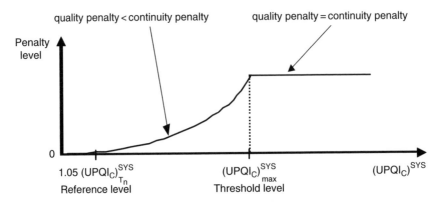

Figure 6.7   Penalty function structure

Analytical relationships that quantify the penalty function shown in Figure 6.7 and that allow the authority to properly balance the penalizing effects can be easily obtained. The penalties are strongly related to the behaviour of the curve representing the penalty function. For example, in the range between the reference level and the threshold level, the higher the slope of the curve around the threshold level is, the higher the penalty levels for violations of the reference level become. The authority can significantly influence the mechanism application effects with a proper choice of curve behaviour for the penalty function.

A similar criterion can be applied for the incentive mechanism. In addition, a mechanism to balance the penalties and incentives can be also considered. As a general criterion in the field of continuity, the incentives are financed by penalties paid by the utilities; when such penalties amount is less than the amount of incentives, the difference should be paid by the users who were granted a higher level of service.

6.3.7   Various countries: power quality contracts

To our knowledge, the first example of a power quality contract was stipulated in the USA [42]. This PQ contract, called the 'Special Manufacturing Contract' (SMC) was stipulated between the Detroit Edison Company and its three largest automotive customers (Chrysler Corporation, Ford Motor Company and General Motors Corporation) in 1994.

The Detroit Edison Company and its customers devised a method of compensating the customers for power quality events if they exceeded the annual permissible level. The PQ events covered by the SMC were interruptions and voltage sags. In the following, we address only the aspects of the PQ contract related to voltage sags.

To characterize the severity of a voltage sag, a new index called the *sag score* was included in this contract. The sag score was defined in terms of the RMS values of the phase voltages V_a, V_b and V_c:

$$\text{Sag Score} = 1 - \frac{V_a + V_b + V_c}{3}. \tag{6.42}$$

In the sag score calculation, only voltage sags characterized by residual voltages below 0.75 p.u. were considered. The strength of this index lies in the simplicity of its computation. However, the index does not take into consideration the duration of the sag event, which indicates its impact on loads. In the presence of interruptions, the sag score computation is zero to prevent overlap with the administration related to voltage interruptions.

For each location, the annual sag score (also called *total sag score*) was obtained by summing the values of the sag scores for the entire year.

The SMC provided for the payment of compensation by Detroit Edison Company to the customer if the sag score total exceeded the annual permissible sag score limit. The sag score limits were defined after an appropriate campaign of measurements and recomputed at the start of each year.[6]

---

[6] In fact, while the SMC was stipulated in 1994, the regulation of voltage sags began in 1998, after the required campaign of measurements used to define the initial severity sag level was completed.

The required payment was obtained by multiplying the sag score total in excess of the sag score limit by a unitary penalty called the *service guarantee payment amount* (SGPA). The SGPA was specified in the contract as the limit of the total annual sag payments.[7]

More recently, in some European countries, customers have begun to negotiate with utilities to receive a specified level of continuity and voltage quality (power quality contracts).

For example, in France, both the transmission and distribution companies offer all customers the possibility to contract for extra quality requirements. In such cases, the customer can ask the operator for customized contractual levels. The customer will have to pay for them, depending on the necessary work required to reach these new levels. In particular, when a customer wants customized levels in his contract, the operator makes a technical and financial proposal that describes the necessary work required on the network and the associated costs that will need to be incurred to reach the levels of quality desired by the customer. If the customer accepts the proposal, the required work will be done at the customer's expense.

It is also possible to demand a power quality contract in Italy. In these contracts, at least three elements must be present:

1. the contractual level of quality;

2. the yearly premium;

3. the penalty for noncompliance.

The contractual level of quality is expressed as a threshold applied to one or more indicators of voltage quality, and the duration of the contract may be no less than one year and no more than four years.

Contracts are totally voluntary, both for customers and for the distribution company (or the transmission system operator). The additional revenues coming from power quality contracts are treated as a service excluded from the company's revenue control. Suppliers can be involved, especially in federating more than one customer interested in quality improvement in the same distribution area; the cost (and the benefits) of the power quality contract can be shared among several customers.

Power quality contracts are still in their initial phase, but they can be seen as an efficient solution for improving voltage quality without imposing excessive costs on the general population of customers. These contracts require that customers who require better voltage quality must be willing to pay for it.

Generally, power quality contracts are rarely monitored by a regulator. Often, where contracts are foreseen, the regulators have no role in the market mechanism for quality. Table 6.7 reports the European countries in which it is foreseen that power quality contracts will be used.

The need to introduce power quality contracts between the different operators involved in the competitive electricity market in order to guarantee specified levels of continuity

---

[7] For example, if the annual sag score is equal to 3.28 and the sag score limit is 3.00, then the sag score excess is 0.28. If the location where the voltage sag was measured has an SGPA of $100.00, then the payment due is $28.00.

Table 6.7 Power quality contracts in Europe [43]

Power quality contract with some ex ante intervention of regulator	France, Italy
Power quality contract with only ex post intervention of regulator	Slovenia
Power quality contract with no intervention of regulator	Czech Republic, Spain, Great Britain, Latvia, Portugal

and voltage quality was also acknowledged by McGranaghan *et al.* in [44]. In this paper, the authors indicated the following main points that each PQ contract should include:

- reliability/power quality concerns to be evaluated;
- performance indices to be used;
- expected level of performance;
- penalties for performance outside the expected level and/or incentives for performance better than the expected level;
- measurement and/or calculation methods to verify performance;
- responsibilities of each party in achieving the desired performance;
- responsibilities of each party in resolving PQ problems.

The requirements of particular PQ contracts and the concerns that must be addressed will depend on the parties involved and the characteristics of the system, so the following different types of PQ contract have been proposed:

1. Contracts between transmission companies and distribution companies. In this type of contract, PQ requirements and responsibilities at the distribution substation interface between the two systems are defined.

2. Contracts between distribution companies and end-use customers. In this type of contract, the PQ requirements at the PCC are defined. In some cases, the end user may be a customer of the distribution company; in other cases, the end customer may be a retail marketer or an energy service company.

3. Contracts between retail marketers or energy service companies and the end-use customer. This type of contract may be much more creative and complicated than the contracts defining the basic distribution company's interface with the customer. The energy service company may offer a whole range of services for improving power quality, efficiency and productivity that dictate the contract requirements.

4. Contracts between distribution companies and independent power producers (IPPs). This type of PQ contract defines the expected power quality that the IPP can expect at

the interface (similar to the contract with end users) and defines the requirements for the IPP in terms of the quality of the generated power. Important areas to consider for the IPP requirements are power fluctuation, harmonics, power factor characteristics and unbalances.

Finally, in [45], a PQ service-pricing approach considering the cost of quality and stop-loss insurance has been proposed.

A product (such as electrical energy) can be characterized by different quality levels. The customer pays two kinds of costs for a product: the purchase cost, which increases as the quality of the product increases, and the use cost, which decreases as the product quality increases. The use cost will directly influence the pricing of the product and, consequently, it will influence the customer's purchase cost. Products with different quality will cause different use costs. Furthermore, for fixed-quality products, different customers have different use costs. The higher the customer's use cost, the larger the potential loss caused by a low-quality product, so the customer may choose the higher-quality product, which will mean a higher initial purchase price. Consequently, because of the lower use cost, some customers are willing to buy higher-quality electricity at a higher cost.

The total quality cost C_T of electrical energy is defined in [45] as the sum of the treatment expense C charged by the distributor for fulfilling the customer's requested PQ level and the customer's loss cost L: $C_T(s) = C(s) + L(s)$. Both C and L costs depend on the PQ level s, and the cost C increases and the cost L decreases as s increases.

The most economical PQ level is that which minimizes the total quality cost; this level will be indicated in the following by the symbol s_T.

In choosing the PQ service, the customer will make the s_T level the PQ level, paying $C(s_T)$ as the treatment expense. The optimal level s_T may be different for customers supplied by the same distribution network; in fact, for a fixed treatment expense function $C(s)$ (given by the distribution company), the loss function $L(s)$ depends on the sensitivity with respect to the power quality of the customer's electrical loads.

In actual operating conditions, the PQ level s fluctuates near the desired PQ level s_T. Customers are willing to endure the risk of loss when the PQ level is higher than the objective PQ level ($s \geq s_T$), but they are not willing to do so when $s < s_T$. So, customers can share the increased risk of loss when $s < s_T$ by buying stop-loss insurance.

By offering a PQ service plus stop-loss insurance, a distributor provides insurance for the customer's extra loss risk. When the real PQ level is lower than the objective PQ level ($s < s_T$), the distributor must compensate the customer for the extra loss, and when $s \geq s_T$ the customer endures the loss. In stop-loss insurance, the distributor compensates the customer according to the customer's willingness to pay.

In this way, when the distributor and the customers sign the PQ service contract, the fee is composed not only of the treatment expense $C(s_T)$, but also of the premium for the stop-loss insurance. The customer chooses the optimal PQ level (from an economical point of view) according to the cost of controlling PQ and selects the amount of protection desired from potential losses by choosing the amount of stop-loss insurance to purchase.

According to the premium calculation theory, the incentive premium for stop-loss insurance is obtained from the probability of violation of the chosen s_T PQ level and by the customer's estimate of the economical damage produced by this violation. In practice, the insurance premium is linked to a risk factor that contemporaneously takes into account both the probability of PQ levels having values lower than the s_T customized value and the economic losses related to the effects of these PQ levels on the customer's equipment and loads.

6.4   Conclusions

In this chapter, the economic aspects of PQ disturbances have been addressed, focusing on the variables that are more sensitive when making the associated cost estimates. Then, some economic mechanisms to use for PQ regulation have been described. These mechanisms, reported in the literature, are based on financial penalties, incentives or both.

Very few of the described mechanisms are actively being used; in particular, among the illustrated mechanisms, the one discussed most extensively is the PQ contract. However, since PQ contracts are voluntary, both for customers and for the distributor (or the transmission system operator), they are infrequently used.

The main conclusion of this chapter is that the economic aspects of PQ disturbances should be the basis by which to regulate interactions between customers and the utility in liberalized markets. However, even though several researchers are working on economic mechanisms for PQ regulation, further studies are needed to clarify all the aspects involved.

References

[1] Emanuel, E., Yang, M. and Pileggi, D.J. (1991) 'The Engineering Economics of Power Systems Harmonics in Subdistribution Feeders. A Preliminary Study', *IEEE Transactions on Power Systems*, **6**(3), 1092–1098.

[2] Pileggi, J., Gentile, T.J., Emanuel, A.E., Gulachenski, E.M., Breen, M., Sorensen, D. and Janczak, J. (1994) Distribution Feeders with Nonlinear Loads in the Northeast U.S.A.: Part II – Economic Evaluation of Harmonic Effects, *IEEE/PES Winter Meeting*, New York, January.

[3] Kawann, C. and Emanuel, A.E. (1996) 'Passive Shunt Harmonic Filters for Low and Medium Voltage: a Cost Comparison Study', *IEEE Transactions on Power Systems*, **11**(4), 1825–1831.

[4] McGranaghan, M. and Roettger, B. (2002) 'Economic Evaluation of Power Quality', *Power Engineering Review*, **22**(2), 8–12.

[5] Baghzouz, Y., Burch, R.F., Capasso, A., Cavallini, A., Emanuel, A.E., Halpin, M., Imece, A., Ludbrook, A., Montanari, G., Olejniczak, K.J., Ribeiro, P., Rios-Marcuello, S., Tang, L., Thaliam, R. and Verde, P. (1998) 'Time-Varying Harmonics. I. Characterizing Measured Data', *IEEE Transactions on Power Delivery*, **13**(3), 938–944.

[6] Baghzouz, Y., Burch, R.F., Capasso, A., Cavallini, A., Emanuel, A.E., Halpin, M., Imece, A., Ludbrook, A., Montanari, G., Olejniczak, K.J., Ribeiro, P., Rios-Marcuello, S., Tang, L., Thaliam, R. and Verde, P. (2002) 'Time-varying harmonics. II. Harmonic Summation and Propagation' *IEEE Transactions on Power Delivery*, **17**(1), 279–285.

[7] IEEE Standard 493 (1997) Recommended Practice for the Design of Reliable Industrial and Commercial Power Systems, December.

[8] IEEE Standard 1346 (1998) Recommended Practice for Evaluating Electric Power System Compatibility With Electronic Process Equipment, May.

[9] Heine, P., Pohjanheimo, P., Lehtonen, M. and Lakervi, E. (2002) 'A Method for Estimating the Frequency and Cost of Voltage Sags', *IEEE Transactions on Power Systems*, **17**(2), 290–296.

[10] IEEE Task Force (1985) 'The Effects of Power System Harmonics on Power System Equipment and Loads', IEEE Transactions on Power Apparatus and Systems, **PAS-104**, 2555–2563.

[11] Fuchs, E.F., Roesler, D.J. and Kovacs, K.P. (1986) 'Aging of Electrical Appliances Due to Harmonics of the Power System's Voltage', *IEEE Transactions on Power Delivery*, **1**(3), 301–307.

[12] IEEE Task Force (1993) 'Effects of Harmonics on Equipment', *IEEE Transactions on Power Delivery*, **8**(2), 672–680.

[13] Anders, G.J. (1997) Rating of Electric Power Cables, *McGraw Hill*, IEEE Press, Piscataway, NJ.

[14] Verde, P. (2000) Cost of Harmonic Effects as Meaning of Standard Limits, *9th International Conference on Harmonics and Quality of Power*, Orlando (USA), October.

[15] Caramia, P., Carpinelli, G., Di Vito, E., Losi, A. and Verde, P. (1996) 'Probabilistic Evaluation of the Economical Damage Due to Harmonic Losses in Industrial Energy Systems', *IEEE Transactions on Power Delivery*, **11**(2), 1021–1031.

[16] Caramia, P., Carpinelli, G., Losi, A., Russo, A. and Verde, P. (1998) A Simplified Method for the Probabilistic Evaluation of the Economical Damage due to Harmonic Losses, *8th International Conference on Harmonics and Quality of Power*, Athens (Greece), October.

[17] Caramia, P., Carpinelli, G., Verde, P., Mazzanti, G., Cavallini, A. and Montanari, G.C. (2000) An Approach to Life Estimation of Electrical Plant Components in the Presence of Harmonic Distortion, *9th International Conference on Harmonics and Quality of Power*, Orlando (USA), October.

[18] Lee, B., Stefopoulos, G.K. and Meliopoulos, A.P.S. (2006) Unified Reliability and Power Quality Index, *12th International Conference on Harmonics and Quality of Power*, Cascais (Portugal), October.

[19] Mine, M.A. (1945) 'Cumulative Damage in Fatigue', *Journal of Applied Mechanics*, **67**, A159–A164.

[20] Allen, P.H. and Tustin, A. (1972) 'The Aging Process in Electrical Insulation: A Tutorial Summary', *IEEE Transactions on Electrical Insulation*, **EI-7**(3), 153–157.

[21] Montanari, G.C. and Pattini, G. (1986) 'Thermal Endurance of Insulating Materials', *IEEE Transactions on Electrical Insulation*, **21**(1), 66–75.

[22] Simoni, L. (1993) Fundamentals of Endurance of Electrical Insulating Materials, *CLUEB*, Bologna (Italy), April.

[23] Cavallini, A., Fabiani, D., Mazzanti, G., Montanari, G.C. and Contin, A. (2000) Voltage Endurance of Electrical Components Supplied by Distorted Voltage Waveforms, *International Symposium on Electrical Insulation*, Anaheim (USA), April.

[24] Mazzanti, G., Passarelli, G., Russo, A. and Verde, P. (2006) The Effects of Voltage Waveform Factors on Cable Life Estimation Using Measured Distorted Voltages, *IEEE/PES General Meeting*, Montreal (Canada), June.

[25] Manson, J. (no date) *Business Model for Investing in Power Quality Solutions, Power Quality Application Guide Background Notes, Section 2 – Costs*, available at: www.lpqi.org.

[26] Pohjanheimo, P. and Lehtonen, M. (2004) Introducing Prob-A-Sag Probabilistic Method for Voltage Sag Management, *11th International Conference on Harmonics and Quality of Power*, Lake Placid (USA), October.

[27] Milanovic, J.V. and Gupta, C.P. (2006) 'Probabilistic Assessment of Financial Losses due to Interruptions and Voltage Sags – Part I: The Methodology', *IEEE Transactions on Power Delivery*, **21**(2), 918–924.

[28] Milanovic, J.V. and Gupta, C.P. (2006) 'Probabilistic Assessment of Financial Losses due to Interruptions and Voltage Sags – Part II: Practical Implementation,' *IEEE Transactions on Power Delivery*, **21**(2), 925–932.

[29] Caramia, P., Carpinelli, G., Rossi, F. and Verde, P. (1994) 'Probabilistic Iterative Harmonic Analysis of Power Systems', *IEE Proceedings of Generation, Transmission and Distribution*, **141**(4), 329–338.

[30] Carpinelli, G., Esposito, T., Varilone, P. and Verde, P. (2001) 'First-order Probabilistic Harmonic Power Flow', *IEE Proceedings of Generation, Transmission and Distribution*, **148**(6), 541–548.

[31] Caramia, P. and Verde, P. (2000) Cost-related Harmonic Limits, IEEE/PES Winter Meeting, *Singapore*, January.

[32] Billinton, R. and Wang, P. (1998) 'Distribution System Cost/Worth Analysis Using Analytical and Sequential Simulation Techniques', *IEEE Transactions on Power Systems*, **13**(4), 1245–1250.

[33] McEachern, A., Mack Grady, W., Moncrief, W.A., Heydt, G.T. and McGranaghan, M. (1995) 'Revenue and Harmonics: an Evaluation of Some Proposed Rate Structure', *IEEE Transactions on Power Delivery*, **10**(1), 474–482.

[34] ANSI/IEEE Standard C57.110–1986 (1986) IEEE Recommended Practice for Establishing Transformer Capability when Supplying Nonsinusoidal Load Current.

[35] McGranaghan, M. (2007) 'Quantifying Reliability and Service Quality for Distribution Systems', *IEEE Transactions on Industry Applications*, **43**(1), 188–195.

[36] Gomez San Roman, T. and Roman Ubeda, J. (1998) 'Power Quality Regulation in Argentina: Flicker and Harmonics', *IEEE Transactions on Power Delivery*, **13**(3), 895–901.

[37] Laspada, H.M. (1999) Regulation of Quality of the Electrical Service in Argentina, *CIRED '99*, Nice (France), June.

[38] Cajamarca, G., Torres, H., Pavas, A., Urrutia, D., Gallego, L. and Delgadillo, A. (2006) A Methodological Proposal for the Estimation of Optimal Power Quality Levels, *Power Systems Conference and Exposition, PSCE 2006*, Atlanta (USA), October/November.

[39] Haghifam, R. and Daneshgar, M. (2001) Developing a New Algorithm to Participate Power Quality Variations in Tariff: an Experience in Iran, *CIRED 2001*, Amsterdam (The Netherlands), June.

[40] Caramia, P., Carpinelli, G., Verde, P., Chiumeo, R., Matrandrea, I. and Tarsia, F. (2006) Indici Globali per la Caratterizzazione della Power Quality nei Sistemi Elettrici, 101 National Conference of AEIT, *Capri* (Italy), September.

[41] EN 50160 (2000) Voltage Characteristics of Electricity Supplied by Public Distribution Systems, March.

[42] Dettloff, A. and Sabin, D. (2000) Power Quality Performance Component of the Special Manufacturing Contracts between Power Provider and Customer, *9th International Conference on Harmonics and Quality of Power*, Orlando (USA), October.

[43] CEER, Electricity Working Group Quality of Supply Task Force (2005) Third Benchmarking Report on Quality of Electricity Supply.

[44] McGranaghan, M., Kennedy, B.W. and Samotyj, M. (1998) Power Quality Contracts in a Competitive Electric Utility Industry, *8th International Conference on Harmonics and Quality of Power*, Athens (Greece), October.

[45] Su, M., Li, G., Zhou, M. and Yang, J. (2007) A Power Quality Service Pricing Approach Considering Treatment Expenses and Stop Loss Insurance, *PowerTech Conference*, Lausanne (Switzerland), July.

Index